METHODS IN PAIN RESEARCH

METHODS & NEW FRONTIERS IN NEUROSCIENCE

Series Editors
Sidney A. Simon, Ph.D.
Miguel A.L. Nicolelis, M.D., Ph.D.

Published Titles

Apoptosis in Neurobiology
Yusuf A. Hannun, M.D., Professor/Biomedical Research and Department Chairman/
  Biochemistry and Molecular Biology, Medical University of South Carolina
Rose-Mary Boustany, M.D., tenured Associate Professor/Pediatrics and Neurobiology,
  Duke University Medical Center

Methods for Neural Ensemble Recordings
Miguel A.L. Nicolelis, M.D., Ph.D., Associate Professor/Department of Neurobiology,
  Duke University Medical Center

Methods of Behavioral Analysis in Neuroscience
Jerry J. Buccafusco, Ph.D., Professor/Pharmacology and Toxicology,
  Professor/Psychiatry and Health Behavior, Medical College of Georgia

Neural Prostheses for Restoration of Sensory and Motor Function
John K. Chapin, Ph.D., MCP and Hahnemann School of Medicine
Karen A. Moxon, Ph.D., Department of Electrical and Computer Engineering,
  Drexel University

Computational Neuroscience: Realistic Modeling for Experimentalists
Eric DeSchutter, M.D., Ph.D., Department of Medicine, University of Antwerp

METHODS IN PAIN RESEARCH

Edited by Lawrence Kruger

Professor of Neurobiology (Emeritus)
University of California, Los Angeles (UCLA)
School of Medicine
Los Angeles, CA

CRC Press

Boca Raton   London   New York   Washington, D.C.

Library of Congress Cataloging-in-Publication Data

Methods in pain research / edited by Lawrence Kruger.
      p. cm. (Methods & new frontiers in neuroscience)
    Includes bibliographical references and index.
    ISBN 0-8493-0035-5 (alk. paper)
    1. Pain—research—Methodology. I. Kruger, Lawrence. II. Methods & new frontiers in
neuroscience series.
    [DNLM: 1. Pain. 2. Research—methods. WL 704 M592 2001]
RB127 .M46 2001
616′.0472′072—dc21
                                           2001025410
                                             CIP

Visit the CRC Press Web site at www.crcpress.com

Methods & New Frontiers in Neuroscience

Series Editors
Sidney A. Simon, Ph.D.
Miguel A.L. Nicolelis, M.D., Ph.D.

Our goal in creating the Methods & New Frontiers in Neuroscience Series is to present the insights of experts on emerging experimental techniques and theoretical concepts that are, or will be, at the vanguard of neuroscience. Books in the series cover topics ranging from methods to investigate apoptosis to modern techniques for neural ensemble recordings in behaving animals. The series also covers new and exciting multidisciplinary areas of brain research, such as computational neuro-science and neuroengineering, and describes breakthroughs in classical fields such as behavioral neuroscience. We want these to be the books every neuroscientist will use in order to get acquainted with new methodologies in brain research. These books can be given to graduate students and postdoctoral fellows when they are looking for guidance to start a new line of research.

Each book is edited by an expert and consists of chapters written by the leaders in a particular field. Books are richly illustrated and contain comprehensive bibli-ographies. Chapters provide substantial background material relevant to the partic-ular subject. Hence, they are not just "methods books." They contain detailed "tricks of the trade" and information as to where these methods can be safely applied. In addition, they include information about where to buy equipment and websites that are helpful in solving both practical and theoretical problems.

We hope that as the volumes become available the effort put in by us, the publisher, the book editors, and individual authors will contribute to the further development of brain research. The extent that we achieve this goal will be deter-mined by the utility of these books.

Preface

The invitation to assemble a book on methodology in pain research proffered by my longtime colleagues, Sid Simon and Miguel Nicholelis, was initially received with considerable uncertainty and hesitation. "Pain" encompasses a rather vast field of research and is hardly a subject that could be reduced to a few formulaic chapters describing laboratory protocols ranging from surgical techniques to automated molecular biology, each of which could encompass an entire volume and still prove inadequate for the needs of investigators seeking methodological guidance in this field of endeavor. Discussions with several friends whose research interests are deeply committed to studying pain persuaded me to recognize that this is no longer an era of publishing lab cookbooks that would be largely out of date by the time they emerge from the printing press. Besides, we are now in an epoch in which new technologies are being developed at a terrifying pace in such areas as high throughput methods, chemical genomics, microarray applications, genotyping, and proteomics. The modern investigator keeps current on such rapidly advancing subjects via the World Wide Web, and protocols generally are supplied by vendors of apparatus and reagents.

But pain research, despite its vastness, is something quite different and special; in part because of its breadth, but largely because it touches upon a subjective report in humans that is poorly understood and yet of fundamental importance and interest to everyone. Pain remains the singular most common complaint of patients world-wide, and this is followed by the next most prevalent affliction — the related phenomenon of itch. Yet pain has not reached the status of a targeted "disease;" it is clearly part of ordinary human experience. It stands apart from other sensations by often being dissociated from a distinctive stimulus condition; i.e., pain can be elicited by apparently innocuous stimuli in pathology or when applied to previously injured tissues. Thus, pain can be described pragmatically as a response to a variety of stimuli and contextual circumstances. Unlike other sensations, it can be altered significantly by widely known drugs, many of which are "controlled substances," imposing serious societal problems and consequent encumbrances for laboratory study. Finally, it should be emphasized that *specific* nerve fibers, now called "nociceptors," were disavowed explicitly over much of the last few decades. Now that laboratory experiments can reveal and enable investigators to parse stimulus and response variables, the field has attracted profoundly expanded interest.

Many of the relevant methods are employed in other areas of neuroscience research and some of the specific techniques are covered in several excellent monographs in the CRC Press series joined by this volume. We have attempted here to select topics and authors to meet the needs of investigators seeking guidance on the "how" and the "means" for studying the idiosyncratic features of pain research. Major areas of contemporary efforts are outlined and explained, and sources are provided for print and digital access to requisite details for designing and performing experiments. The complex ethical and political problems arising from studying pain in animals and in human subjects are alluded to only briefly, in the introductory chapter. Such issues are obviously relevant and extremely important, but we have evaded the thornier facets because they cannot be cast easily aside with brief commentary. One can only hope that the study of a problem that offers the promise of relieving human suffering is judged as sufficiently worthy to justify intelligent use of animals models (largely "manufactured" (man-made) mice with manipulated genes).

The authors have been selected as recognized experts representing much of the cutting edge of current methodology. Selecting the topics to be represented proved more difficult and in attempting to round out the expanse of appropriate coverage, the chapter authors have been remarkably thoughtful, cooperative, and prompt, rendering the task of producing a timely volume a tolerable and achievable goal. They get special thanks for responding to editorial comments, suggestions, and criticisms with cheer, tolerance, and promptness.

The editor is much indebted to various colleagues, too numerous to list individually, for suggestions and feedback. The editors of this series were persuasive in fostering a pragmatic approach and time schedule as well as specific advice. Special thanks go to Sid Simon of Duke University and Barbara Norwitz of CRC Press for the "hand-holding" that was needed periodically and tendered with thoughtfulness and sensitivity. Expert assistance with literature searches and various mundane editorial tasks was gratefully received from Shauna Mulvihill and Howard Kim at UCLA. Special acknowledgement belongs to the several people at CRC Press who guided this book into print and gracefully endured my tormenting commentaries about the extraordinary election debacle and the young Cuban refugee who dominated the news in South Florida last year. Barbara Norwitz, with the able assistance of Tiffany Lane, counseled and nurtured the editorial process with comfortable grace. Suzanne Lassandro deserves special thanks for expert and efficient handling of production. The copy editing and crucial final stages were gently guided with skill, excellence, and warm spirits by Mimi Williams, rendering the editorial task remarkably painless, perhaps in recognition of what pain research is all about — or should be. The attractive cover was designed by Dawn Boyd, and last but not least, the cheering section was provided by my family and especially my wife, Ginny.

Lawrence Kruger

About the Editor

Lawrence Kruger has been active in various aspects of pain-related research for more than 4 decades as Professor of Anatomy and more recently as Research Professor of Neurobiology at the UCLA School of Medicine and Brain Research Institute. For the past 25 years he has held a joint appointment in anesthesiology and served as an advisor to the department's pain clinic.

Dr. Kruger earned his Ph.D. in Physiology from Yale University, New Haven, Connecticut. He has completed postdoctoral training in physiology and anatomy from Johns Hopkins, Institut Marey, College de France, and Oxford University.

Dr. Kruger was among the founding members of the International Society for the Study of Pain (IASP) and the Society for Neuroscience. He has been a member of numerous journal editorial boards, and served as founding editor of *Somatosensory and Motor Research* for more than a decade. He has contributed reviews and edited two recent books dealing with pain. He was supported by a prestigious NIH Jacob Javits Neuroscience Investigator Award until recently assuming emeritus status and closing his laboratory to devote his energies largely to writing on diverse areas of neuroscience, principally on the emergence of pain research.

Professor Kruger's pain research has ranged from early studies of thalamic and trigeminal nociceptive-specific neuron discharge properties, to anatomical studies of pain pathways encompassing the tracing of peripheral patterns in innervation labeled by peptide and other molecular markers of pain endings.

Contributors

A. Vania Apkarian, Ph.D.
Department of Physiology
Northwestern University
  Medical School
Chicago, Illinois

Lars Arendt-Nielsen, Ph.D., Dr. Sci.
Center for Sensory-Motor Interaction
Laboratory for Experimental
  Pain Research
Aalborg University
Denmark

Janice L. Arruda, B.S.
Department of Anesthesiology
Dartmouth Hitchcock Medical Center
Lebanon, New Hampshire

Gary J. Bennett, Ph.D.
Department of Anesthesiology
  and Faculty of Dentistry
McGill University
Montreal, Canada

Susan M. Carlton, Ph.D.
Marine Biomedical Institute
Department of Anatomy
  and Neurosciences
University of Texas Medical Branch
Galveston, Texas

Richard E. Coggeshall, M.D.
Department of Anatomy and Neuroscience
  and Marine Biomedical Institute
University of Texas Medical Branch
Galveston, Texas

Karen D. Davis
University of Toronto
and
Toronto Western Research Institute
Toronto, Ontario, Canada

Joyce A. DeLeo, Ph.D.
Departments of Anesthesiology
  and Pharmacology/Toxicology
Dartmouth Hitchcock
  Medical Center
Lebanon, New Hampshire

Gerald F. Gebhart, Ph.D.
Department of Pharmacology
The University of Iowa
Iowa City, Iowa

Michael S. Gold, Ph.D.
Department of Oral and Craniofacial
  Biological Sciences
University of Maryland Dental School
Baltimore, Maryland

Igor D. Grachev, M.D., Ph.D.
Departments of Neurosurgery
  and Neuroscience and Physiology
SUNY Upstate Medical University
Syracuse, New York

Thomas Graven-Nielsen, Ph.D.
Center for Sensory-Motor Interaction
Laboratory for Experimental
  Pain Research
Aalborg University
Denmark

Michael D. Hayward, Ph.D.
Vollum Institute
Oregon Health Sciences University
Portland, Oregon

Beth R. Krauss
Department of Neurosurgery
SUNY Upstate Medical University
Syracuse, New York

Lawrence Kruger, Ph.D.
Departments of Neurobiology
 and Anesthesiology
Brain Research Institute
UCLA School of Medicine
Los Angeles, California

Alan R. Light, Ph.D.
Department of Cellular
 and Molecular Physiology
University of North Carolina
Chapel Hill, North Carolina

Malcolm J. Low, M.D., Ph.D.
Vollum Institute
Oregon Health Sciences University
Portland, Oregon

Nigel T. Maidment, Ph.D.
Department of Psychiatry
 and Biobehavioral Sciences
Brain Research Institute
UCLA School of Medicine
Los Angeles, California

Yoshizo Matsuka, D.D.S., Ph.D.
Division of Oral Biology and Medicine
UCLA School of Dentistry
Los Angeles, California

Jeffrey S. Mogil, Ph.D.
Department of Psychology
 and Neuroscience Program
University of Illinois at Urbana-
 Champaign
Champaign, Illinois

Timothy J. Ness, M.D., Ph.D.
Department of Anesthesiology
University of Alabama at Birmingham
Birmingham, Alabama

John K. Neubert, D.D.S., Ph.D.
Division of Oral Biology and Medicine
UCLA School of Dentistry
Los Angeles, California
and
Pain and Neurosensory
 Mechanisms Branch
National Institutes of Health
Bethesda, Maryland

Märta Sergerdahl, M.D., Ph.D.
Department of Anaesthesia and
 Intensive Care
Huddinge University Hospital
Sweden

Igor Spigelman, Ph.D.
Division of Oral Biology and Medicine
and
Dental and Brain Research Institutes
UCLA School of Dentistry
Los Angeles, California

Peter Svensson, D.D.S., Ph.D., Dr. Sci.
Center for Sensory-Motor Interaction
Laboratory for Experimental
 Pain Research
Aalborg University
Denmark
and
Department of Prosthetic Dentistry
 and Stomatognathic Physiology
Orofacial Pain Clinic
Royal Dental College
Aarhus University
Denmark

Sarah M. Sweitzer, Ph.D.
Department of Pharmacology/
  Toxicology
Dartmouth Hitchcock Medical Center
Lebanon, New Hampshire

Andrew Todd, Ph.D.
Spinal Cord Group
Institute of Biomedical
  and Life Sciences
University of Glasgow
Glasgow, Scotland

Nikolaus M. Szeverenyi, Ph.D.
Departments of Radiology and
  Neuroscience and Physiology
SUNY Upstate Medical University
Syracuse, New York

You Wan, Ph.D.
Department of Psychology
  and Neuroscience Program
University of Illinois at Urbana-
  Champaign
Champaign, Illinois

Sonya G. Wilson, B.S.
Department of Psychology
  and Neuroscience Program
University of Illinois at Urbana-
  Champaign
Champaign, Illinois

Table of Contents

1 The Idiosyncratic Problems Associated with Pain Research

Lawrence Kruger

CONTENTS

1.1 INTRODUCTION

Pain research earns a distinctive niche for several reasons, but none more persuasive than its practical importance in human welfare, for pain is the most common clinical complaint and the pervasive source of much suffering. Pain provides an essential mechanism for protecting organisms from destructive influences and those rare cases of congenital insensitivity to pain are susceptible to a variety of harmful events that usually curtail the longevity of those unfortunate individuals. Thus, pain cannot be regarded as less useful or desirable than vision or any other sense, for it is of key importance in sustaining life. Yet it must be considered quite separately from other sensory systems because its key role is not that of exploration, but rather to guide warning and avoidance behavior. Indeed, pain is a reaction that frequently defies description in terms of a given stimulus condition and thus requires precise definition that is not obfuscated by the important subjective issues of suffering.

To the extent that pain research can relieve human suffering it is a noble enterprise, but this must be tempered by concern for and development of sensible and sensitive policies to avoid cruelty or suffering in experimental subjects. No other area of scientific endeavor is scrutinized and politicized with such passion. It is, therefore, the responsibility of pain researchers to define their aims and to weigh carefully the ethical, legal, and political consequences of pain research with an enthusiasm that matches their desire to understand and control pain for its obvious human benefit.

0-8493-0035-5/01/$0.00+$1.50
© 2001 by CRC Press LLC

1

Recognition of pain as a scientific discipline is relatively recent, deriving largely from the formation of pain clinics and large-scale organizations, i.e., the International Society for the Study of Pain (IASP) and the American Pain Society (APS). But pain is rarely included in the medical curriculum as a distinct subject or as a clinical entity, and not surprisingly, sources of funding for research support usually must also be found within other contexts. The idiosyncrasies of pain research spring from the equivocal and ambiguous status of a subject historically subsumed within other scientific disciplines.

1.2 DEFINING PAIN

The definition of pain has been elusive for a variety of reasons deriving from the idiosyncratic history of pain research,[3,4] but principally from the failure to distinguish sharply between stimulus and response. The customary description of pain as a stimulus, especially in clinical usage, has contributed massively to the obfuscation. Clearly, pain is an experiential response to a variety of stimulus conditions, at times including stimuli that are ordinarily innocuous under normal circumstances, as in the painful experience of gentle air movement eliciting paroxysmal pain in trigeminal neuralgia, light contact of a wound margin or a burn blister. Modern studies of the distribution and molecular specialization of peripheral nociceptor neurons suggest that these nerves serve effector functions throughout life even if they are rarely, or ever, activated to elicit pain. The response aspects are also evident in the subjective components of pain that are profoundly influenced by learning and experience. Thus, it should be emphasized that pain is enormously multifaceted or multidimensional and cannot be successfully defined as a unitary phenomenon.

The semantic issues have become inscrutable encumbrances and are difficult to modify and modernize. The recognition of pain as a signal portending the threat of injury is most easily understandable when it evokes behavioral (usually affective) and autonomic responses capable of providing protection for the organism and the reestablishment of balance or homeostasis. This is less obvious for prolonged, chronic, persistent, or long-term pain syndromes; these are often maladaptive and actually systemically debilitating, aside from the profound psychological impact of episodic or prolonged suffering. Several chapters in this volume wrestle with the issue of measuring and describing pain as a sensory experience. The term analgesia is widely used to denote reduced painful perception of a noxious stimulus, although etymologically an absence of pain is implied; a condition generally achieved only by local or general anesthesia, i.e., absence of sensation. Strict usage would require use of the term hypalgesia or hypoalgesia but current practice precludes attempting to impose semantic rules. Unfortunately, considerable ambiguity derives from linguistic looseness. Thus, an allodynic response where a light tactile stimulus elicits pain is sometimes loosely referred to in stimulus terms. Clinical pain usually refers to chronic pain, implying that it is prolonged or persistent and generally outlasting the duration of a recognizable stimulus. Further confusion derives from the hedonic spectrum extending from pain to pleasure associated with some noxious stimulus conditions. Responses to intense scratching of an itch or even sadomasochistic practices exemplify the range of responses associated with stimuli that are generally conceded to possess a painful component.

1.3 ETHICAL AND LEGAL ISSUES

The ethical and legal issues pervading pain research are inextricably intertwined and it seems pointless to deal with them separately. Anthropocentric moral values are seriously derided by the animal rights movement and it should be acknowledged directly and openly that throughout history, man has exploited animals for self-serving reasons — especially to provide food and clothing. Most recently, there has been a surge of manufacturing animal models for research, literally man-made genetically engineered animals that are not found in nature and that cannot be attributed to the influence of a deity. Such animal research models, largely mice, have become a key tool for examining the molecular basis of many fundamental biological mechanisms, including those underlying pain. While most biomedical research is energetically designed to avoid pain and suffering, the study of pain obviously is especially susceptible to criticism, careful scrutiny, and legal regulation. Imposing pain upon animals (even those manufactured for research) to serve human welfare is not invoked lightly or without care and planning. It is regulated appropriately by law at federal and state levels, and locally by serious, dedicated committees of scientists, veterinarians, and lay persons at each research institution. Guidelines have been established by IASP[8] and periodic thoughtful discussions can be found in the journal *Pain*.

The elimination of testing altogether, or the substitution of alternatives to animals in research has been welcomed by some manufacturers, e.g., the cosmetics industry, because they become unconstrained by stringent requirements for testing the deleterious and irritant properties of agents applied to the skin. There is room for disagreement concerning the desirability of failing adequately to protect humans in order to defend animals, and encouraging manufacturers to abrogate their responsibilities under the guise of humane efforts should not be greeted with applause or impunity.

Improving the human condition, and especially the alleviation of the suffering resulting from some painful conditions, is accepted as a worthy and perhaps noble goal, one that inevitably requires studying the behaviors associated with pain in man and other animals. This does not imply that other models should not be exploited and, indeed, studies of inflammatory mechanisms in the sensory neurons of the invertebrate sea-slug, *Aplysia*,[1,2] have revealed valuable information of direct relevance to the largely intractable sequelae of neuropathic pain. But ultimately, pain must be studied in a complete, behaving organism, not a tissue culture dish.

It should be emphasized that pain should not be equated with suffering and that experimental protocols must be designed to narrowly limit the magnitude and duration of noxious stimuli to avoid misery. Failure to do so would not be approved by institutional oversight committees and would be subject to serious legally mandated sanctions. One of the responsibilities of pain researchers is to reach out to society, including the anti-vivisection and animal rights advocates, and to convince these people that understanding pain and alleviating its unfortunate consequences in all animals, but especially in humans, is worthwhile and deserving of their support.

We have omitted wrestling with the ethical problems resulting from the use of placebos and the withholding of therapeutic aid to control groups, because this cannot be explored practically in a methods book of useful size. However, it should be emphasized that experimental design is of paramount importance in clinical pain studies and that most modern studies reflect the sophistication of that emphasis. Electronic (i.e., digital) search tools have become an essential and mandatory requirement for exploiting the full range of modern and future experimental innovations.

A larger ethical and legal issue surrounds the use of analgesic agents for pain control; many fall in the realm of controlled substances because they are susceptible to abuse. The euphorigenic effect of narcotic drugs useful in pain control, especially opiates, has resulted in their widespread use despite the dangers of addiction and physically dangerous dependence. While some addictive drugs, notably alcohol, are usually legally obtainable, the illegal status of opiate drugs has fostered a vastly influential multi-billion dollar criminal enterprise. The economic impact and power of the international illegal drug industry is such that it cannot be ignored by society as a whole, but especially by the many pain researchers who elect to study addictive drugs. This is not the place for meaningful editorial comment on the failure of modern society to address this problem satisfactorily; suffice it to admonish researchers to be wary of the variety of dangers that can derive from studying potentially addictive analgesics. Failure to achieve adequate pain relief, especially in terminally ill cancer patients when attributable to reticence to prescribe and deliver controlled substances is not easy to rationalize, but problems requiring political solutions have rarely been successfully solved by scientific organizations. Efforts on behalf of such goals have not been ignored by national medical societies and the IASP, and it is useful to track their publications for news of their political efforts.

1.4 METHODOLOGICAL ADVANCES

Achieving a balanced view of current methodology has been attempted by discussing what topics would be most useful with colleagues currently active in pain research. A full list of those consulted would be cumbersomely long and yet perhaps incomplete. To whatever extent this summary succeeds in achieving balance, it is to the credit of numerous workers in the field whose ideas guided the outline of this volume. Obviously, an all-inclusive approach would be impractical and lengthy, forcing emphasis on specific applications in pain studies and placing constraints on covering topics that are too large and broad, especially if employed more commonly in other fields and suitably available from sources that might prove more practical. Examples include broad topics such as receptor research, extending from detection and binding to pharmacokinetics, or narrower methods not commonly employed in pain studies, e.g., karyology for examining chromosomal properties. The new vast field of DNA microarray technologies has accelerated emphasis on expression analysis, leading most recently to advances in arraying antibodies in gelpack-based chips to achieve protein array identification of, among various possibilities, proteins that bind to other proteins, to protein kinases, or to drug-like molecules bearing a fluorescent tag. The possibilities seem almost infinite in applying analytical biochemistry to the core questions in cellular chemistry, but for the most part these methods remain at the fringe of pain research for the moment. Thus, the bias in chapter selection largely reflects the range of principal current activity. The fads of today will inevitably change within the next decade to reflect the rapidity of technological development.

Examining neuronal function is covered in a variety of ways that involve a few areas of overlap. These include chapters on membrane properties with emphasis on putative nociceptive sensory ganglion cells (Gold, Chapter 9) to a broad survey of various electrophysiological recording methods for studying neuronal impulse discharge properties (Spigelman et al., Chapter 8). Recent advances of relevance in clinical research include surveys of the variety of extracellular sampling methods used in experiments on animal models and humans (Spigelman et al., Chapter 7), and methods of sampling chemical and electrical properties in a more global manner that permits study in human subjects (Davis, Apkarian et al., and Graven-Nielsen et al., Chapters 12 through 14).

In sheer numbers, rodents are the most widely used animal models and most of the experimental methods described in this volume have been applied principally to the laboratory rat. In the last few years the choice has been shifting rapidly toward the lab mouse (*Mus musculus*), the rodent of choice for genetic studies because of the reliability of commercially available inbred strains, their rapid reproduction in large numbers, and most importantly, as a tool for genetic studies. Thus, great emphasis is being placed on behavioral assessment of nociception and extensive studies have emerged that attempt to evaluate the efficacy of analgesic agents, especially those possibly applicable to human subjects. The range of those mouse strains commonly employed in pain research is reviewed by the expert team of Mogil and his colleagues (Chapter 2), but with the admonition that "mice are *not* small rats." The ability to study strain variants is now being supplanted by studies of transgenic manipulations, including gene deletions (knockouts) and gene overexpression. The initial sequence map of the mouse genome, based on several strains, is complete and with it the revelation of single nucleotide polymorphisms (SNPs) that ultimately will provide a useful SNP database for understanding genetic diversity in many species, including humans. By the time this book is widely available to readers, the techniques of mutagenesis in mice are likely to have grown exponentially. The variety and limitations of gene manipulation and strain assessment tools are discussed in admirably broad scope and within the specific context of opioids by Hayward and Low (Chapter 3) as a means of emphasizing its practical application in pain research. It is easy to predict that the enormous number of genes shared by man and mouse will provide the impetus for studies of mouse mutagenesis and the temporal control of gene activation and inactivation. There is also a new technology based on literally shooting biologically active particles (especially DNA) into tissue or cells. This involves using a "gene gun" for "bolistic" particle delivery, a contraction of "biological" and "ballistic" and best followed with electronic search tools. This most rapidly growing sector of pain research in the immediate future clearly will be dominantly dependent upon gene-modified or selected strains of mice.

We initially discussed including chapters on several topics but they have grown enormously in technical diversity and variations to the point where print publication seems too quickly outdated and less effective than web search for specific methods, instruments, and kits. A prominent example derived from the important stride in cloning the nociceptor-specific capsaicin receptors.[5] Consulting one of the experts on receptor identification, detection, binding, gene expression, etc. (John A. Maggio, University of Cincinnati) proved convincingly that even citing the several dedicated specialized volumes would be less suitable than guiding readers to the web sites of commercial providers of reagents, kits, and instruments. They also regularly update their technical web sites for methods and most maintain readily accessible hotlines.

Quantification of tissue changes correlating with noxious stimulus application or observation of pain-related behaviors is relatively recent in pain research. The principal anatomical methods are outlined and explained in sufficient detail to enable practical bench procedures with little prior specialized experience except for developing skills with microtomy. Chapter 10 by Carlton and Todd includes protocols as well as references for the most widely employed experimental procedures specific to particular molecular markers employing antibodies, *in situ* hybridization, and the range of techniques in current pain research. It does not attempt to explain some of the new fluorescence methods that enable measurement of dynamic cellular processes, e.g., FRET (fluorescence resonance energy transfer) for study of protein interactions or LRET (luminescence resonance energy transfer) for detecting conformational changes, etc., because these technologies have not yet significantly penetrated pain studies. Most of the quantification in recent years has relied on neuronal and axonal counts or area measures, and because this has been a subject of some controversy and claimed discrepancies, a brief account of the principles of stereology and its practical application has been provided in a valuable updated account by Coggeshall (Chapter 11).

Over a century of studying pain elicited from accessible integumentary regions (cf., skin, cornea, and teeth) has led to long-established methods of stimulus control and standardization. In recent years, new methods have been devised for studying visceral pain evoked by controlled stimuli; these are recounted with detailed expertise in Chapter 5 by Ness and Gebhart. Neuroinflammatory mechanisms, especially those involved in chronic pain syndromes, have been examined in detail only in the past decade with the identification of cytokine mediators and still obscure immune mechanisms. An account of modern approaches from a leading laboratory is provided in Chapter 6 by Sweitzer, Arruda, and DeLeo.

The gradual decline in use of radioisotopes for labeling and detecting biomolecules and the utilization of fluorescent and luminescent technologies are developing in the hands of commercial enterprise and new kits are marketed at an astonishing pace. It should be emphasized that fluorescence methods have provided tools for quantification and for real-time dynamic studies of physiological processes ranging from polymerase chain reaction (PCR) or other nucleic acid amplification products to fluorescent resonance energy transfer (FRET) to detect the assembly, dissociation, or conformational rearrangements of proteins; chemiluminescent determination of chemical reactions (e.g., assay of ATP bioenergetics); optical detection of synaptic vesicle dynamics; changes in membrane permeability; and the use of lipophilic calcium indicators enabling study of Ca^{2+} concentration transients at membrane surfaces as well as the longer established means for monitoring global cytosolic calcium dynamics. Attempts to summarize and update the rapid development of optical techniques and new reagents would be out of date before the volume could appear in print and although some general primers are becoming available, the availability of instruments and new reagents is largely driven by commercial sources. Such information is often best accessed from web sites of manufacturers, e.g., http://www.probes.com/handbook/sections/0002.html. New links (URLs) and the new reagents announced each month would seem to preclude the need for preparing a concise summary in print form at this time.

Subjective magnitude scaling methods can be exploited for human psychophysical studies and these are a useful and important means for assessing successful pain therapy, recognizing that analgesics have little effect on pain detection *thresholds*. Measuring the magnitude of the *response* to suprathreshold stimuli has been crucial for assessing clinical therapeutic effectiveness, but behavioral assessment is far more difficult in animal experiments, and magnitude scaling of a pain response has been sufficiently elusive to require more modern accounts of methodological advances, as outlined in Chapters 2 and 4.

Ultimately, it must be acknowledged that in a strict sense, "pain" can only be properly studied in human subjects who can report and estimate the quality and magnitude of the wide range of stimulus conditions that elicit pain. For practical purposes this is dealt with briefly and with emphasis on objective methods of examining nociception in Chapter 14 by Graven-Nielsen and colleagues, and represents one of the leading clinical pain services in Europe with an active research program. There are more than a few major monographs dealing specifically and often solely with clinical studies referred to in their bibliographies and additional sources can be found in the Wall and Melzack *Textbook of Pain*,[7] several monographs, (e.g., Reference 6), and, of course, most comprehensively by a web search through PubMed or some other convenient search engine. Finally, it should be noted that there are other books in the CRC Press *Methods and New Frontiers in Neuroscience* series that provide valuable resources for in-depth surveys of specific relevant subjects, such as computational and behavioral analysis, as well as advanced techniques in general use for electrophysiology and functional imaging.

1.4.1 ELECTRONIC SOURCES

Electronic publishing has dramatically altered the manner in which the proliferation of scientific information can be explored but despite the convenience and strength of powerful new search mechanisms, usage can be complex and difficult. The major commercial publishers of scientific journals have erected barriers in order to establish new ways to charge readers for accessing journal articles and there are services available on an annual fee basis such as Current Contents from the Institute for Scientific Information (ISI) which provides the **Web of Science** citation indices. University library systems, or sometimes consortiums of multiple institutions, have devised literature search tools (e.g., *Melvyl* and *Orion* in the University of California system) and fee arrangements have been negotiated with publishers to enable direct, no-cost access for their constituency (e.g., the *California Digital Library* in the University of California system). For the present there are different rules and mechanisms at various venues. At the time of this writing a plethora of new sites and services have been instituted or announced and by the time this volume appears in print there is certain to be further innovation. We can only hope to present some selected highlights as exemplars of the latest developments in search tools.

The National Institutes of Health (NIH) and the National Library of Medicine (NLM) have recently instituted a **PubMed** search engine and a new and somewhat controversial repository for peer-reviewed primary research articles, **PubMed Central**, still in its infancy and with minimal content, but growing rapidly (www.pubmedcentral.nih.gov). **Cite Track** alerts readers to new material in PubMed or HighWire, matching the reader's preselected search choices by author, keywords, etc. A European counterpart **E-BioSci** is also in the making (www.embo.org), and we shall soon have initiatives from commercial publishers such as **BioMed Central** (www.biomedcentral.com) that also will deposit its articles in **PubMed Central**. Stanford University's non-profit electronic publishing venture **High Wire Press** is one of the largest free-access information repositories (www.highwire.org), and a new site enabling bibliographic linking, **CrossRef** is due to be launched this year (www.crossref.org).

The biotech industry provides a web site (www.bio.com) and a technical information URL **TechDox**, without advertising and promotions (www.techdox.com); also a comprehensive site for access to thousands of vendors and sources of reagents, instruments, and services can be found at www.sciquest.com, including SciCentral to search for other databases. A forum of opinions and evaluations on equipment, products, reagents, kits, etc. is now available from Biowire at www.biowire.com; this site also has a section called "Voodoo Hints" which provides a place for important lab hints that don't seem to make it into materials and methods sections.

Informatics sites are offered by some universities (e.g., University of Michigan and Virginia Polytech) at www.bioinformatics.med.umich.edu and www.bioinformatics.vt.edu; Michigan also plans a proteomics site. Information on the rapidly growing field of genomics is updated daily by the National Human Genome Research Institute (www.nhgri.nih.gov), and further information is forthcoming from commercial sources such as Celera Genomics Group (www.celera.com). The public working draft version of the sequence of the human genome has recently become available via three complementary (but unique) links providing assembled views of the human genome, together with browsing tools for examining the variety of annotations of the sequence. These sites are updated continuously: University of California at Santa Cruz, http://genome.ucsc.edu; National Center for Biotechnology Information (NCBI), http://www.ncbi.nlm.nih.gov/genome/guide/ and click "Map Viewer"; and European Bioinformatics Institute (EBI), http://www.ensembl.org/.

With the recent completion of the entire genetic code of man and the publication of the sequence for the animal model of choice for gene testing (the mouse) by the time this book appears in print, it will be abundantly evident that we have entered a new age of "bioinformatics." This marriage between computer science and biology is of enormous consequence in the expected "tsunami of information" about to explode on the scene. There are already more than 50 private and publicly traded companies offering bioinformatics products and services. The bioinformatics databases maintained by the National Center for Biotechnology Information (NCBI) can be accessed at www.ncbi.nlm.nih.gov; the biweekly newsletter *BioInform* can be found at www.bioinform.com and the recent report *Trends in Commercial Bioinformatics* can be logged onto free at www.oscargruss.com/reports.htm. The astronomical number of biochemical events and interactions that will be studied in the next decade thrusts major efforts in pharmacology and medical diagnostics into a new "in silico" era of unimaginable dimensions.

Most scientific societies maintain useful web sites and these are usually found by typing in their initials (e.g., Society for Neuroscience, www.sfn.org or International Society for Computational Biology, www.iscb.org) but it is usually more practical and fruitful to search under specific subjects when hunting for methodological advances and procedures. Electronic (i.e., digital) search tools have become an essential, mandatory requirement for exploiting the full range of modern and future experimental innovations.

REFERENCES

1. Clatworthy, A.L. and Grose, E., Immune-mediated alterations in nociceptive sensory function in *Aplysia californica, J. Exp. Biol.*, 202, 623, 1999.
2. Clatworthy, A.L. and Walters, E.T., Comparative analysis of hyperexcitability and synaptic facilitation induced by nerve injury in two populations of mechanosensory neurones of *Aplysia californica, J. Exp. Biol.*, 190, 217, 1994.
3. Kruger, L. and Kroin, S., A brief historical survey of concepts in pain research, in *Handbook of Perception*, Vol. VIB, E.C. Carterette and M.P. Friedman, Eds., Academic Press, New York, 1978, 159–179.
4. Perl, E. and Kruger, L., Nociception and pain: evolution of concepts and observations. Pain and touch in *Handbook of Perception and Cognition*, 2nd ed., L. Kruger, Ed., Academic Press, 1996, 179–211.
5. Tominaga, M., Caterina, M.J., Malmberg, A.B., Rosen, T.A., Gilbert, H., Skinner, K., Raumann, B.E., Basbaum, A.I., and Julius, D., The cloned capsaicin receptor integrates multiple pain-producing stimuli, *Neuron*, 21, 531, 1998.
6. Turk, D.C. and Melzack, R., Eds., *Handbook of Pain Assessment,* Guilford Press, New York, 1992.
7. Wall, P.D. and Melzack, R., *Handbook of Pain,* 3rd ed., Churchill Livingstone, Edinburgh, 1994.
8. Zimmermann, M., Ethical guidelines for investigations of experimental pain in conscious animals, *Pain*, 16, 109, 1983.

2 Assessing Nociception in Murine Subjects

Jeffrey S. Mogil, Sonya G. Wilson, and You Wan

CONTENTS

2.1 INTRODUCTION

Pain research has always been hampered by the subjective nature of the phenomenon. In addition to the complexities of the independent variables being applied, pain researchers have also faced interpretational challenges related to the dependent variables being measured. The existence of pain in laboratory animals can never be ascertained with certainty, merely inferred from behaviors (e.g., withdrawal, licking, immobility, vocalization). Physiological signs (e.g., autonomic changes, EEG, evoked potentials) have proven to be even less reliable as indices of pain, since they

TABLE 2.1

Choice of Species in Pain Research Published in the Last Two Years[a]

Species	Publications in last two years[a]	Publications since 1966[b]	% Publications in last two years
Human[c]	11,223	102,949	10.8
Rat	718	5609	12.8
Mouse	281	1671	16.8
Cat	39	1035	3.8
Dog	44	669	6.6
Rabbit	19	590	3.2
Monkey	19	370	5.1

[a] Number of published pain studies using one of these species, from a search of the National Library of Medicine's MEDLINE database (PubMed, http://www.ncbi.nlm.nih.gov/PubMed/) with the Medical Subject Heading (MeSH) search terms Pain[MeSH] and *SpeciesName*[MeSH] and an entry date limit of 2 years, conducted on March 16, 2001. Species with 10 or more relevant publications in the previous 2 years were included.
[b] Number of published pain studies since 1966, the span of time covered by the MEDLINE database. Search performed as above, but with no date limit.
[c] Includes clinical studies.

can be produced by arousing and/or threatening stimuli in the absence of pain.[161] A further problem with pain research in animals concerns the generalization of standard nociceptive assays to human clinical pain states. Although it must be conceded that even simple reflex withdrawal assays accurately predict the clinical efficacy of a wide range of analgesic manipulations,[146] the development of novel assays that more closely model features of human pain has remained a high priority.[142]

The most common choice of model species for non-human pain research has been the laboratory rat. However, recent years have seen an increase in the use of the laboratory mouse (*Mus musculus*); almost 16% of the pain research studies in the mouse since 1966 have been published in the last 2 years, the highest ratio of any species (see Table 2.1). The likely reason for the accelerating use of mice in pain research, and in biobehavioral research in general, is their utility for genetic studies. The large number of extant, commercially available inbred strains and the small size of this rodent have made them the species of choice for gene mapping efforts.[139] More tellingly, though, the embryonic stem cell technology at the heart of current transgenic manipulations — knockouts — is at present restricted to the mouse. Of the 235 published articles on murine pain research in the last 2 years, 50 used or reviewed findings from knockout mice.

The desire to test knockout mice has encouraged many investigators to adapt their ongoing laboratory procedures to this species. Many techniques used in pain research are affected by the relatively small size of mice compared to rats (average 10-week-old ICR male mouse, 37 g; average 10-week-old Sprague Dawley male rat, 330 g), notably radioimmunoassay and single-cell recording. It may not appear obvious that behavioral nociceptive testing would greatly differ between rats and mice. However, there are, in fact, a great deal of quantitative and qualitative differences in behavioral responses on standard nociceptive assays between these two species. Simply put, mice are *not* small rats. Although they belong to the same rodent family, *Muridae*, the *Mus* and *Rattus* genera diverged from a common ancestor over 10 million years ago.[65]

There exists, to our knowledge, no comprehensive review of algesiometric techniques in the mouse, and between us, we have 15 years of hands-on experience performing a large number of nociceptive assays on unanesthetized mice of a variety of genetic backgrounds. In addition to species differences, we will focus on known and suspected intraspecies genetic and environmental factors affecting the accuracy and reliability of nociceptive testing.

2.2 GENERAL CONSIDERATIONS: HOUSING, HANDLING, AND HABITUATION

Mice and rats have differing general behavioral characteristics that impact on virtually all laboratory testing situations, especially the assessment of nociceptive sensitivity. For example, it is known that unlike rats, mice do not engage in social play when housed in same-sex groups.[123] Although isolation housing is thus a larger stressor in rats, producing permanent deficits in habituation,[39] group housing is generally preferred for both species. The social interactions of male mice more commonly involve dominance displays, including fighting. This species difference has clear implications for nociceptive testing. Stress-induced antinociception (SIA) associated with conspecific defeat is a well-known phenomenon,[97] and group-housed male mice fight frequently, especially those of advanced ages. Obviously, the presence of defeat SIA in some subjects at the start of an experiment will confound the accuracy of baseline measurements. Further, chronic opioid release due to defeat SIA may render such subjects opioid-tolerant before any experimental manipulations occur.[97] Most mice (except for some strains like BALB/c) will fight minimally with their littermates, but it is exceedingly difficult to ensure littermate housing in mice purchased commercially. There is little that can be done about this problem, except to test male mice as young adults. We try to restrict our experimental subjects to 6 to 8 weeks of age.

Standard practice in biobehavioral experiments with laboratory rats is to handle them extensively prior to any testing, in order to render them gentle and presumably non-stressed for the experiment itself. This strategy is not only useless in the laboratory mouse, but actually makes matters worse. In our experience, the vast majority of mouse strains become increasingly agitated with repeated handling attempts, reflected in an emotion-induced increase in body temperature[17] and plasma glucose levels.[145] Thus, SIA associated with human handling is a larger worry for nociceptive testing in mice vs. rats, and the problem will increase over time. The mouse is, therefore, a particularly unsuitable subject for experimental designs involving repeated trials. This issue is especially acute in transgenic knockout studies, where

TABLE 2.2
Dimensions Characterizing Nociceptive Assays

Dimension	Levels
Stimulus etiology	Nociceptive, chemical/inflammatory, neuropathic
Stimulus modality	Spontaneous (chemical)
	Evoked (thermal, mechanical)
Stimulus intensity	Mild → severe
Activated primary afferents	Aδ, C
Location	Cutaneous, subcutaneous, visceral, nervous system
Duration	Acute, subacute, tonic, chronic
Response type	Threshold, suprathreshold
Response characteristics	Reflexive (flexion/extension, writhe, flinch);
	organized (vocalization, recuperative, escape)
Highest level of processing	Spinal, supraspinal

very limited numbers of the *same* mice are tested over and over again on multiple behavioral assays (e.g., open field, rotarod) before nociceptive testing is attempted. The fact that both wild-type and knockout mice are subjected to the same procedures is not able to fully ameliorate this problem, since many targeted genes and their chromosomal neighbors (see 2.8 below) have effects on stress and emotionality.

One factor appearing not to differ greatly between mice and rats is the need to habituate these animals to the testing room and/or observation chamber before experimental data are to be obtained. Both species exhibit marked neophobia (i.e., fear of novel stimuli), which decreases over time and trial number, although in a highly strain-dependent manner.[73,154] Mice are generally more active than rats, and have far more room to maneuver in comparison to their body size in standard observation chambers. Thus, mice will take a longer time to stop exploring the chamber, and this cessation is necessary for many standard nociceptive assays (e.g., Hargreaves' test of paw withdrawal, von Frey fiber testing) to be accurately applied. Partial solutions to this problem include scaling down the size of the chamber when using mice, and habituation on the day or days preceding testing. But even so, in our experience most mouse strains must be habituated for several hours on the testing day itself before any data collection requiring a motionless subject can begin.

2.3   WHICH NOCICEPTIVE ASSAY(S) TO USE?

In recent years it has become increasingly appreciated that as pain is not a unitary phenomenon, nociceptive assays can be dissociated from one another on the basis of what type of pain they presume to model. Nociceptive assays can be characterized on many dimensions (see Table 2.2). Anatomical, ontogenetic, neurochemical, pharmacological, and genetic evidence has been marshaled to suggest functional dissociation of mechanisms underlying assays differing on one or more dimensions.[105,110] For example, a large body of evidence purports that the supraspinally organized paw licking/shaking response in the hot-plate test is mediated and modulated by different neural circuitry than that underlying reflexive withdrawal of the tail from a thermal stimulus.[9,44,67,78,118,175] Similarly, acute pain models (e.g., hot-plate test, tail-withdrawal test) have been dissociated from tonic pain models (e.g., formalin test).[32,131]

Some of these proposed dissociations are suspect because they conflate differences on one dimension with differences on others. For example, the tail-withdrawal test and the formalin test differ as follows: The former involves a spinally reflexive response of the undamaged tail at nociceptive threshold to a high-intensity but acute, escapable, thermal stimulus applied to a restrained animal; whereas the latter involves the measurement of recuperative behaviors (e.g., licking) directed at the inflamed hind paw of an unrestrained animal presumably experiencing modest but prolonged, inescapable, suprathreshold, chemically induced nociception. Thus, the observed ability of a compound to produce antinociception on the formalin test but not the tail-withdrawal test may reflect any number of factors other than stimulus duration.

Because of increasing concern over the generalization of data obtained with any particular method, more and more published investigations feature the use of multiple nociceptive assays. This is especially true for transgenic knockout mice, for which a number of claims have been made linking targeted genes with particular types of pain. For example, Zimmer's laboratory demonstrated an increased sensitivity to escape jumping on the hot-plate test, but no change in latencies on the tail-withdrawal test, in mutant mice lacking preproenkephalin.[78] An analogous demonstration of altered supraspinal but not spinal nociception was reported by this group for tachykinin-1 mutants,[175] although this conclusion was directly contradicted by findings from Basbaum's laboratory using independently derived tachykinin-1 knockout mice[18] (see 5.4 below for further discussion of spinal/supraspinal distinctions). This latter group has demonstrated the specific role in inflammatory and neuropathic nociception, respectively, of the signal transduction molecules protein kinase A, type I regulatory subunit and protein kinase C, gamma.[92,93] As a final example, a modulatory role of the κ-opioid receptor specific to visceral nociception was gleaned from the altered sensitivity of these mutants to the abdominal constriction (writhing) test, but not the hot-plate, tail-withdrawal, tail-pressure, or formalin tests.[140]

Our laboratory has recently collected data from a common set of 11 inbred mouse strains on a wide variety of common nociceptive assays in this species.[101,109,110] Although the primary purpose of these studies is to provide a database of strain sensitivities as an aid to gene mapping and interpretation of transgenic experiments (see 2.8 below), the observed genetic correlation among assays in this strain set can be used to identify fundamental pain types. Nociceptive assays featuring the same high-responding and low-responding strains must be mediated by similar genes, and thus similar physiological mechanisms. Using this logic, we determined that the fundamental dimension underlying the genetic correlation or dissociation of the nociceptive assays considered was stimulus modality: thermal, chemical, or mechanical.[110] No support was found for genetic distinctions between nociceptive vs. neuropathic assays, acute vs. tonic/chronic assays, or spinal vs. supraspinal assays. We will thus use the stimulus modality classification to order the subsequent sections of this chapter. Although we found no strong evidence for the existence of distinct genes mediating inflammatory vs. neuropathic hypersensitivity states, these phenomena nonetheless represent major foci of current pain research.

However, many inflammatory algogens (e.g., carrageenan, mustard oil, complete Freund's adjuvant, zymosan) and neuropathy models (e.g., Bennett and Xie's[12] chronic constriction injury, Kim and Chung's[72] spinal nerve ligation injury, and Seltzer et al.'s[135] partial sciatic ligation injury), although quite likely producing some spontaneous nociception, are more commonly studied with respect to their induced hypersensitivity to evoked thermal and/or mechanical stimuli. Therefore, present discussion will be restricted to evoked thermal and mechanical assays (used of their own accord and/or to evaluate hypersensitivity), and chemical assays of ongoing, spontaneous nociception commonly used in the mouse. Electrical assays (e.g., the flinch/jump test, flexion reflex, tail-shock test) will not be discussed, since they are virtually never used in the mouse.

2.4   DOES STIMULUS INTENSITY MATTER?

Of the various parameters that can be manipulated within the application of any nociceptive assay, stimulus intensity is usually the first considered. Many studies have shown that in threshold assays (e.g., tail-withdrawal test, paw-pressure test), response latency is not a function of stimulus intensity, but rather of heat/pressure transfer (for heat, expressed in $kcal/s/m^2$).[54] For example, Ness and Gebhart[112] demonstrated that the threshold cutaneous tail temperature producing a tail-withdrawal response in the Sprague Dawley rat was 42.6°C, invariant of the voltage applied and the resultant latency. Analgesics may increase this threshold, and injury may decrease it, and thus such models are valid for conducting pain research. By contrast, all chemical nociceptive assays in common use are suprathreshold. Stimulus intensity may have important consequences in such models as well.

2.4.1   NOCICEPTION OF VARIED INTENSITY MAY BE DIFFERENTIALLY PROCESSED

Some data suggest that the exact stimulus intensity chosen may qualitatively affect the nociceptive processing of the stimulus. Yeomans and colleagues[171] have shown that in rats, as in humans, low skin-heating rates produced by exposure to a 50 V bulb (0.9°C/s; mean withdrawal latency, 13.4 s) evoked capsaicin-sensitive, and therefore C-fiber-mediated, nociceptive responses, whereas high heating rates produced by exposure to a 100 V bulb (6.5°C/s; mean withdrawal latency, 2.9 s) evoked responses that were capsaicin insensitive. Higher stimulus intensities thus recruited Aδ nociceptors. The high and low heating rate versions also displayed quantitatively and qualitatively different sensitivities to inhibition by opioid agonists.[90,171] In an intriguing study suggesting that choice of stimulus intensity may be critical to the neurochemical mediation of the nociception produced, Cao and colleagues[18] observed that tachykinin-1 knockout mice differed from their wild-type littermates only at intermediate intensities of thermal and chemical stimulation, for example, at a hot-plate temperature of 55.5°C but not 52.5°C or 58.5°C. Arguing against a qualitative difference between processing of differently intense thermal stimuli are data from Elmer and colleagues,[40] who showed that the stimulus-effect curves of 8 inbred mouse strains on the hot-plate test and abdominal constriction test differed in position, but were of equivalent slope; that is, the same strains were sensitive and resistant to nociception regardless of its intensity. We have obtained similar findings on the tail-withdrawal test[101] and abdominal constriction test (unpublished data).

2.4.2 PRACTICAL IMPACT OF NOCICEPTIVE INTENSITY

These issues aside, the choice of stimulus intensity in nociceptive assays has tremendous practical impact on the sensitivity and reliability of the test. For example, it is well-known that common thermal assays are unable to detect antinociception from non-steroidal antiinflammatory drugs (NSAIDs) like aspirin. This may be related to the specific mechanism of action of NSAIDs (e.g., efficacy against chemical but not thermal nociception), but it has been noted more recently that antinociception from these compounds, and also weak partial or mixed opioid agonists, can be detected in mice using more sensitive warm plate (50 to 51°C) and/or increasing-temperature hot-plate assays.[6,61,116,144] Milder stimuli also render nociceptor sensitization less of a potential problem, but increase the possibility of reflex habituation.[19] The real trade-off for increased sensitivity is increased susceptibility to confounds. For example, writhing responses in the abdominal constriction test, which has more than adequate sensitivity to detect NSAID antinociception, are also inhibited by antipsychotics, antihistamines, and cholinesterase inhibitors at high doses.[54] This test is also exquisitely sensitive to SIA; we have noted, for instance, that the SIA produced by a mere 30-s swim in room temperature water can completely inhibit mice responding on this test [unpublished data, see also Reference 130].

We have noticed a progressive *decrease* in the stimulus intensity of commonly used nociceptive assays, especially thermal ones, in the past few years. As the focus of pain research changes from investigating the neural circuitry underlying antinociception to that underlying hypersensitivity (i.e., hyperalgesia and/or allodynia following inflammatory or neuropathic injury), ceiling effects imposed by ethical but arbitrary cut-off latencies become less of a problem, but floor effects related to minimum reflex latencies become more worrisome. A bulb intensity producing a baseline tail- or paw-withdrawal latency of 2.0 s is ideal for measuring the antinociceptive action of morphine (with a standard cut-off latency of 10 to 15 s), but is unable to clearly display the hyperalgesic effect of nerve ligation. The apparent solution, therefore, is to lower the bulb intensity to produce a baseline latency of, say, 8.0 s.

While this alteration in protocol is no doubt advantageous for the demonstration of hyperalgesia, we and others have shown that SIA-related confounds are concurrently produced. d'Amore and colleagues[29] strikingly demonstrated in rats that tail-withdrawal latencies to a low-intensity stimulus (≈ 4.0 s) were elevated by up to 200% by a restraint stressor, whereas responses to a high-intensity stimulus (≈ 1.5 s) were stable with repeated testing and unaffected by stress. We demonstrated in mice that the stress associated with intracerebroventricular (i.c.v.) injection of artificial cerebrospinal fluid — a procedure requiring a prior indwelling cannula implantation in rats, but in mice achieved by injecting directly through the skull under light anesthesia[85] — produces measurable SIA on the abdominal constriction, hot-plate, and tail-withdrawal tests.[103] We and others believe that this injection-related SIA was responsible for confounding initial studies of the effect of the orphan opioid ligand, orphanin FQ/nociceptin.[96,127] In these studies, and many that followed, vehicle-treated and peptide-treated mice were compared, and the relatively decreased thermal latencies of the latter group were used as evidence of a hyperalgesic effect of orphanin FQ/nociceptin. Accounting for the SIA produced by the i.c.v. injection by taking baseline data and including uninjected groups, we were able to show that this apparent hyperalgesia was in fact a reversal of SIA, and thus an anti-opioid action.[103,104] The injection-related SIA is only a measurable confound at mild noxious stimulus intensities (e.g., < 47.5°C tail-withdrawal test) and is strain dependent.[107] We believe this phenomenon was never reported before because the use of mild stimulus intensities is such a recent development.

Finally, unpublished data from our laboratory have revealed the existence of measurable SIA (on mildly noxious assays) from an intraperitoneal (i.p.) injection of saline in some genetic backgrounds. We conclude that SIA is a ubiquitous feature of nociceptive testing, whose impact can only be minimized by the use of highly noxious stimuli. Since such stimuli are unsuitable for many research purposes, investigators need to be aware of its existence so that appropriate controls can be incorporated into experimental protocols.

2.5 THERMAL ASSAYS OF NOCICEPTION

The most common stimulus used for pain research in any species is acute thermal stimulation. Although acute thermal pain is virtually non-existent as a clinical entity ("Doc, it hurts when I put my hand on a hot stove!"), the heat-evoked flexion reflex has proven to be a surprisingly good predictor of the analgesic potential of pharmacological compounds,[30,146] and nicely matches subjective pain ratings in normal humans.[22] The specificity and validity of such tests in animals as models of human pain have been repeatedly questioned, however.[132] One problem with the application of such tests in mice is that virtually nothing is known about the functional properties of sensory neurons in this species. The one study of which we are aware found that noxious heat stimuli activated Aδ-fibers in the hairy skin of the mouse at an average threshold of 42.5°C and C-fibers at an average threshold of 37.6°C.[76] These data are very similar to those from the rat, but may not be entirely relevant to thermal nociceptive assays in common use, where the heat stimulus is applied to the glabrous skin of the paw, or to the tail.

2.5.1 THE TAIL-WITHDRAWAL TEST

The tail-withdrawal test measures sensitivity to cutaneous thermal stimulation via the flexor withdrawal reflex. Two versions are common: the classic radiant heat "tail-flick" test of D'Amour and Smith[30] and the hot water "tail-immersion" test of Ben-Bassat et al.[11] A cold version of the latter has been developed, using water/ethylene glycol (or other solutions remaining liquid at temperatures < 0°C).[120] We have found mouse strain sensitivities on this test to correlate very highly with the hot water version, although featuring more variability.[101] The major advantage of these tests compared to all others is their relative stability with multiple repeated measurements; we have observed no significant effect of repeated measures even when mice are tested once every 10 min over a 2-h period.

We are aware of no published dissociation between the radiant heat and hot water versions of the tail-withdrawal assay. However, we strongly recommend the latter for use in mice, especially in genetic experiments. First, it has been clearly demonstrated that the exact location of the tail stimulated by the bulb can affect resultant latencies, antinociceptive potencies, and tolerance development.[114,174] Such variability does not greatly affect the hot-water test, since by convention the distal half of the tail is stimulated by virtually all investigators. Second, both homemade and commercial radiant heat units feature divergent time–temperature profiles, contributing to variability among laboratories. By contrast, immersion in 49°C hot water will cause equivalent heat transfer at all locations. Third, whereas the intensity of the noxious stimulus can be succinctly described in the hot water test (common parameters range from 46 to 52°C), the intensity of the noxious stimulus in the radiant heat version can only be indirectly assessed by considering the resultant baseline latency values (commonly, 2 to 8 s), which are highly strain dependent. Fourth, it is necessary to restrain mice more forcefully in the radiant heat version of the test, since precise aiming of the stimulus requires a more completely motionless subject; this extra restraint may impose additional SIA. Finally, the radiant heat version of the test is affected by the light reflectance of the tail, introducing a confound when comparing variously pigmented mice, or young mice to old. One investigation showed that the well-known strain difference in thermal latencies between black C57BL/6 mice (low) and gray DBA/2 (high) mice could be abolished by painting their tails black with permanent marker.[157] In contrast, we have shown that on the hot-water tail-withdrawal test, this strain difference is robust and unaffected by tail color (unpublished data).

Two important parameters affecting both versions of the test are restraint and tail temperature. Restraint, absolutely necessary for positioning the mouse with respect to the noxious stimulus, can be achieved in two ways. Some investigators habituate mice in Plexiglas™ tubes, with their tails protruding through an opening, and restrain them thus for the duration of the experiment.[50] Others remove animals from their home cages immediately prior to each test, and restrain them temporarily only for the duration of the test itself. We typically use the latter approach, using a cloth/cardboard pocket that is voluntarily entered by all but the most agitated mice. Although such pockets can be washed between experiments, their use necessitates introducing naïve mice to the odors of previously tested mice, a confound since rodents (strain-dependently) emit alarm substances[3] that can produce SIA.[43] We conducted an experiment to directly compare these two approaches on the 48°C hot-water tail-withdrawal test; the results are shown in Fig. 2.1. As can be seen, pocket restraint produced consistently lower latencies, even after the habituation required with Plexiglas restraint was achieved. Thus, the use of the Plexiglas method was associated with extra restraint SIA, and resulted in overestimation of baseline sensitivity that could confound subsequent measurements of antinociception or hyperalgesia. Worse still, since pain inhibitory mechanisms are known to interact,[75] the preexistence of SIA might alter the expression of the phenomenon being studied. It should be noted, though, that obtaining consistent and low tail-withdrawal values using restraining pockets is a learned skill, requiring considerable practice.

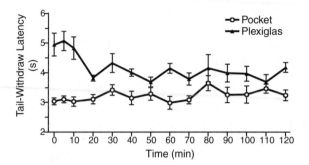

FIGURE 2.1 A comparison of repeated baseline 48°C hot-water tail-withdrawal latencies in Swiss Webster (Hsd:ND4; Harlan Sprague Dawley, Inc.) mice acutely restrained in cloth/cardboard pockets while being tested (Pocket group) vs. restrained for the duration of the experiment in Plexiglas holders (Plexiglas group). A repeated measures ANOVA revealed significant main effects of group and repeated measure, and a significant interaction (all $p < 0.005$). The Plexiglas group yielded a significantly higher area under the time × latency curve (AUC) compared to the Pocket group ($AUC_{0-30 \, min}$; $p = 0.003$; $AUC_{0-120 \, min}$; $p = 0.01$), indicative of additional ongoing SIA.

Tjolsen and Hole[152] have pointed out a serious confound related to the use of the tail-withdrawal test: the role of tail skin temperature. The tail is the most important thermoregulatory organ in the rodent, and as such will respond differently to thermal changes than other tissues. This group has observed significant relationships among ambient temperature, tail skin temperature, vasomotor tone, and radiant heat tail-withdrawal latencies, and recommends measuring tail temperature at every latency determination.[98,152] They point out, in addition, that stress and many analgesic drugs can affect body temperature and blood flow parameters.[91] However, other investigators have questioned ambient temperature as a confound,[20] and one investigation in mice clearly dissociated these phenomena.[88]

2.5.2 THE HOT-PLATE TEST

This assay was originally described by Woolfe and MacDonald,[169] although the version most often used today is as modified by Eddy and Leimbach.[38] In this test, the animal is confined by Plexiglas walls to a metal or porcelain surface heated to a specified temperature (commonly 50 to 56°C), and the latency is measured to the performance of an endpoint response considered indicative of nociception. Unlike the tail-withdrawal test, where only one obvious behavior is commonly observed, it is a matter of continuing debate exactly which behaviors should be properly considered nociceptive endpoints on the hot-plate test. Mice placed on a hot plate will exhibit one or more of the following: freezing, exploring, forepaw licking, grooming, hind paw lifting/guarding, hind paw licking, hind paw fluttering (also known as shaking or stamping), and vertical jumps. The use of forepaw responses as endpoints, although continuing to appear in the literature, is regarded by most experts as confounded since this behavior is a component of grooming,[54] and/or a response to warmth rather than noxious heat.[42] Most investigators use one or more of hind paw lifting, hind paw licking, hind paw fluttering, and escape jumping as hot-plate endpoints. Disagreement also exists over whether the first observation of *any* of these behaviors should terminate the test.

These issues have been seriously considered from an ethological perspective in the rat.[21,42] One investigator concluded that in this species, sensitivity and reliability were maximized by the use of hind paw lick or jump, whichever occurred first.[21] However, this investigator also alluded to the fact that species differences existed here, such that mice were more apt to engage in hind paw fluttering. Espejo and Mir[42] recommend using hind paw flutter or lick, whichever occurs first, but not hind paw lift or jump. Belknap and colleagues[10] have pointed out that frequencies of particular hot-plate behaviors are strain dependent in the mouse, with DBA/2 mice showing much lower propensity to lick their hind paw than C57BL/6 or C3H/He mice. Our experience with a large number of mouse strains is that the best hot-plate endpoint in this species is hind paw flutter or lick, whichever occurs first. Hind paw lifting is too subtle to score accurately, and is unanimously followed within a few seconds by a more obvious response. We submit that it is better to err conservatively by a few seconds than to terminate the test with an endpoint that may or may not be nociceptive. We rarely see jumping in naïve mice of most strains, except at latencies greatly exceeding those producing scorable hind paw behaviors, rendering the use of this endpoint scientifically and ethically questionable. In contrast, mice given multiple exposures to the hot plate are particularly apt to jump (presumably as an attempt to escape) many seconds *before* hind paw licking or fluttering, leading us to conclude that escape jumping may be a learned response (see below).

It should be noted, of course, that even hind paw flutter and licking behaviors are highly vulnerable to confound by any experimental manipulation affecting motor coordination or muscle tone (but see Reference 122).[54,161] Other limitations of the hot-plate assay include its extreme unsuitability for repeated testing. Repeated exposures to a noxious[41,47,48,81,121] or sham (i.e., room or body temperature)[7,47,61,99] hot plate can markedly decrease latencies and reduce sensitivity to antinociceptive agents. This phenomenon, termed behavioral tolerance, has been explained in terms of (1) learning of the correct behavior that results in removal from the plate,[48] and/or (2) habituation/tolerance to the testing situation, presuming the existence of test-related SIA.[7,99] Although this phenomenon has also been documented on the tail-withdrawal test in the rat,[47,98,99] we have not observed it on that test in the mouse (unpublished data). It has also been demonstrated that repeated hot-plate testing can induce a non-opioid form of SIA, altering subsequent tail-withdrawal latencies.[56] A largely unappreciated additional confound of the hot-plate test concerns the contact of the hind paw with urine on the plate, which will often facilitate heat transfer and cause premature endpoint behaviors. Thus, the advantages of the hot-plate test's supraspinal organization and lack of restraint are more than offset by disadvantages. It should be noted that the hot-plate test features higher inter- and intra-strain variation in mice compared to the tail-withdrawal test.[109]

Two new versions of the hot-plate test have been introduced over the past decade. An increasing-temperature test — similar to the original test proposed by Woolfe and MacDonald,[169] but featuring Peltier-controlled heating and cooling from and to a room temperature or body-temperature holding point — is reported to feature less variability, require no prior habituation, and to be sensitive to NSAID antinociception in mice. Both rats and mice displayed nociceptive thresholds in the range of 47 to 48°C, regardless of the rate of temperature increase. We have been largely unable to demonstrate NSAID antinociception using this test, however (unpublished data), and have found that the long time needed for the plate to cool down represents a practical limitation of its use. Jasmin and colleagues[66] developed a cold-plate test in rats (0 to 10°C) to quantify cold allodynia, a prevalent feature of neuropathy in humans. Uninjured rats displayed licking behaviors at temperatures < 3°C, but we have been unable to observe such behaviors in intact mice, even with prolonged contact on a 0°C surface (unpublished data).

2.5.3 HARGREAVES' TEST OF PAW WITHDRAWAL

A useful combination of the tail-withdrawal and hot-plate tests is the paw-withdrawal assay of Hargreaves and colleagues.[55] In this test, animals are habituated to Plexiglas chambers atop a transparent glass floor. A radiant heat source located under the floor is aimed at its target location, usually the plantar surface of one or the other hind paw, but sometimes the tail instead, and the latency to a withdrawal from the stimulus is measured. A very similar assay has been developed by Yeomans and Proudfit,[172] with the major difference being that the dorsal surface of the hind paw is stimulated from above. It is unclear whether the paw withdrawal is a reflex and/or a supraspinally organized behavior, especially since the withdrawal is often followed by licking or shaking behaviors. The obvious advantage of this test over either the tail-withdrawal or hot-plate assays is that it allows independent assessment of treatment effects on either side of the body. Thus, in addition to baseline controls, when using the paw-withdrawal test one can simultaneously test the hind paw ipsilateral and contralateral to an inflammatory or neuropathic injury. Baseline measurements should be taken nonetheless, since nociceptive sensitivity may be asymmetric and injuries may often produce contralateral hyperalgesia.[77] In extensive experiments with a number of mouse strains, we have found this assay resistant against systematic latency changes with repeated testing at > 5-min intervals, although it does feature a high degree of intra-individual variability [see Reference 109, unpublished data]. However, such variability has been shown to be reduced in rats when the glass floor is heated to hind paw skin temperature, 30°C, which avoids having the glass act as a heat sink.[34] As might be expected, a particularly high degree of genetic correlation ($r = 0.85$) was observed between strain sensitivities on the paw-withdrawal and hot-plate tests.[110]

We have found that latencies on the paw-withdrawal test are decreased when the beam is aimed near the toes, and increased when aimed near the heel. Extensive habituation is required to render mice motionless enough to collect data, but sleeping or grooming mice may exhibit altered latencies. The buildup of urine and feces on the glass floor, which can't be cleaned without disturbing the mice, can also represent a major confound of this test.

2.5.4 SPINAL VS. SUPRASPINAL THERMAL ASSAYS

Irwin and colleagues[64] first noted that the flexor withdrawal reflex is organized segmentally, and can be demonstrated in the spinalized animal. The tail-withdrawal assay is also a feature of many experiments on anesthetized animals. By contrast, a spinal-ized or anesthetized subject will be wholly unable to perform the suprasegmentally organized nocifensive responses scored in the hot-plate test. For this reason, it is commonly assumed that the tail-withdrawal and hot-plate tests represent spinal and supraspinal assays, respectively, and that any experimental manipulation affecting one test but not the other necessarily implies a spinal vs. supraspinal dissocia-tion.[9,44,67,78,118,175] While this conclusion is certainly possible, it should be borne in mind that tail-withdrawal latency changes can reflect alterations in any number of supraspinal sites that participate in ascending transmission of nociceptive information to the cortex[166] and/or descending modulation of nociception.[8,112] Indeed, it has been convincingly demonstrated that the tail-withdrawal assay is affected by avoidance learning (but see Reference 19).[74] In contrast, the fact that supraspinal organization is required to perform hot-plate test endpoints does not exclude the involvement of spinal circuitry in that test. Interneurons involved in spinal reflexes are also involved in the performance of supraspinally organized movements.[132] Finally, the observation of divergent effects of an experimental manipulation on the tail-withdrawal vs. the hot-plate test may be due to any number of additional factors dissociating these assays, including stimulus intensity, restraint SIA, learning, and order-of-testing effects (since in many studies animals are exposed to one test and then the other).

2.6　MECHANICAL ASSAYS OF NOCICEPTION

Mechanical assays of nociception are used far less commonly in rodents than in humans for practical reasons — rats and especially mice just won't sit still long enough to be poked in a particular location with any accuracy — but may represent a more valid model of human pain. It is unquestionably true that mechanical allodynia (i.e., pain from a normally innocuous stimulus, for example, clothing brushing against the skin) is a far more relevant clinical entity than thermal hyperalgesia.[14] It is also becoming more and more apparent that the physiological mechanisms underlying mechanical and thermal hypersensitivity are dissociable clinically,[153] electrophysiologically,[51,137] biochemically,[95] anatomically,[89] neurochemically,[31] pharmacologically,[86] and geneti-cally.[110] Thus, it is interesting from a scientific perspective to simultaneously evaluate the effects of experimental manipulations on both thermal and mechanical hyper-sensitivity, and an increasing number of investigators are doing so. As for thermal nociception, the receptive properties of mouse sensory neurons to mechanical stimuli have virtually never been studied in the mouse. In the one pertinent study of which we are aware, using constant force stimuli via calibrated von Frey monofilaments, the threshold stimulus causing firing of high-threshold mechanosensitive Aδ-fibers (sural and saphenous nerve) ranged from 4.0 to 5.6 mN.[76] Polymodal C-fibers with high mechanical thresholds were also found in the mouse, with thresholds similar to the rat.

2.6.1　THE VON FREY FIBER TEST OF MECHANICAL SENSITIVITY

In this very old assay introduced by von Frey[159] and modified by Weinstein,[163] sets of nylon monofilaments (also called fibers or hairs) are applied to the hind paw, exerting a blunt, static pressure whose maximal logarithmic force is determined by the thickness of the fiber. The scored "positive" response is a brisk withdrawal of the paw, often followed by a sustained retraction and/or licking, probably indicative of supraspinal organization. Fibers can be applied to either the dorsal (hairy) or ventral (glabrous) surfaces of the hind paw, the latter associated with lower withdrawal thresholds.[168] For the more common ventral application, mice are enclosed in a Plexiglas chamber atop a wire mesh floor. What is unclear to us is whether the application of a suprathreshold fiber to an intact mouse is a noxious stimulus, or just an annoying one; this test is variably referred to as one of mechanical nociception or one of mechanical sensibility. The most accurate and reliable thresholds are yielded by the psychophysically based up-down method of Dixon[35] as modified by Chaplan and colleagues,[23] but many other protocols are used.

Like Hargreaves' test of thermal paw withdrawal, the von Frey test is advantageous in that mechanical thresholds can be determined simultaneously for both ipsilateral and contralateral hind paws. We have noted in mice (in possible contrast to stable values with repeated testing in rats[23]) that variability and thresholds decrease with repeated applications over days, not becoming stable for at least three sessions.[109] Published confounds of the von Frey test include the sensitivity of the fibers to temperature and relative humidity,[5,87] and variability due to careless application,[87] and to the precise location stimulated.[134] As pointed out by Levin and colleagues,[87] many users misreport the numbers marked on commercial fiber sets (e.g., Weinstein-Semmes algesiometer; Stoelting, Inc.) as grams, when in fact these numbers represent the logarithm of 10 times the force in milligrams required to bow the filament. In our experience, the calibrated values of individual fibers need to be periodically confirmed, since the force they exert can be altered with repeated use. Kauppila and colleagues[70] have demonstrated that an important confounding factor when comparing thresholds of ipsilateral and contralateral hind paws is weight bearing. After injury, unequal weight bearing may result in false positive findings of allodynia. The problem was ameliorated by testing suspended rats who were not bearing their weight on their hindlimbs, but such a modification introduces restraint SIA as a confound.[70] It should also be noted that paw withdrawal can be elicited by non-painful stimuli like those associated with the startle response.[133] The time- and labor-intensive nature of this test may be reduced significantly by the recent availability of automated von Frey systems in which the force of a single fiber is slowly increased until a withdrawal response occurs, but such systems have yet to be extensively compared to the classical test.

2.6.2 The Tail-Clip Test

When it is desirable to apply a clearly noxious mechanical stimulus, or investigate mechanical hyperalgesia rather than mechanical allodynia, the tail-clip (or tail-pinch) test of Haffner[53] could be considered. In its common form in rodents, artery clips or binder clips are applied to the base of the tail. Depending on the pressure applied, mice will either immediately (< 2 s) attempt to attack (i.e., bite) the clip, show a delayed response in doing so (possibly indicative of antinociception), or largely ignore the clip for an extended period of time. Although one may attempt using a set of calibrated clips to determine the nociceptive threshold, most investigators simply report the latency to attack a suprathreshold clip.

Although this test is simply applied and scored, we have found it exceedingly unreliable. We tested 11 mouse strains (those reported in Reference 109) using a variety of binder clip sizes (unpublished data). We mostly observed bimodal responses: either immediate and frantic attempts to remove the clip, or total disregard for its presence (often exceeding 60 s) except for mere curiosity evidenced by sniffing. Some of this bimodality was due to the existence of responder and non-responder strains, but bimodal responses were also observed within-strain. In our hands the test also featured extreme alterations in attack latency with repeated tests. As such, we concluded that the tail-clip test was poorly suited to measure either antinociception, because non-responding animals would be mistaken as anti-nociceptive, or hyperalgesia, because the only reliable responses are too close to the zero floor latency. Also, arguing for skepticism over the reliability of this assay are data from various investigators estimating the threshold in the rat anywhere from 10 to 178 g, and in the mouse from 71 to 1070 g.[147]

2.6.3 THE RANDALL-SELITTO PAW-PRESSURE TEST

This test[126] is similar to pressure algometer tests in humans, in which a blunt applicator is applied with increasing force to tissue. Rodents are restrained, and their hind paw, either ipsilateral and contralateral to an injury, is carefully placed in contact with the applicator. The unnatural position rodents are forced to adopt in this test makes paw withdrawal a poor endpoint, since it would occur long before the stimulus becomes noxious. Thus, other less well-accepted nociceptive endpoints, variously struggling or vocalization, must be used. Although developed specifically for the rat, a number of reports exist claiming the successful use of this method in the mouse. The obvious confound here is the restraint necessary to correctly position the animal. The small size of the mouse, combined with its relative inability to be habituated to handling, poses extreme challenges for the successful use of this test, unconfounded by SIA, in this species.

2.7 CHEMICAL ASSAYS OF NOCICEPTION FEATURING SPONTANEOUS BEHAVIORS

Clinical pain states in humans feature prolonged, suprathreshold pain, and thus chemical assays producing ongoing nociception possess attractive face validity. These assays also yield drug half-maximal antinociceptive dose (AD_{50}) estimates far more similar to clinically relevant doses in humans than do either thermal or mechan-ical assays.[54] The substances most commonly used as algogens (e.g., acids, formalde-hyde) are virtually never encountered by organisms in the wild, but it is thought that pain due to trauma and inflammation is mediated by chemical algogens released from damaged cells.[36] A wide variety of substances, both exogenous and endogenous, have been demonstrated to be painful when injected into humans, and produce nociceptive behaviors in animals when injected into the skin (e.g., capsaicin, bradykinin, prostaglandins, formalin) or peritoneal cavity (e.g., non-isotonic saline, acetic acid, phenylquinones, magnesium sulfate, acetylcholine). As mentioned earlier, we restrict our present discussion to those chemical assays featuring strong behavioral evidence of spontaneous nociception rather than the induction of a hypersensitivity state.[165]

2.7.1 THE ABDOMINAL CONSTRICTION (WRITHING) TEST

This assay has the distinction of being to our knowledge the only major nociceptive test largely developed, and still used, on mice. The technique dates back to the mid-1950s, when two groups described a syndrome of writhing — lengthwise stretches of the torso accompanied (usually) by concave arching of the back — produced by phenylquinone or acetic acid, which was sensitive to abolition by a wide range of analgesics including NSAIDs.[58,138,156] Writhing, thought to be reflexive in nature, is produced by a caudally directed wave of abdominal wall muscle constrictions and elongations, and is often followed by a characteristic hindlimb extension. The technique was modified to use 0.6% (60 mg/kg) acetic acid,[79] and this algogen and concentration remain the most commonly used, if not the most ideal. Its sensitivity to weak analgesic agents whose activities cannot otherwise be detected in animals has probably accounted for the wide use of the abdominal constriction assay, despite a number of serious drawbacks (see below).

Abdominal constrictions from acetic acid start to appear within 5 min of injections, peak from 5 to 10 min post-injection, and decline thereafter. Most investigators score this test for 30 min, although the full duration of writhing can be far longer. The large number of substances that produce writhing differ in their time course, but not in the qualitative nature of the writhing behavior itself. Thus, it has been proposed that short-acting substances act directly on nociceptors whereas delayed-acting substances like acetic acid may work indirectly, by releasing endogenous mediators.[26] Although acetic acid injection produces acute peritoneal inflammation,[115] inflammation is not in fact required for writhing to occur. For example, 200 mg/kg magnesium sulfate produces writhing within a small window of time, 2 to 5 min, likely too short in duration for inflammation to develop.[26] In support of this contention, magnesium-sulfate writhing has been shown to be unaffected by antiinflammatory drugs (including serotonin and histamine receptor antagonists) and unaccompanied by prostaglandin release.[52]

As described earlier, one limitation of the abdominal constriction test derives directly from one of its strengths. The test is not only sensitive enough to pick up the actions of weak analgesics, but also those of weak stressors. The confounding influence of SIA may help to explain what is probably the major drawback of the test: the large number of non-responders. For example, Collier and colleagues[26] reported that 0.5% acetic acid produced no writhes whatsoever in 16% of mice. We have obtained similar results, although the frequency of non-responders can be reduced somewhat by using higher acetic acid concentrations (e.g., 0.9%). If a particular mouse shows no, or very few, writhes, it is very difficult to decide whether that mouse was highly antinociceptive or a non-responder to acetic acid. To compensate for this, of course, experiments using this assay require larger group sample sizes. The opposite problem can occur as well; Collier et al.[26] reported that 23% of mice displayed at least one writhe after a sham injection. Finally, the test has been severely criticized for its inhibition by non-analgesic drugs, including neuroleptics, anticholinergics, and antihistamines,[119] although only at doses producing obvious sedation.[54]

The abdominal constriction test has been largely employed as an assay of visceral nociception, which represents a major clinical problem and may be differentially mediated than somatic nociception.[94] However, this test is clearly not viscerospecific, as it also involves the parietal peritoneum, which belongs to the abdominal wall and is of somatic embryological origin.[83] A number of attempts have been made to develop a truly viscerospecific assay, but adapting such assays to the mouse has proven problematic. Many models feature fine surgery or intubations, great technical challenges in the mouse. For example, in rats mechanical distention of the colon produces reflexive contractions of the abdominal musculature, and appears to be a valid and reliable assay of visceral nociception.[113] We have made repeated and unsuccessful attempts to adapt this model; in our hands, the only observable distention-induced behavior in the mouse is freezing (Mayer and Mogil, unpublished data). Another problematic cross-species adaptation of a visceral assay is cyclophosphamide cystitis, which in the rat produces an easily scorable behavioral syndrome featuring lacrimation, piloerection, assumption of a peculiar head immobility and rounded back posture, and various brief abdominal "crises."[82] Olivar and Laird[117] also observed occasional "crises" in the mouse, but no lacrimation or postural changes. Working independently, we were unable to observe *any* of these behaviors in over 10 different mouse strains, but instead noted dose-dependent hypolocomotion and evidence of referred pain (Bon, Lichtensteiger, and Mogil, submitted).

2.7.2 THE FORMALIN TEST

Originally developed by Dubuisson and Dennis[37] in cats and rats, and adapted for use in mice,[62,111,136,148] the formalin test is currently the most popular chemical assay of nociception (see References 124 and 151 for excellent reviews). It involves the subcutaneous injection of a dilute solution of formalin, itself a 37% w/w solution of formaldehyde, into the dorsal or ventral surface of the hind paw, inducing a variety of recuperative behaviors that last for approximately 1 h (see below). This limited duration is considered an advantage of the formalin test over the more prolonged time course of other models of inflammatory pain. In a comparison of the effects of various algogens known to cause inflammation and/or hyperalgesia, including yeast, carrageenan, serotonin, kaolin, platelet-activating factor, and mustard oil, only formalin and acetic acid produced obvious evidence of spontaneous nociception in the rat.[165] Bee venom is now known to be an effective and more ecologically valid alternative.[84] A wide range of formalin concentrations can be used in the mouse, 0.02 to 5%, with injection volumes typically 20 to 25 μl. Such doses produce relatively greater amounts of inflammation in mice relative to rats.[151] It has been suggested that mechanisms underlying high (5%) and low (1%) concentration responses may be dissociable.[170]

The most intriguing property of the formalin test is its biphasic nature in a large number of species including rats and mice. Subcutaneous formalin produces two distinct phases of behavioral responses *and* firing of Aδ and C dorsal horn convergent neurons.[33,57,125] The early or acute phase (0 to 5 min post-injection) is thought to reflect direct activation of nociceptors, whereas the late or tonic phase (approximately 15 to 60 min) has been variously and controversially attributed to central sensitization[25,155] and/or ongoing inflammation-related afferent input.[28,149] The interphase period (5 to 15 min) is hypothesized to be due to active inhibition at the supraspinal[46] or spinal[59] level. Thus, an attractive feature of this assay is that acute and tonic pain can be modeled using a single noxious stimulus. A large number of studies have demonstrated anatomical,[158] neurochemical,[136] and pharmacological[63] dissociations between the two phases of the formalin test. The late phase in particular is known to be sensitive to a number of analgesic types including NSAIDs, κ-opioid agonists, and gabapentin that are less effective or ineffective against the early phase and other acute models of nociception.[124] Although most now believe that ongoing inflammation is required for the expression of the late phase of the formalin test, it is clear that the intensity of the behavioral response is independent from the extent of the inflammation (but see Reference 170).[16,106,164,165]

Debate continues over the appropriate nociceptive endpoint(s) in the formalin test. In the rat, many investigators[2,24,162] favor some modification of the weighted-scores technique originally proposed by Dubuisson and Dennis[37] (0 = no favoring, 1 = no weight bearing, 2 = elevation, 3 = licking/biting/shaking) but others report the incidence of hind paw flinches/jerks/shakes and/or the total time spent licking/biting the affected paw. A major species difference appears to be the lack of flinching behavior in the mouse, especially in the late phase. In a stepwise multiple regression study designed to glean the optimal scoring strategy for the formalin test in mice, we[143] concluded that a measure of the total time spent licking/biting was the most adequate index, in agreement with the suggestions of earlier investigators.[61,111]

A number of investigations in rodents have identified environmental factors that greatly affect outcomes on the formalin test. Novelty SIA, which is minimized by habituation to the observation chambers, has been noted frequently as a confound.[1,111] In addition, ambient temperature,[128] ambient pressure,[13] and site of injection[4] are known to be important modifiers. Ambient temperature produces striking effects on late-phase responding in mice, with 3-fold higher licking durations seen after 1% formalin injections at > 23°C compared to 20 to 21°C.[128] Finally, we have noticed that within our laboratory, the subcutaneous injection of formalin into the hind paw by different investigators yields greatly differing licking durations, indicative of the important effect of injection accuracy. One easy way to control for confounding bad injections is to routinely measure hind paw edema at the cessation of testing.

Given its several advantages over other common nociceptive assays, one might have expected the formalin test to largely replace the use of abdominal constriction, hot-plate, and tail-withdrawal assays in modern nociception experiments. However, compared to these alternatives the formalin test is remarkably labor-intensive, requiring extensive habituation and scoring time. A number of automated scoring systems have been developed, based either on image processing[69] or force transduction[68] technologies. However, such systems have not been shown to work in mice. It is likely that force transduction paradigms may never be successfully adapted to this species, since they largely detect flinches, an extremely rare behavior in the mouse. We have had success with a semi-automated formalin test setup, in which mice are videotaped from below a glass floor in a quiet room (obviating stressors associated with human presence during the test), and the tapes are later scored with Observer™ software (Noldus, Inc.). Since formalin responding in the late phase occurs in bouts of several minutes each, separated by many minutes of inactivity, the videotape can be fast-forwarded through such periods with attendant scoring time savings.

2.8 STRAIN-DEPENDENCE OF NOCICEPTIVE SENSITIVITY IN THE MOUSE

Pain researchers considering switching their model species from rat to mouse are faced with considerations in addition to species differences in the application of nociceptive assays. Mice feature prodigious intraspecies variability, and the laboratory strains in common use differ greatly with respect to a number of nociception-related traits.[100] This issue is of great importance if transgenic knockout mice are to be studied, since they are unanimously derived from embryonic stem (ES) cells originally isolated from one or more substrains of the 129 mouse strain.[141] Since 129 mice are poor breeders, and there are practical advantages to be gained in knockout mouse development by using a contrasting (i.e., dark) coat color, a large proportion of existing knockout mice of relevance to pain have been placed on a mixed 129 × C57BL/6 genetic background. This standard practice yields two interpretational confounds on the results of knockout experiments in addition to the better-known confound of developmental compensation.[102,108] First, in contrast to what is commonly assumed, wild-type and knockout mice are *not* coisogenic, differing only at the targeted gene. In fact, genes on the same chromosome as the transgene — that is, genes linked to the transgene — will be 129-derived in knockout mice, but C57BL/6-derived in wild-type mice. The region containing these hitchhiking donor genes[27] may include most of the chromosome in (129 × C57BL/6)F_2 hybrids, and will still be ≈ 16 cM long (containing up to 300 genes) even after 12 generations of backcrossing. The problem, then, is that any phenotypic difference observed between knockout and wild-type mice may actually be due to differential 129 and C57BL/6 alleles at one of these neighboring genes.[49] A second problem concerns the default choice of the C57BL/6 strain as the genetic background for the induced mutation, since it is well established that the effects of mutations are largely dependent on epistatic interactions with other genes.[15,80,129,150]

These problems are rendered more serious for particular experiments if (a) 129 and C57BL/6 strains differ importantly on the trait of interest, since they will possess allelic differences at a number of genes that may be located near a targeted mutation; and (b) the C57BL/6 strain is somehow a poor representative of mice in general with respect to the trait, since it would possess many rare alleles at genes acting epistatically with the targeted mutation. Unfortunately, both of these situations are true in pain research. We have recently collected much data on the sensitivity of a common set of 11 inbred mouse strains to 19 different nociceptive and 9 different antinociceptive phenotypes.[45,71,84a,101,109,160,167] In general, the 129 strain is resistant to nociception and sensitive to antinociception, whereas the C57BL/6 strain is sensitive to nociception and highly resistant to antinociception. In fact, the 129 and C57BL/6 strains (specifically, the 129/J and C57BL/6J substrains) differ significantly from each other on

22 of 28 pain traits, rendering the hitchhiking donor gene confound a possible explanation of a large number of existing transgenic studies of pain.[102] Perhaps more troubling is the fact that C57BL/6 mice are clear outliers on so many pain-related traits, and as such a poor representative of even laboratory "mousedom." They likely do, therefore, possess many unique alleles at pain-relevant genes that may be interacting with induced mutations, altering observed phenotypes accordingly. We have argued previously that this strain is the worst possible choice of genetic background on which to place transgenic or chemically induced spontaneous mutations, if the resultant mice are to be tested for nociceptive sensitivity. A far better choice is the DBA/2 strain, which we have found to be an average responder on a wide variety of pain-related traits.[109]

Even if one is simply working with outbred (i.e., non-defined, genetically heterogeneous) mice, subject population differences, known as vendor effects, may affect results of pain experiments.[60,173] For example, we recently observed that 49°C hotwater tail-withdrawal latencies in male Swiss Webster mice obtained from Simonsen, Inc. (Gilroy, CA) exceed those in male Swiss Webster or ICR (CD-1®) mice from Harlan Sprague Dawley, Inc. (Indianapolis, IN) by ≈ 33% (Mogil et al.[110a]). In contrast to inbred strains, differences between outbred populations may be due to genetic and/or environmental factors, and can change over time. Unfortunately, at the present time very little data exist comparing responses of outbred mouse populations that would allow informed choice of subjects for nociception experiments.

2.9 CONCLUSIONS

It seems likely that the next few years will bring a further acceleration of the use of murine subjects in pain research. In addition, technologies such as transgenesis have facilitated the combining of cutting-edge molecular advancements with nociceptive testing in live, behaving subjects. Thus, the optimization of algesiometric techniques in mice will help make the most of this golden opportunity to advance our knowledge of pain, hopefully leading to novel and effective therapies.

ACKNOWLEDGMENTS

This work was supported by PHS grants DA11394 and DE12735.

REFERENCES

1. Abbott, F.V., Franklin, K.B.J., and Connell, B., The stress of a novel environment reduces formalin pain: possible role of serotonin, *Eur. J. Pharmacol.*, 126, 141, 1986.
2. Abbott, F.V., Franklin, K.B.J., and Westbrook, R.F., The formalin test: scoring properties of the first and second phases of the pain response in rats, *Pain*, 60, 91, 1995.
3. Abel, E.L., Response to alarm substance in different rat strains, *Physiol. Behav.*, 51, 345, 1992.
4. Aloisi, A.M., Decchi, B., and Carli, G., Importance of the site of injection in the formalin test, *Med. Sci. Res.*, 23, 601, 1995.

5. Andrews, K., The effect of changes in temperature and humidity on the accuracy of von Frey hairs, *J. Neurosci. Meth.*, 50, 91, 1993.

6. Ankier, S.I., New hot plate tests to quantify antinociceptive and narcotic antagonist actions, *Eur. J. Pharmacol.*, 27, 1, 1974.

7. Bardo, M.T. and Hughes, R.A., Exposure to a nonfunctional hot plate as a factor in the assessment of morphine-induced analgesia and analgesic tolerance in rats, *Pharmacol. Biochem. Behav.*, 10, 481, 1979.

8. Basbaum, A.I. and Fields, H.L., Endogenous pain control systems: brainstem spinal pathways and endorphin circuitry, *Annu. Rev. Neurosci.*, 7, 309, 1984.

9. Baumeister, A.A., Nagy, M., Hebert, G., Hawkins, M.F., Vaughn, A., and Chatellier, M.O., Further studies of the effects of intranigral morphine on behavioral responses to noxious stimuli, *Brain Res.*, 525, 115, 1990.

10. Belknap, J.K., Lamé, M., and Danielson, P.W., Inbred strain differences in morphine-induced analgesia with the hot plate assay: a reassessment, *Behav. Genet.*, 20, 333, 1990.

11. Ben-Bassat, J., Peretz, E., and Sulman, F.G., Analgesimetry and ranking of analgesic drugs by the receptacle method, *Arch. Int. Pharmacodyn. Ther.*, 122, 434, 1959.

12. Bennett, G.J. and Xie, Y.-K., A peripheral mononeuropathy in rat that produces disorders of pain sensation like those seen in man, *Pain*, 33, 87, 1988.

13. Berge, O.-G., Garcia-Cabrera, I., and Furset, K., Hyperbaric exposure and morphine alter the pattern of behavior in the formalin test, *Pharmacol. Biochem. Behav.*, 40, 197, 1991.

14. Bonica, J.J., Causalgia and other reflex sympathetic dystrophies, in *The Management of Pain*, 2nd ed., J.J. Bonica, Ed., Lea and Febiger, Philadelphia, PA, 1990.

15. Bowers, B.J., Owen, E.H., Collins, A.C., Abeliovich, A., Tonegawa, S., and Wehner, J.M., Decreased ethanol sensitivity and tolerance development in γ-protein kinase C null mutant mice is dependent on genetic background, *Alcohol.: Clin. Exp. Res.*, 23, 387, 1999.

16. Brown, J.H., Kissel, J.W., and Lish, P.M., Studies on the acute inflammatory response. I. Involvement of the central nervous system in certain models of inflammation, *J. Pharmacol. Exp. Ther.*, 160, 231, 1968.

17. Cabanac, A. and Briese, E., Handling elevates the colonic temperature of mice, *Physiol. Behav.*, 51, 95, 1992.

18. Cao, Y.Q., Mantyh, P.W., Carlson, E.J., Gillespie, A.-M., Epstein, C.J., and Basbaum, A.I., Primary afferent tachykinins are required to experience moderate to intense pain, *Nature*, 392, 390, 1998.

19. Carstens, E., Hindlimb flexion withdrawal evoked by noxious heat in conscious rats: magnitude measurement of stimulus-response function, suppression by morphine, and habituation, *J. Neurophysiol.*, 70, 621, 1993.

20. Carstens, E., Is skin temperature a significant confounding factor in the tail-flick test?, *APS J.*, 2, 112, 1993.

21. Carter, R.B., Differentiating analgesic and non-analgesic drug activities on rat hot plate: effect of behavioral endpoint, *Pain*, 47, 211, 1991.

22. Chan, C.W. and Dalliere, M., Subjective pain sensation is linearly correlated with the flexion reflex in man, *Brain Res.*, 479, 145, 1989.

23. Chaplan, S.R., Bach, F.W., Pogrel, J.W., Chung, J.M., and Yaksh, T.L., Quantitative assessment of tactile allodynia evoked by unilateral ligation of the fifth and sixth lumbar nerves in the rat, *J. Neurosci. Meth.*, 53, 55, 1994.

24. Coderre, T.J., Fundytus, M.E., McKenna, J.E., Dalal, S., and Melzack, R., The formalin test: a validation of the weighted-scores method of behavioral pain rating, *Pain*, 54, 43, 1993.

25. Coderre, T.J., Vaccarino, A.L., and Melzack, R., Central nervous system plasticity in the tonic pain response to subcutaneous formalin injection, *Brain Res.*, 535, 155, 1990.

26. Collier, H.O.J., Dinneen, L.C., Johnson, C.A., and Schneider, C., The abdominal constriction response and its suppression by analgesic drugs in the mouse, *Br. J. Pharmacol.*, 36, 313, 1968.

27. Crusio, W.E., Gene-targeting studies: new methods, old problems, *Trends Neurol. Sci.*, 19, 186, 1996.

28. Dallel, R., Raboisson, P., Clavelou, P., Saade, M., and Woda, A., Evidence for a peripheral origin of the tonic nociceptive response to subcutaneous formalin, *Pain*, 61, 11, 1995.

29. d'Amore, A., Chiarotti, F., and Renzi, P., High-intensity nociceptive stimuli minimize behavioral effects induced by restraining stress during the tail-flick test, *J. Pharmacol. Toxicol. Meth.*, 27, 197, 1992.

30. D'Amour, F.E. and Smith, D.L., A method for determining loss of pain sensation, *J. Pharmacol. Exp. Ther.*, 72, 74, 1941.

31. Davies, M.F., Kingery, W.S., Poree, L.R., Guo, T.Z., and Maze, M., The α_{2A} adrenoceptor (α_{2A} AR) is critical to the development of thermal, but not mechanical neuropathic hyperalgesia, *Soc. Neurosci. Abstr.*, 23, 1538, 1997.

32. Dennis, S.G. and Melzack, R., Comparison of phasic and tonic pain in animals, in *Advances in Pain Research and Therapy*, Vol. 3, J.J. Bonica, Ed., Raven Press, New York, 1979, 747.

33. Dickenson, A.H. and Sullivan, A.F., Subcutaneous formalin-induced activity of dorsal horn neurones in the rat: differential response to an intrathecal opiate administered pre or post formalin, *Pain*, 30, 349, 1987.

34. Dirig, D.M., Salami, A., Rathbun, M.L., Ozaki, G.T., and Yaksh, T.L., Characterization of variables defining hind paw withdrawal latency evoked by radiant thermal stimuli, *J. Neurosci. Meth.*, 76, 183, 1997.

35. Dixon, W.J., Efficient analysis of experimental observations, *Annu. Rev. Pharmacol. Toxicol.*, 20, 441, 1980.

36. Dray, A., Inflammatory mediators of pain, *Br. J. Anaesth.*, 75, 125, 1995.

37. Dubuisson, D. and Dennis, S.G., The formalin test: a quantitative study of the analgesic effects of morphine, meperidine, and brain stem stimulation in rats and cats, *Pain*, 4, 161, 1977.

38. Eddy, N.B. and Leimbach, D., Synthetic Analgesics. II. Diethienylbutenyl- and dithienylbutylamines, *J. Pharmacol. Exp. Ther.*, 107, 385, 1953.

39. Einon, D.F., Humphreys, A.P., Chivers, S.M., Feild, S., and Naylor, V., Isolation has permanent effects upon the behavior of the rat, but not the mouse, gerbil, or guinea pig, *Dev. Psychobiol.*, 14, 343, 1981.

40. Elmer, G.I., Pieper, J.O., Negus, S.S., and Woods, J.H., Genetic variance in innate nociception and its relationship to the potency of morphine-induced analgesia in thermal and chemical tests, *Pain*, 75, 129, 1997.

41. Espejo, E.F. and Mir, D., Differential effects of weekly and daily exposure to the hot plate on the rat's behavior, *Physiol. Behav.*, 55, 1157, 1994.

42. Espejo, E.F. and Mir, D., Structure of the rat's behaviour in the hot plate test, *Behav. Brain Res.*, 56, 171, 1993.

43. Fanselow, M.S., Odors released by stressed rats produce opioid analgesia in unstressed rats, *Behav. Neurosci.*, 99, 589, 1985.

44. Fasmer, O.B., Berge, O.-G., Tveiten, L., and Hole, K., Changes in nociception after 6-hydroxydopamine lesions of descending catecholaminergic pathways in mice, *Pharmacol. Biochem. Behav.*, 24, 1441, 1986.

45. Flores, C.M., Wilson, S.G., and Mogil, J.S., Pharmacogenetic variability in neuronal nicotinic receptor-mediated antinociception, *Pharmacogenetics*, 9, 619, 1999.
46. Franklin, K.B.J. and Abbott, F.V., Pentobarbital, diazepam, and ethanol abolish the interphase diminution of pain in the formalin test: evidence for pain modulation by GABA$_A$ receptors, *Pharmacol. Biochem. Behav.*, 46, 661, 1993.
47. Gamble, G.D. and Milne, R.J., Repeated exposure to sham testing procedures reduces reflex withdrawal and hot-plate latencies: attenuation of tonic descending inhibition?, *Neurosci. Lett.*, 96, 312, 1989.
48. Gebhart, G.F., Sherman, A.D., and Mitchell, C.L., The influence of learning on morphine analgesia and tolerance development in rats tested on the hot plate, *Psychopharmacologia*, 22, 295, 1971.
49. Gerlai, R., Gene-targeting studies of mammalian behavior: is it the mutation or the background genotype?, *Trends Neurol. Sci.*, 19, 177, 1996.
50. Grotto, M. and Sulman, F.G., Modified receptacle method for animal analgesimetry, *Arch. Int. Pharmacodyn.*, 165, 152, 1967.
51. Guilbaud, G., Benoist, J.M., Neil, A., Kayser, V., and Gautron, M., Neuronal response thresholds to and encoding of thermal stimuli during carrageenin-hyperalgesic-inflammation in the ventro-basal thalamus of the rat, *Exp. Brain Res.*, 66, 421, 1987.
52. Gyires, K. and Torma, Z., The use of the writing test in mice for screening different types of analgesics, *Arch. Int. Pharmacodyn.*, 267, 131, 1984.
53. Haffner, F., Experimentelle prufung schmerzstellender mittel, *Deutsch. Me. Worchenschr.*, 55, 731, 1929.
54. Hammond, D.L., Inference of pain and its modulation from simple behaviors, in *Advances in Pain Research and Therapy: Issues in Pain Management*, C.R. Chapman and J.D. Loeser, Eds., Raven Press, New York, 1989, 69.
55. Hargreaves, K., Dubner, R., Brown, F., Flores, C., and Joris, J., A new and sensitive method for measuring thermal nociception in cutaneous hyperalgesia, *Pain*, 32, 77, 1988.
56. Hawranko, A.A., Monroe, P.J., and Smith, D.J., Repetitive exposure to the hot-plate test produces stress induced analgesia and alters β-endorphin neuronal transmission within the periaqueductal gray of the rat, *Brain Res.*, 667, 283, 1994.
57. Heapy, C.G., Jamieson, A., and Russell, N.J.W., Afferent C-fibre and A-delta activity in models of inflammation, *Br. J. Pharmacol.*, 90, 164P, 1987.
58. Hendershot, L.C. and Forsaith, S., Antagonism of the frequency of phenylquinone-induced writhing in the mouse by weak analgesics and non-analgesics, *J. Pharmacol. Exp. Ther.*, 125, 237, 1959.
59. Henry, J.L., Yashpal, K., Pitcher, G.M., and Coderre, T.J., Physiological evidence that the 'interphase' in the formalin test is due to active inhibition, *Pain*, 82, 57, 1999.
60. Ho, I.K., Loh, H.H., and Way, E.L., Morphine analgesia, tolerance and dependence in mice from different strains and vendors, *J. Pharm. Pharmacol.*, 29, 583, 1977.
61. Hunskaar, S., Berge, O.-G., and Hole, K., A modified hot-plate test sensitive to mild analgesics, *Behav. Brain Res.*, 21, 101, 1986.
62. Hunskaar, S., Fasmer, O.B., and Hole, K., Formalin pain in mice, a useful technique for evaluating mild analgesics, *J. Neurosci. Meth.*, 14, 69, 1985.
63. Hunskaar, S. and Hole, K., The formalin test in mice: dissociation between inflammatory and non-inflammatory pain, *Pain*, 30, 103, 1987.
64. Irwin, S., Houde, R.W., Bennett, D.R., Hendershot, L.C., and Seevers, M.H., The effects of morphine, methadone and meperidine on some reflex responses of spinal animals to nociceptive stimulation, *J. Pharmacol. Exp. Ther.*, 101, 132, 1951.
65. Jaeger, J.-J., Tong, H., and Denys, C., The age of *Mus-Rattus* divergence: paleontological data compared with the molecular clock, *C. R. Acad. Sci. Paris*, 302, Ser. II, 917, 1986.

66. Jasmin, L., Kohan, L., Franssen, M., Janni, G., and Goff, J.R., The cold plate as a test of nociceptive behaviors: description and application to the study of chronic neuropathic and inflammatory pain models, *Pain*, 75, 367, 1998.

67. Jensen, T.S. and Yaksh, T.L., Comparison of antinociception action of morphine in the periaqueductal gray, medial and paramedial medulla in rat, *Brain Res.*, 363, 99, 1986.

68. Jett, M.F. and Michelson, S., The formalin test in rat: validation of an automated system, *Pain*, 64, 19, 1996.

69. Jourdan, D., Ardid, D., Bardin, L., Bardin, M., Neuzeret, D., Lanphouthacoul, L., and Eschalier, A., A new automated method of pain scoring in the formalin test in rats, *Pain*, 71, 265, 1997.

70. Kauppila, T., Kontinen, V.K., and Pertovaara, A., Weight bearing of the limb as a confounding factor in assessment of mechanical allodynia in the rat, *Pain*, 74, 55, 1998.

71. Kest, B., Wilson, S.G., and Mogil, J.S., Sex differences in supraspinal morphine analgesia are dependent on genotype, *J. Pharmacol. Exp. Ther.*, 289, 1370, 1999.

72. Kim, S.H. and Chung, J.M., An experimental model for peripheral neuropathy produced by segmental spinal nerve ligation in the rat, *Pain*, 50, 355, 1992.

73. King, D.L. and Appelbaum, J.R., Effect of trials on "emotionality" behavior of the rat and mouse, *J. Comp. Physiol. Psychol.*, 85, 186, 1973.

74. King, T.E., Joynes, R.L., and Grau, J.W., Tail-flick test: II. The role of supraspinal systems and avoidance learning, *Behav. Neurosci.*, 111, 754, 1997.

75. Kirchgessner, A.L., Bodnar, R.J., and Pasternak, G.W., Naloxazone and pain-inhibitory systems: evidence for a collateral inhibition model, *Pharmacol. Biochem. Behav.*, 17, 1175, 1982.

76. Koltzenburg, M., Stucky, C.L., and Lewin, G.R., Receptive properties of mouse sensory neurons innervating hairy skin, *J. Neurophysiol.*, 78, 1841, 1997.

77. Koltzenburg, M., Wall, P.D., and McMahon, S.B., Does the right side know what the left is doing?, *Trends Neurosci.*, 22, 122, 1999.

78. Konig, M., Zimmer, A.M., Steiner, H., Holmes, P.V., Crawley, J.N., Brownstein, M.J., and Zimmer, A., Pain responses, anxiety and aggression in mice deficient in pre-proenkephalin, *Nature*, 383, 535, 1996.

79. Koster, R., Anderson, M., and de Beer, E.J., Acetic acid for analgesic screening, *Fed. Proc.*, 18, 412, 1959.

80. Kustova, Y., Sei, Y., Morse III, H.C., and Basile, A.S., The influence of a targeted deletion of the IFNγ gene on emotional behaviors, *Brain Behav. Immun.*, 12, 308, 1998.

81. Lai, Y.-Y. and Chan, S.H.H., Shortened pain response time following repeated algesiometric tests in rats, *Physiol. Behav.*, 28, 1111, 1982.

82. Lanteri-Minet, M., Bon, K., de Pommery, J., Michiels, J.F., and Menetrey, D., Cyclophosphamide cystitis as a model of visceral pain in rats: model elaboration and spinal structures involved as revealed by the expression of c-Fos and Krox-24 proteins, *Exp. Brain Res.*, 105, 220, 1995.

83. Lanteri-Minet, M., Isnardon, P., de Pommery, J., and Menetrey, D., Spinal and hindbrain structures involved in visceroception and visceronociception as revealed by the expression of Fos, Jun and Krox-24 proteins, *Neuroscience*, 55, 737, 1993.

84. Lariviere, W.R. and Melzack, R., The bee venom test: a new tonic-pain test, *Pain*, 66, 271, 1996.

84a. Lariviere, W.R. and Mogil, J.S., Transgenic studies of pain: mutation or background genotype, *J. Pharmacol. Exp. Ther.*, in press.

85. Laursen, S.E. and Belknap, J.K., Intracerebroventricular injections in mice: some methodological refinements, *J. Pharmacol. Meth.*, 16, 355, 1986.
86. Lee, S.H., Kayser, V., Desmeules, J., and Guilbaud, G., Differential action of morphine and various opioid agonists on thermal allodynia and hyperalgesia in mononeuropathic rats, *Pain*, 57, 233, 1994.
87. Levin, S., Pearsall, G., and Ruderman, R.J., Von Frey's method of measuring pressure sensibility in the hand: an engineering analysis of the Weinstein-Semmes pressure aesthesiometer, *J. Hand Surg.*, 3, 211, 1978.
88. Lichtman, A.H., Smith, F.L., and Martin, B.R., Evidence that the antinociceptive tail-flick response is produced independently from changes in either tail-skin temperature or core temperature, *Pain*, 55, 283, 1993.
89. Lima, D., Avelino, A., and Coimbra, A., Differential activation of c-*fos* in spinal neurones by distinct classes of noxious stimuli, *Neuroreport*, 4, 747, 1993.
90. Lu, Y., Pirec, V., and Yeomans, D.C., Differential antinociceptive effects of spinal opioids on foot withdrawal responses evoked by C fiber or A*d* nociceptor activation, *Br. J. Pharmacol.*, 121, 1210, 1997.
91. Lund, A., Tjolsen, A., and Hole, K., The apparent antinociceptive effect of desipramine and zimelidine in the tail flick test in rats is mainly caused by tail skin temperature, *Pain*, 38, 65, 1989.
92. Malmberg, A.B., Brandon, E.P., Idzerda, R.L., Liu, H., McKnight, G.S., and Basbaum, A.I., Diminished inflammation and nociceptive pain with preservation of neuropathic pain in mice with a targeted mutation of the Type I regulatory subunit of cAMP-dependent protein kinase, *J. Neurosci.*, 17, 7462, 1997.
93. Malmberg, A.B., Chen, C., Tonegawa, S., and Basbaum, A.I., Preserved acute pain and reduced neuropathic pain in mice lacking PKCγ, *Science*, 278, 279, 1997.
94. McMahon, S.B., Are there fundamental differences in the peripheral mechanisms of visceral and somatic pain?, *Behav. Brain Sci.*, 20, 381, 1997.
95. Meller, S.T., Thermal and mechanical hyperalgesia: a distinct role for different excitatory amino acid receptors and signal transduction pathways?, *APS J.*, 3, 215, 1994.
96. Meunier, J.-C., Mollereau, C., Toll, L., Suaudeau, C., Moisand, C., Alvinerie, P., Butour, J.-L., Guillemot, J.-C., Ferrara, P., Monsarrat, B., Mazarguil, H., Vassart, G., Parmentier, M., and Costentin, J., Isolation and structure of the endogenous agonist of opioid receptor-like ORL$_1$ receptor, *Nature*, 377, 532, 1995.
97. Miczek, K.A., Thompson, M.L., and Shuster, L., Opioid-like analgesia in defeated mice, *Science*, 215, 1520, 1982.
98. Milne, R.J. and Gamble, G.D., Habituation to sham testing procedures modifies tail-flick latencies: effects on nociception rather than vasomotor tone, *Pain*, 39, 103, 1989.
99. Milne, R.J., Gamble, G.D., and Holford, N.H.G., Behavioral tolerance to morphine analgesia is supraspinally mediated: a quantitative analysis of dose-response relationships, *Brain Res.*, 491, 316, 1989.
100. Mogil, J.S., The genetic mediation of individual differences in sensitivity to pain and its inhibition, *Proc. Natl. Acad. Sci. USA*, 96, 7744, 1999.
101. Mogil, J.S. and Adhikari, S.M., Hot and cold nociception are genetically correlated, *J. Neurosci.*, 19, RC25 (1–5), 1999.
102. Mogil, J.S. and Grisel, J.E., Transgenic studies of pain, *Pain*, 77, 107, 1998.
103. Mogil, J.S., Grisel, J.E., Reinscheid, R.K., Civelli, O., Belknap, J.K., and Grandy, D.K., Orphanin FQ is a functional anti-opioid peptide, *Neuroscience*, 75, 333, 1996.
104. Mogil, J.S., Grisel, J.E., Zhang, G., Belknap, J.K., and Grandy, D.K., Functional antagonism of μ-, δ- and κ-opioid antinociception by orphanin FQ, *Neurosci. Lett.*, 214, 131, 1996.

105. Mogil, J.S., Kest, B., Sadowski, B., and Belknap, J.K., Differential genetic mediation of sensitivity to morphine in genetic models of opiate antinociception: influence of nociceptive assay, *J. Pharmacol. Exp. Ther.*, 276, 532, 1996.

106. Mogil, J.S., Lichtensteiger, C.A., and Wilson, S.G., The effect of genotype on sensitivity to inflammatory nociception: characterization of resistant (A/J) and sensitive (C57BL/6) inbred mouse strains, *Pain*, 76, 115, 1998.

107. Mogil, J.S., Nessim, L.A., and Wilson, S.G., Strain-dependent effects of supraspinal orphanin FQ/nociceptin on thermal nociceptive sensitivity in mice, *Neurosci. Lett.*, 261, 147, 1999.

108. Mogil, J.S. and Wilson, S.G., Nociceptive and morphine antinociceptive sensitivity of 129 and C57BL/6 inbred mouse strains: implications for transgenic knock-out studies, *Eur. J. Pain*, 1, 293, 1997.

109. Mogil, J.S., Wilson, S.G., Bon, K., Lee, S.E., Chung, K., Raber, P., Pieper, J.O., Hain, H.S., Belknap, J.K., Hubert, L., Elmer, G.I., Chung, J.M., and Devor, M., Heritability of nociception. I. Responses of eleven inbred mouse strains on twelve measures of nociception, *Pain*, 80, 67, 1999.

110. Mogil, J.S., Wilson, S.G., Bon, K., Lee, S.E., Chung, K., Raber, P., Pieper, J.O., Hain, H.S., Belknap, J.K., Hubert, L., Elmer, G.I., Chung, J.M., and Devor, M., Heritability of nociception. II. "Types" of nociception revealed by genetic correlation analysis, *Pain*, 80, 83, 1999.

110a. Mogil, J.S., Chesler, E.J., Wilson, S.G., Juraska, J., and Sternberg, W.F. Sex differences in thermal nociception and morphine antinociception in the rodent depend on genotype, *Neurosci. Biobehav. Rev.*, 24, 375, 2000.

111. Murray, C.W., Porreca, F., and Cowan, A., Methodological refinements to the mouse paw formalin test. An animal model of tonic pain, *J. Pharmacol. Meth.*, 20, 175, 1988.

112. Ness, T.J. and Gebhart, G.F., Centrifugal modulation of the rat tail flick reflex evoked by graded noxious heating of the tail, *Brain Res.*, 386, 41, 1986.

113. Ness, T.J. and Gebhart, G.F., Visceral pain: a review of experimental studies, *Pain*, 41, 167, 1990.

114. Ness, T.J., Jones, S.L., and Gebhart, G.F., Contribution of the site of heating to variability in the latency of the rat tail flick reflex, *Brain Res.*, 426, 169, 1987.

115. Northover, B.J., The permeability of plasma proteins of the peritoneal blood vessels of the mouse, and the effect of substances that alter permeability, *J. Path. Bacter.*, 85, 361, 1963.

116. O'Callaghan, J.P. and Holtzman, S.G., Quantification of the analgesic activity of narcotic antagonists by a modified hot-plate procedure, *J. Pharmacol. Exp. Ther.*, 192, 497, 1975.

117. Olivar, T. and Laird, J.M.A., Cyclophosphamide cystitis in mice: behavioural characterisation and correlation with bladder inflammation, *Eur. J. Pain*, 3, 141, 1999.

118. Pastoriza, L.N., Morrow, T.J., and Casey, K.L., Medial frontal cortex lesions selectively attenuate the hot plate response: possible nocifensive apraxia in the rat, *Pain*, 64, 11, 1996.

119. Pearl, J., Stander, H., and McKean, D.B., Effects of analgesics and other drugs on mice in phenylquinone and rotorod tests, *J. Pharmacol. Exp. Ther.*, 167, 9, 1969.

120. Pizziketti, R.J., Pressman, N.S., Geller, E.B., Cowan, A. and Adler, M.W., Rat cold water tail-flick: a novel analgesic test that distinguishes opioid agonists from mixed agonist-antagonists, *Eur. J. Pharmacol.*, 119, 23, 1985.

121. Plone, M.A., Emerich, D.F., and Lindner, M.D., Individual differences in the hot plate test and effects of habituation on sensitivity to morphine, *Pain*, 66, 265, 1996.

122. Plummer, J.L., Cmielewski, P.L., Gourlay, G.K., Owen, H., and Cousins, M.J., Assessment of antinociceptive drug effects in the presence of impaired motor performance, *J. Pharmacol. Meth.*, 26, 79, 1991.

123. Poole, T.B. and Fish, J., An investigation of playful behavior in *Rattus norvegicus* and *Mus musculus* (Mammalia), *J. Zool.*, 175, 61, 1975.

124. Porro, C.A. and Cavazzuti, M., Spatial and temporal aspects of spinal cord and brainstem activation in the formalin pain model, *Prog. Neurobiol.*, 41, 565, 1993.

125. Puig, S. and Sorkin, L.S., Formalin-evoked activity in identified primary afferent fibers: systemic lidocaine suppresses phase-2 activity, *Pain*, 64, 345, 1995.

126. Randall, L.O. and Selitto, J.J., A method for measurement of analgesic activity on inflamed tissue, *Arch. Int. Pharmacodyn.*, 111, 409, 1957.

127. Reinscheid, R.K., Nothacker, H.-P., Bourson, A., Ardati, A., Henningsen, R.A., Bunzow, J.R., Grandy, D.K., Langen, H., Monsma, F.J., and Civelli, O., Orphanin FQ: a novel neuropeptide which is a natural ligand of an opioid-like G protein-coupled receptor, *Science*, 270, 792, 1995.

128. Rosland, J.H., The formalin test in mice: the influence of ambient temperature, *Pain*, 45, 211, 1991.

129. Rozmahel, R., Wilschanski, M., Matin, A., Plyte, S., Oliver, M., Auerbach, W., Moore, A., Forstner, J., Durie, P., Nadeau, J., Bear, C., and Tsui, L.C., Modulation of disease severity in cystic fibrosis transmembrane conductance regulator deficient mice by a secondary genetic factor, *Nature Genet.*, 12, 280, 1996.

130. Rubinstein, M., Mogil, J.S., Japon, M., Chan, E.C., Allen, R.G., and Low, M.J., Absence of opioid stress-induced analgesia in mice lacking β-endorphin by site-directed mutagenesis, *Proc. Natl. Acad. Sci. USA*, 93, 3995, 1996.

131. Ryan, S.M., Watkins, L.R., Mayer, D.J., and Maier, S.F., Spinal pain suppression mechanisms may differ for phasic and tonic pain, *Brain Res.*, 334, 172, 1985.

132. Schomburg, E.D., Restrictions on the interpretation of spinal reflex modulation in pain and analgesia research, *Pain Forum*, 6, 101, 1997.

133. Schomburg, E.D., Spinal sensorimotor systems and their supraspinal control, *Neurosci. Res.*, 7, 265, 1990.

134. Schouenborg, J., Weng, H.-R., and Holmberg, H., Modular organization of spinal nociceptive reflexes: a new hypothesis, *News Physiol. Sci.*, 9, 261, 1994.

135. Seltzer, Z., Dubner, R., and Shir, Y., A novel behavioral model of causalgiform pain produced by partial sciatic nerve injury in rats, *Pain*, 43, 205, 1990.

136. Shibata, M., Ohkubo, T., Takahashi, H., and Inoki, R., Modified formalin test: characteristic pain response, *Pain*, 38, 347, 1989.

137. Shir, Y. and Seltzer, Z., A-fibers mediate mechanical hyperesthesia and allodynia and C-fibers mediate thermal hyperalgesia in a new model of causalgiform pain disorders in rats, *Neurosci. Lett.*, 115, 62, 1990.

138. Siegmund, E., Cadmus, R., and Lu, G., A method for evaluating both non-narcotic and narcotic analgesics, *Proc. Soc. Exp. Biol. Med.*, 95, 729, 1957.

139. Silver, L.E., *Mouse Genetics: Concepts and Applications*, Oxford University Press, Oxford, U.K., 1995.

140. Simonin, F., Valverde, O., Smadja, C., Slowe, S., Kitchen, I., Dierich, A., Le Meur, M., Roques, B.P., Maldonado, R., and Kieffer, B.L., Disruption of the κ-opioid receptor gene in mice enhances sensitivity to chemical visceral pain, impairs pharmacological actions of the selective κ-agonist U-50,488H and attenuates morphine withdrawal, *EMBO J.*, 17, 886, 1998.

141. Simpson, E.M., Linder, C.C., Sargent, E.E., Davisson, M.T., Mobraaten, L.E., and Sharp, J.J., Genetic variation among 129 substrains and its importance for targeted mutagenesis in mice, *Nature Genet.*, 16, 19, 1997.

142. Sternbach, R.A., The need for an animal model of chronic pain, *Pain*, 2, 2, 1976.

143. Sufka, K.J., Watson, G.S., Nothdurft, R.E., and Mogil, J.S., Scoring the mouse formalin test: a validation study, *Eur. J. Pain*, 2, 351, 1998.

144. Suzuki, T., Narita, M., Misawa, M., and Nagase, H., Pentazocine-induced biphasic analgesia in mice, *Life Sci.*, 48, 1827, 1991.

145. Tabata, H., Kitamura, T., and Nagamatsu, N., Comparison of effects of restraint, cage transportation, anaesthesia and repeated bleeding on plasma glucose levels between mice and rats, *Lab. Anim.*, 32, 143, 1998.

146. Taber, R.I., Predictive value of analgesic assays in mice and rats, *Adv. Biochem. Psychopharmacol.*, 8, 191, 1974.

147. Takagi, H., Inukai, T., and Nakama, M., A modification of Haffner's method for testing analgesics, *Jpn. J. Pharmacol.*, 16, 287, 1966.

148. Takahashi, H., Ohkubo, T., Shibata, M., and Narese, S., A modified formalin test for measuring analgesia in mice, *Jpn. J. Oral Biol.*, 26, 543, 1984.

149. Taylor, B.K., Peterson, M.A., and Basbaum, A.I., Persistent cardiovascular and behavioral nociceptive responses to subcutaneous formalin require peripheral nerve input, *J. Neurosci.*, 15, 7575, 1995.

150. Threadgill, D.W., Dlugosz, A.A., Hansen, L.A., Tennenbaum, T., Lichti, U., Yee, D., LaMantia, C., Mourton, T., Herrup, K., Harris, R.C., Barnard, J.A., Yuspa, S.H., Coffey, R.J., and Magnuson, T., Targeted disruption of mouse EGF receptor: effect of genetic background on mutant phenotype, *Science*, 269, 230, 1995.

151. Tjolsen, A., Berge, O.-G., Hunskaar, S., Rosland, J.H., and Hole, K., The formalin test: an evaluation of the method, *Pain*, 51, 5, 1992.

152. Tjolsen, A. and Hole, K., The tail-flick latency is influenced by skin temperature, *APS J.*, 2, 107, 1993.

153. Treede, R.-D., Meyer, R.A., Raja, S.N., and Campbell, J.N., Peripheral and central mechanisms of cutaneous hyperalgesia, *Prog. Neurobiol.*, 38, 397, 1992.

154. Trullas, R. and Skolnick, P., Differences in fear motivated behaviors among inbred mouse strains, *Psychopharmacology*, 111, 323, 1993.

155. Vaccarino, A.L. and Chorney, D.A., Descending modulation of central neural plasticity in the formalin pain test, *Brain Res.*, 666, 104, 1994.

156. Vander Wende, C. and Margolin, S., Analgesic tests based upon experimentally induced acute abdominal pain in rats, *Fed. Proc.*, 15, 494, 1956.

157. Vetulani, J., Castellano, C., Lason, W., and Oliverio, A., The difference in the tail-flick but not hot-plate response latency between C57BL/6 and DBA/2J mice, *Pol. J. Pharmacol. Pharm.*, 40, 381, 1988.

158. Voege-Sipahi, J., Ramsaroop, N.K., and Cox, V.C., The differential effects of lateral midbrain lesions on the two phases of formalin pain, *Neuroreport*, 3, 587, 1992.

159. von Frey, M., Verspatete schmerzempfindungen, *Z. Gesamte. Neurol. Psychiat.*, 79, 324, 1922.

160. Wan, Y., Malpeli, J., Han, J., and Mogil, J.S., Genetic mediation of electroacupuncture analgesia in mice: sensitivity of inbred strains and transgenic beta-endorphin knockouts, *Soc. Neurosci. Abstr.*, 25, 932, 1999.

161. Watkins, L.R., Algesiometry in laboratory animals and man: current concepts and future directions, in *Advances in Pain Research and Therapy: Issues in Pain Management*, C.R. Chapman and J.D. Loeser, Eds., Raven Press, New York, 1989, 249.

162. Watson, G.S., Sufka, K.J., and Coderre, T.J., Optimal scoring strategies and weights for the formalin test in rats, *Pain*, 70, 53, 1997.

163. Weinstein, S., Tactile sensitivity of the phalanges, *Percept. Mot. Skills*, 14, 351, 1962.

164. Wheeler-Aceto, H. and Cowan, A., Neurogenic and tissue-mediated components of formalin-induced edema: evidence for supraspinal regulation, *Agents Actions*, 34, 264, 1991.

165. Wheeler-Aceto, H., Porreca, F., and Cowan, A., The rat paw formalin test: comparison of noxious agents, *Pain*, 40, 229, 1990.

166. Willis, W.D. and Westlund, K.N., Neuroanatomy of the pain system and of the pathways that modulate pain, *J. Clin. Neurophysiol.*, 14, 2, 1997.

167. Wilson, S.G., Melton, K.A., Wickesburg, R.E., and Mogil, J.S., Strain-dependent antinociception from the cannabinoid receptor agonist, WIN, 55,212-2, *Soc. Neurosci. Abstr.*, 25, 924, 1999.

168. Woolf, C.J. and Swett, J.E., The cutaneous contribution to the hamstring flexor reflex in the rat: an electrophysiological and anatomical study, *Brain Res.*, 303, 299, 1984.

169. Woolfe, G. and MacDonald, A.D., The evaluation of analgesic action of pethidine hydrochloride (Demerol), *J. Pharmacol. Exp. Ther.*, 80, 300, 1944.

170. Yashpal, K. and Coderre, T.J., Influence of formalin concentration on the anti-nociceptive effects of anti-inflammatory drugs in the formalin test in rats: separate mechanisms underlying the nociceptive effects of low- and high-concentration formalin, *Eur. J. Pain*, 2, 63, 1998.

171. Yeomans, D.C., Pirec, V., and Proudfit, H.K., Nociceptive responses to high and low rates of noxious cutaneous heating are mediated by different nociceptors in the rat: behavioral evidence, *Pain*, 68, 133, 1996.

172. Yeomans, D.C. and Proudfit, H.K., Characterization of the foot withdrawal response to noxious radiant heat in the rat, *Pain*, 59, 85, 1994.

173. Yoburn, B.C., Kreuscher, S.P., Inturrisi, C.E., and Sierra, V., Opioid receptor upregulation and supersensitivity in mice: effects of morphine sensitivity, *Pharmacol. Biochem. Behav.*, 32, 727, 1989.

174. Yoburn, B.C., Morales, R., Kelly, D.D., and Inturrisi, C.E., Constraints on the tailflick assay: morphine analgesia and tolerance are dependent upon locus of tail stimulation, *Life Sci.*, 34, 1755, 1984.

175. Zimmer, A., Zimmer, A.M., Baffi, J., Usdin, T., Reynolds, K., Konig, M., Palkovits, M., and Mezey, E., Hypoalgesia in mice with a targeted deletion of the tachykinin 1 gene, *Proc. Natl. Acad. Sci. USA*, 95, 2630, 1998.

3 Techniques for Mutagenesis of the Murine Opioid System *in Vivo*

Michael D. Hayward and Malcolm J. Low

CONTENTS

3.1 INTRODUCTION

Gene-targeting and transgenic manipulations constitute a developing technology that is of enormous value in studies of both opioid receptors and their respective ligands. Induced mutations of the mouse can be produced by four different methods: random mutagenesis by a chemical agent, infection with viral-based gene constructs, trans-genesis, or targeted mutagenesis. Many targeted mutant mice and transgenic mice are of interest to those wishing to study the opioid system. Null mutants of all three opioid receptors and their corresponding endogenous ligands have been generated and are reviewed below. Some of these mice are available from the Jackson Labo-ratory, which maintains a depository of mutant mice in addition to its collection of inbred mouse strains. An up-to-date list of these mice and their availability can be found on their web-site at http://www.jax.org/tbase.

The opioid system is a powerful modulator of pain and pleasure. Membrane-bound receptors that bind opioid drugs have been identified and characterized pharmacologically, and the genes that encode these receptors have been cloned and mapped to their chromosomal loci. A family of endogenous opioid peptides binds these receptors with affinities specific to individual receptors. Currently, experiments using genetically manipulated mice are unambiguously identifying the physiological processes controlled by the opioid peptides and their cognate receptors. A major strength of this technique is that experiments are no longer subject to the limitations of pharmacological specificity (e.g., an agonist's affinity for more than one receptor). Technology for transgenesis and targeted mutagenesis in mice has progressed to a point where we can now examine single gene changes *in vivo*, using biologically relevant paradigms. Genetically modified mice, mostly those harboring a null allele for an opioid gene, behave consistently with several hypotheses based on earlier pharmacological data, but unexpected results have been observed as well. We review here what those experiments have found and identify important experiments that have not been done. In addition, we review the many techniques available to manip-ulate the mouse genome and point out caveats and pitfalls of these technologies, particularly as they apply to the opioid-mediated behaviors of analgesia and addiction.

The three opioid receptor genes that have been identified and cloned are the MOP (mu, MOR, OPR-3), KOP (kappa, KOR, OPR-2), and DOP (delta, DOR, OPR-1).[1-5] These receptors are 60% similar at the amino acid level, contain 7 membrane spanning domains, and share a similar coupling to intracellular signaling path-ways.[6-9] The corresponding endogenous opioid peptides are beta-endorphin and endomorphins 1 and 2, which bind to the mu receptor; enkephalin, which binds to the delta receptor; and dynorphin, which binds to the kappa receptor. This distinction may not be absolute, however, as beta-endorphin only has a slightly higher affinity for the mu receptor than for the delta receptor.[10] Presumably, the colocalization of peptide and receptor also provides specificity in the endogenous opioid system. For example, experiments using electrically evoked release of beta-endorphin in brain slices suggest that, functionally, beta-endorphin uses presynaptic mu receptors in neocortical slices.[11] The most recently discovered of the opioid peptides are endo-morphins 1 and 2, two naturally occurring peptides isolated from bovine cortex but, as of yet, uncloned.[12] Endomorphins 1 and 2 have a higher affinity for the mu receptor than beta-endorphin ($K_i = 0.36$ and 0.69 nM for endomorphins 1 and 2, respectively, while $K_i = 4.4$ nM for beta-endorphin). Consequently, there are at least two different endogenous peptides that bind to the mu receptor with high affinity.

The endogenous opioid system is usually considered a regulator of nociception, but it also can be considered part of a general physiological stress regulatory system working in parallel with the autonomic nervous system. Since the endogenous opioids can function as neurotransmitters, modulators of neurotransmission, or neurohormones, they are capable of influencing a diverse array of behaviors. Evidence suggests that the endogenous opioids can modulate endocrine, cardiovascular, gastrointestinal, and immune functions in addition to their effects through the central and peripheral nervous systems on nociception, behavioral reinforcement, feeding, and locomotor activity.[9,13–15] Most of what we know of the opioid system is based on pharmacological experiments with synthetic drugs and peptides. For example, we know that morphine can produce analgesia, decrease body temperature, depress respiration, dilate peripheral blood vessels, and decrease intestinal motility.[10] Unfortunately, it is difficult to dissect out relevant physiological processes from pharmacological actions when administering exogenous drugs. Elimination of the endogenous peptides stands as a definitive test for their specific roles in many of these physiological processes.

3.2 TRANSGENESIS

Transgenesis involves the introduction of a cloned deoxyribonucleic acid (DNA) construct that is randomly integrated within the genome. This technique is useful for studies using reporter genes fused to the promoter of a gene of interest in order to precisely examine expression patterns.[16] It is also useful for experiments involving dominant-negative proteins, antisense molecules, and overexpression or ectopic expression of a particular gene. Inducible transgenes have recently been developed, in which basal expression is low or nonexistent but can be turned on by the addition or removal of an exogenous factor. Inducible transgenes have been used in co-injection with targeted constructs to design inducible null mutants, which will be discussed later. Use of the transgenic technique does not lend itself easily to introducing mutations into specific genetic loci.

Transgenic constructs require their own promoter due to the nature of their freestanding design. It is important to use a well-characterized promoter unless the experiments are being used to directly examine the promoter for elements conferring characteristics such as tissue-specific expression or inducibility. Boundary elements or viral enhancer sequences can be useful to maximize expression levels. It is important to include elements in transgenes that are required for proper transcription, processing, and translation; elements often included in commercially available expression vectors for transfection of cell cultures. For example, it is vital to include the several bases just upstream of the translation initiation site known as the Kozak sequence[17] in order to ensure proper translation. Additionally, a polyadenylation signal is required for proper ribonucleic acid (RNA) processing. Inclusion of introns has also been shown to enhance expression levels.[18] However, with all of these elements the site of integration for the transgene can be the most important factor. Transgenes integrate randomly into the genome so the site of integration is not controlled. In a later section on cis-DNA regulatory element (Cre) recombinase we will describe a technology to potentially generate transgenes that integrate into characterized loci. The surrounding site of integration can influence expression of a transgene such that it is expressed in unintended tissues or at unexpected levels.

The primary precaution one can take is to generate several lines and compare them. An additional reason to analyze multiple lines is that transgenes often integrate as concatamers of multiple copies. Expression levels of transgenes are often related to gene dosage, so different lines of mice will have different expression levels. Not uncommonly, transgenes will integrate at more than one site resulting in independently sorting alleles, which is detectable by non-Mendelian inheritance. The definitive way to determine if this has occurred is by Southern blot analysis. Restriction endonuclease digestion with a rare cutter will result in more than one band being recognized by an appropriate probe. In short, the key to successful experiments with transgenic mice is the generation and analysis of multiple lines.

3.2.1 INDUCIBLE TRANSGENES

Advances in inducible gene expression in culture have led to several possible methods for inducing transgene expression *in vivo*. The first and most commonly applied method uses the *E. coli* tetracycline resistance system. The tetracycline-controlled transactivator protein (tTA) is composed of the repressor from the *E. coli* tetracycline resistance operon (tetR) and is fused to the activation domain of the virion protein 16 (VP16) from herpes simplex virus.[19] A concatamerized *tet* operator sequence and minimal promoter from the human cytomegalovirus (CMV) will drive expression of a transgene when tTA binds the promoter. In the presence of tetracycline the tTA cannot bind the operator, and expression does not occur. Upon removal of tetracycline the transactivator undergoes a steric conformational change, binds the operator, and expression is induced. This method was successfully used to drive the expression of a luciferase reporter in transgenic mice by a constitutively active tTA transgene crossed to mice harboring the luciferase reporter construct containing the *tet* operator and CMV promoter.[20] The basal state of these mice requires the presence of tetracycline and so has been called tet-off. A modified version of the tTA was developed (reverse tetracycline-controlled trans-activator protein, rtTA) that bound to the *tet* operator sequences in the presence of the tetracycline analogue, doxocycline, i.e., tet-on.[21] The tet-on system has the important advantage that mice do not need to be maintained on tetracycline to prevent the transgene from expressing. However, in the tet-on system, the promoter has a high basal activity in the absence of tetracycline and so the tet-off system remains the most used one. In addition, the rtTA also requires approximately a 100-fold higher dose of antibiotic. The difference in dose dependence of doxocycline for tTA and rtTA has been used for the simultaneous expression and repression of two different constructs.[22]

One drawback of the use of tetracycline is its slow pharmacokinetics, which may not induce expression as quickly as desired. Other forms of inducible promoters that more rapidly induce transciption utilize the family of steroid hormone receptors and have been adapted to transgenics. For example, the ecdysone receptor from *Drosophila melanogaster* has been used as a transactivator since it has no activity in mammals and neither does its ligand 20-OH ecdysone.[23] The transactivators in this system are the ecdysone receptor (EcRE) fused with VP16 and the retinoic acid receptor (RXR) fused to VP16, which are both required for heterodimerization before they can induce transcription. Both of these transgenes had to be injected into the same zygote for transgenic mice. The target gene contains four copies of the ecdys-one response element (EcRE) and a CMV minimal promoter. The advantage of this system is that the drug is only required to turn on expression, similar to the tet-on system. Compared to the tet-on system, there was very little basal expression without the drug.[23]

Another system uses a mutant estrogen receptor estrogen-binding domain (EBD) that is incapable of binding estrogen but binds 4-OH-tamoxifen.[24] This system has the advantage of producing a single receptor transgene unlike the EcRE and RXR. However, the mutation in the EBD also reduces the affinity for tamoxifen by 100-fold,[25] which may interfere with pregnancy. The glucocorticoid receptor (GR) has also been adapted for ligand-inducible gene expression using a mutant GR that only binds the synthetic analog dexamethasone (GR^{dex}) but not naturally occurring glucocorticoids.[26] This system may not interfere with *in utero* studies but clearly dexamethasone has central nervous system effects by itself, notably on the hypothalamic-pituitary-adrenal axis. Use of this system has not been demonstrated in a transgenic mouse yet.

Transgenic mice have been made with a mutant version of the human progest-erone receptor (hPR), which binds the synthetic analog mifeprestone (RU486) but not progesterone.[27] Expression of growth hormone was induced in the liver of these mice following RU486 treatment. The constructs used for this system contained the mutant hPR fused to the yeast transcriptional activator GAL4. A target construct contained four hPR binding sites as well as a liver-specific promoter.[28,29] This system also demonstrated extremely low basal activity and robust expression upon induction with RU486. Dose-dependent amounts of expression also could be demonstrated with different concentrations of RU486. There are concerns for reproduction with *in vivo* treatment of RU486, the same as with tamoxifen, although mutated hPR is sensitive enough to be activated by levels of RU486 which do not have significant physiological activity.[29] Interestingly, this system was also adapted for an adenoviral expression system and was used to induce expression of human growth hormone targeted to the liver by viral delivery.[30]

The ligand-binding domain methods that utilize steroid hormones represent an alternative to the tetracycline system and have the advantage of rapid changes in transcription, low basal expression, and no requirement for long-term treatment with a drug. The advantage of the tetracycline systems over these is the relatively minor side effects with tetracycline compared to the doses of steroids necessary. The EcRE system uses a hormone that has no physiological activity but requires a tricky dual injection of two transgenes. Techniques for generating inducible transgenes are still quite time intensive and cumbersome as two transgenes are required, the trans-activator-containing transgene and the target gene containing the elements for induc-ible transcription. Thus, two lines of transgenic mice have to be crossed for the system to be operational.

3.2.2 TRANSGENICS IN THE OPIOID SYSTEM

There are few reports of using transgenic mice for experiments involving the opioid system. A transgenic mouse was made with the human proenkephalin A promoter using chloramphenicol acetyltransferase (CAT) as a reporter. Expression of the reporter showed neuronal expression with regional specificity and was inducible *in vivo* by haloperidol and footpad injection with Freud's adjuvant.[31] Although regional specificity and inducibility were retained, a further study with several versions of this promoter failed to identify critical enhancer elements for brain expression even though some enhancer regions were found to be critical for expression in the testes.[32] Another proenkephalin transgenic mouse used the bacterial gene beta-galactosidase (*lacZ*) as a reporter and also demonstrated neuronal expression with regional specificity.[33] Several models of stress including immune-stress were able to induce the expression of the reporter.[34] Increased sensitivity to morphine analgesia was analyzed in lines of transgenic mice overexpressing the mu receptor under the control of the tyrosine hydroxylase promoter.[32] A transgenic analysis of the promoters for the opioid receptors could be used to examine the tissue-specific distribution of the receptors by using different portions of the promoter to drive the expression of a reporter gene such as *lacZ*. This was recently done with the kappa receptor, and an analysis of expression during development was described.[35] The size of these receptor genes, including their large introns, suggests a complicated control that could also contain intronic enhancer regions or alternatively spliced forms. Experiments similar to this used the proopiomelanocortin (POMC) gene, which encodes beta-endorphin, to a region necessary for expression only in the pituitary while a larger portion of the promoter is necessary for hypothalamic expression.[36,37] Many more experiments examining opioid function could be done, including more promoter analyses, ectopic expression, and inducible expression of opioid peptides.

3.3 CONSTITUTIVE TARGETED MUTATIONS

Targeted genetic manipulations of mice use transfection of embryonic stem cells (ES) to produce a specifically altered allele (a knockout).[38] Targeted mutagenesis generates null or hypomorphic mutations by inserting a drug resistance gene (e.g., *neo* for ampicillin resistance) into an exon, deleting a portion of the exon or shifting the translation frame so that no functional gene product can be made (for methods, see Reference 18). These constructs do not require mammalian promoters and *in vitro* selection cassettes normally contain all the elements necessary for their expression. Constructs that target a mutation require genomic regions flanking the drug resistance cassettes. The genomic fragments are the targeting arms that are necessary for homologous recombination at a single site in the genome. Constructs used for targeted recombination in ES cells can be generally described as replacement vectors or insertion vectors.[38,39] As the name implies, a replacement vector will replace sequences in the targeted allele while insertion vectors will insert into the allele adding new sequences and duplicating others. For null mutation constructs the replacement type vector is used most often.

3.3.1 SOURCE OF GENOMIC DNA FOR CONSTRUCTS

The genomic library from which the targeting arms are acquired is preferably isogenic to the parental mouse strain of the ES cells. Virtually all reliable and available ES cells are derived from the 129 strain of mice. One important aspect of this reality is that 129 substrains are actually quite genetically diverse and many different substrains have been used for the derivation of stem cells.[40] These differences are reflected at the nucleotide level, especially within noncoding stretches that make up the majority of genomic DNA. Thus, genomic fragments used for the targeting construct that do not originate from the same subspecies of 129 mice as the ES cells could result in decreased recombination efficiency. ES cells have been established from more than only the 129 strain of mice. For example, cells were established using the C57BL/6 strain and were successfully used to generate null mutants.[41] An ES cell line for the BALB/c line has been used for making transgenics since BALB/c zygotes are too fragile to withstand microinjection.[42] Although there are a few exceptions, 129 cells are usually the ES cells of choice, so it is important that the targeting construct be made from genomic DNA that is isogenic to the ES cells.

3.3.2 TARGETED MUTANTS OF THE OPIOID SYSTEM

Functionally null mutations of all of the cloned members of the opioid system have been produced. Mice null for the mu opioid receptor have been generated independently in at least five labs.[43–47] Mice lacking the kappa opiate receptor[48] and the delta receptor[49] have also been published. Beta-endorphin was the first member of the opioid system to be mutated,[50] and a null mutant of enkephalin has also been reported.[51] A null mutation in the prodynorphin gene has resulted in viable and fertile mice in the absence of dynorphin peptides throughout development and additional studies are ongoing.[52,52a,53] In addition to the single gene mutations, double null mutant mice that lack both beta-endorphin and enkephalin[54] and double null mutants for each pair of mu, delta, and kappa receptors are viable and healthy.[55]

3.3.2.1 Complete Loss of Function

Null mutants allow one to infer a gene's involvement in a specific physiological process by examining that process in the absence of the gene product. The clearest result is the absence of a physiological process when a single gene's product is not translated. This is clear evidence of the necessity for that protein in the process. Null mutants of the opioid system have demonstrated the complete loss of many physiological processes thought to require the opioid peptides or receptors, thus confirming the importance of those proteins. For example, environmental stressors can produce an antinociception, mediated in part by endogenous opioid pain-inhibition mechanisms, which is called stress-induced analgesia (SIA). The relative importance of opioid and nonopioid pathways in SIA depends on the type and magnitude of the stressor.[56] Mice lacking beta-endorphin were produced by a unique point-mutation strategy that introduced a premature stop codon at the N-terminus of the beta-endorphin coding region within the POMC gene.[57] These mice no longer demonstrated the opioid-mediated SIA measured using the abdominal constriction assay when mice were stressed by a room temperature swim.[50] Contrary to some expectations, mice null for enkephalin had intact opioid and nonopioid forms of SIA.[51] It remains to be shown whether mice lacking the mu receptor no longer produce opioid SIA. This experiment would be of considerable value as it is still not certain whether all beta-endorphin-mediated behaviors are transduced through the mu receptor because beta-endorphin has nearly equivalent affinities for both the mu and delta receptors.[6] Thus, SIA is a physiological process that can be examined in the absence of a gene product thought to control that action.

Mu-receptor mutant mice were analyzed by saturation radioligand binding on brain membranes and, predictably, there was nearly no detectable mu-receptor binding by [^{3}H]DAMGO ([D-Ala^2, N-McPhe^4, Gly^5-ol]enkephalin) in homozygous mutants.[43,45,58] The prototypical opiate, morphine, is a relatively selective agonist for mu receptors. In fact, mu receptors were defined by their affinity for morphine.[10] However, at high enough concentrations morphine will interact with delta and kappa receptors so some analgesic effects of morphine could be mediated through non-mu opioid receptors.[59,60] The mu-receptor mutant mice had an extraordinary loss of morphine-induced analgesia by both the hot-plate and the tail-withdrawal assays.[43,45] At doses below 56 mg/kg there was essentially no analgesic action of morphine and at greater doses there was only a trend toward modest analgesia. However, at such heroic doses it is likely that other opioid receptors are involved.[45] Thus, the mu-receptor mutant mice confirmed that the mu receptor is the only receptor directly responsible for the analgesic actions of morphine to thermal stimuli at clinically relevant doses.

A complete loss of morphine analgesia suggests that the MOP-1 gene must encode any mu-receptor subtype. Targeting different exons produced the different mu-receptor null mutants. Intriguingly, targeting of the first exon produced a strain of mu-receptor mutant mice that lacked morphine analgesia but responded almost normally to fentanyl, heroin, 6-acetylmorphine, and morphine-6-glucuronide (M6G).[46] It was also reported that [^{3}H]-M6G binding could be detected in the exon 1 mu-receptor null mutant mice.[46] The authors suggested that a form of the mu opioid receptor was retained in this exon 1 null mutant mouse. The mu-specific antagonists beta-funaltrexamine and naloxonazine but not the delta-specific antagonist naltrindole or the kappa-specific antagonist nor-binaltrophamine (norBNI) were able to block the analgesia produced by M6G.[46] Interestingly, this group compared their mice to the exon 2 null mutants from Matthes et al.[43] that lacked analgesia in response to the aforementioned ligands. However, at a recent meeting of the International Narcotics Research Conference another comparison of the two mouse strains was presented which described some modest amounts of analgesia remaining in both strains, but to a greater extent in the exon 1 null mutants.[61]

One possible explanation for these unexpected findings is the existence of an uncloned receptor. Alternatively, there may be a mu-receptor splice variant that was not eliminated by the construct used to generate the mutant mice. A transcript encoded by exons 2 and 3 could be detected by reverse transcriptase PCR in the brains of the exon 1 knockout mouse.[46] It may not be surprising that a receptor lacking exon 1 of the mu receptor binds mu receptor agonists, as a chimeric receptor containing the TM1 and 2 of the kappa receptor with the remaining portion of the mu receptor binds DAMGO with high affinity.[62] Additionally, we have found that a chimeric receptor containing exon 1 of the mu receptor but exons 3 and 4 of the kappa receptor binds the kappa ligand U50,488H with an affinity similar to that of wild-type kappa receptors (Fig. 3.1). Thus, if an alternative transcription start site

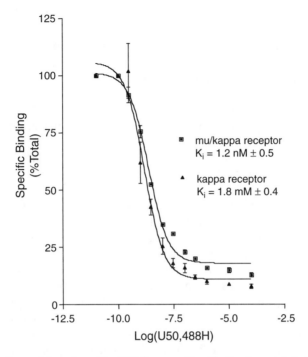

FIGURE 3.1 Competition of specific [³H] diprenorphine binding by unlabeled U50,488H in membranes prepared from Cos-7 cells transfected with either a mouse kappa cDNA or a chimeric mu/kappa receptor cDNA containing the first exon of the mu receptor and the third and fourth exons of kappa. The cDNAs were cloned into pcDNA 3.1 (InVitrogen) and cell membranes were isolated as described.[140] Each curve was plotted using data from one representative experiment. The curves were obtained by fitting a logistic equation to the data using nonlinear regression analysis, and the K_i was calculated based on the IC_{50} value for the high-affinity site using the equation of Cheng and Prussof where IC_{50} is the concentration of unlabeled ligand that inhibits 50% of the labeled ligand binding, and K_D is defined as the dissociation constant for [³H]diprenorphine.

was not removed in the first exon knockout mouse it is possible that an alternative mu opioid receptor could be generated. In fact, it appears that there are as many as 18 alternatively spliced forms of the mouse mu opioid receptor,[63,64] and a new transcriptional start site has been recently reported.[65] The presence of several alternatively spliced forms of a receptor may frustrate plans to easily design a truly null mutant mouse but targeted deletion techniques can be used to selectively disrupt alternatively spliced forms of a gene and determine their physiological relevance. For example, the new putative alternative transcriptional start and alternative promoter of the mu receptor could be eliminated. The originally identified transcriptional start site could be left intact and reveal the relevance of these particular alternatively spliced forms.

The opioid system is equally well-known for its roles in antinociception and behavioral reinforcement. Mu agonists produce reinforcing effects that can lead to the development of sensitization, drug craving, and conditioned withdrawal. The mesolimbic dopamine pathway to the nucleus accumbens (NA) originates from the ventral tegmental area (VTA) and this pathway appears to be involved in the rewarding properties of several drugs of abuse. Experiments have shown that dopamine is released in the NA following administration of both opioids and psychostimulants.[66-70] These data support the hypothesis that mu agonists are reinforcing because they increase dopamine release in the NA. Consistent with these behavioral and biochemical data, dense binding of mu ligands and mu receptor mRNA expression are detected in the VTA.[9] Additional support for the idea that the mu receptor is necessary for reinforcement comes from the mu-receptor null mice, which did not prefer morphine in conditioned place preference tests.[43] Thus, the reinforcing property of mu agonists may be due to the expression of mu receptors in the VTA.

Most predictions of the pharmacology in delta-receptor null mutant mice were also fulfilled. The delta-receptor mutant mice did not exhibit [D-Pen,D-Pen] enkephalin (DPDPE) or deltorphin II (DELT II) spinal analgesia while morphine analgesia was intact.[49] These data suggest that the delta 1 receptor with higher affinity to DPDPE and the delta 2 receptor with higher affinity to DELT II are both products of the same gene. Unexpectedly, supraspinal analgesia produced by the same delta receptor-specific compounds and the nonpeptide delta agonist BW373U69 actually produced significantly more analgesia in delta null mutant mice than in wild-type mice. The supraspinal analgesia produced by DPDPE and DELT may be mediated by mu receptors in the delta receptor null mutants. The supraspinal analgesia was naltrexone reversible but beta-funaltrexamine, a mu receptor-specific antagonist, failed to reverse the antinociception. There have even been suggestions that all thermal, supraspinal opioid analgesia in the mouse is mediated solely via mu receptors.[71,72] The recent evidence from this null mutant mouse suggests that a delta or mu receptor system is unregulated in the delta null mutant.

Kappa-receptor mutant mice, as expected, lack analgesia produced by U50,488H when the animals were tested by both the hot-plate and tail-withdrawal assays.[48] Analgesia produced by morphine was not altered when tested by either assay, demonstrating that the kappa receptor is not required for mu receptor analgesia. Taken together, these studies suggest that although delta- and kappa-mediated analgesia may require a functional mu receptor, morphine analgesia does not require either a functional delta or kappa receptor.

Multiple opioid pathways could be involved in the rewarding properties of morphine since kappa agonists blocked reward to morphine.[73] In fact, pharmacological experiments have shown that kappa agonists were aversive[74-77] and led to a decrease in dopamine levels in the NA.[78,79] Experiments with the kappa receptor mutant mice demonstrated that the rewarding potential of morphine was unaltered.[48] Surprisingly, signs of physical withdrawal in morphine-dependent mice were reduced in the absence of the kappa opiate receptor.[48] This was also contrary to expectations as the kappa peptide agonist dynorphin A-(1-17) attenuated physical signs of morphine withdrawal in wild-type mice.[80] In fact, the kappa receptor antagonist norBNI potentiated the weight loss that occurred during withdrawal, although other signs of opiate withdrawal were not significantly altered.[80] Thus, it appears that kappa receptor expression does not affect the rewarding properties of morphine, but it can modulate physical withdrawal from morphine. It would be helpful, however, to know if kappa-receptor ligands are still aversive to kappa-receptor null mutant mice.

3.3.2.2 Changes in Basal Function

Although a complete loss of a physiological process in a null mutant is unambiguous, more subtle phenotypes can also be identified. These phenotypes often demonstrate a homeostatic role of that gene product. For example, although SIA was preserved in mice lacking enkephalin, a basal nociceptive phenotype was discovered. Mice lacking enkephalin had normal tail-flick responses but decreased jump latencies on a 55°C hot-plate test (i.e., they were hyperalgesic).[51] This finding is in contrast to the beta-endorphin-deficient mice which had no or minimal alteration in their basal response to thermal stimuli.[81] One of the mu-receptor null mutant mice had altered basal nociception in the tail-flick and hot-plate tests[45] but not for mechanical nociception.[82] Another study found no difference in the basal nociception of the mu-receptor null mutants for the tail flick or paw licks on the hot-plate, although there was a difference in the number of jumps on a hot-plate.[83] There is evidence from these null mutant mice that the endogenous opioid system not only plays a role in a mouse's response to its environment but that it also has a homeostatic role in setting basal nociception.

3.3.2.3 Revealing Interactions through Epistasis

Null mutants of one receptor may have consequences for other receptors, especially considering the recent evidence that the delta and kappa opioid receptors form heterodimers[84] and suggestions that mu and delta receptors functionally interact.[85] A small decrease in kappa binding was detected using [3H]CI-977 in a few brain regions of mu-receptor null mutant mice.[58] A decrease in delta binding was also detected in the mu-receptor null mutant mice in regions that are normally abundant with mu-receptor binding.[58] Notably, [3H]DELT II was decreased in the nucleus accumbens and amygdala of the mu-receptor null mutant. Interestingly, in the kappa-receptor null mutant there was an increase in delta receptor binding in the nucleus accumbens but no change in mu receptor binding.[86] Kappa and mu receptors mediate opposing actions on dopamine in the nucleus accumbens as part of the mesolimbic dopamine pathway involved in reward to drugs of abuse.[67,79] Therefore, the changes in receptor binding detected in the two null mutant mice suggest that these opposing pathways may be linked by delta receptors.

The behavioral pharmacology to delta and kappa agonists in the mu-receptor null mutant has suggested that much of what was thought to be central analgesia through delta receptors may, in fact, require the mu receptor. The delta-selective agonist BuBu [Tyr-D-Ser(*O*-tert-butyl)-Gly-Phe-Leu-Thr(*O*-tert-butyl)] did not induce any analgesia measured by the tail-flick test but did in the hot-plate assay.[43,87] Another delta-selective agonist, DPDPE injected I.C.V., produced nearly no analgesic effect as assessed by both the hot-plate and tail-withdrawal assays.[73,88] Analgesia produced by I.C.V. injection of DPDPE in mu-receptor null mutants was also ineffective for mechanical withdrawal as measured by von Frey filaments.[82] In another study, the delta agonists

DPDPE and DELT II produced approximately 30 to 40% less antinociception in mu-receptor null mutant mice compared to wild-type mice as measured by the tail-flick assay.[83] There was no difference between the genotypes in analgesia as measured by the hot-plate. Agonists for delta and mu receptors were also able to stimulate GTPγs hydrolysis and inhibit adenylate cyclase normally in mu-receptor null mutant mice, suggesting that these receptors are present and functional.[83] A complement to these studies is the finding that some supraspinal analgesia produced by the delta receptor agonists DPDPE and BW373U69 persist in the delta-receptor null mutant mice.[49] These results were intriguing because substantial evidence indicates that there are interactions between the mu receptor and other opioid receptor types. Morphine analgesia can be modulated by specific delta agonists,[60,85] and analgesia produced by mu agonists can also be modulated by kappa agonists.[59] Data from the mu- and delta-receptor null mutants support the idea that opioid receptor interaction is an important feature of opioid analgesia and of physiological relevance.

Loss of an endogenous ligand may have consequences on the expression or function of its receptor. We have examined both the expression and function of all three opioid receptors in the beta-endorphin mutant mice and have found no apparent change in the distribution or amount of opioid receptor expression in brain and spinal cord.[89,90] Analgesic dose-response curves to morphine administered I.P. and measured by abdominal constriction and hot-plate tests were normal in the mutant mice.[50] We have also found that analgesic tolerance to morphine I.P. develops normally in beta-endorphin-deficient mice.[91] Further tests for analgesia have suggested that a differential change in sensitivity to mu agonists occurs in the beta-endorphin-deficient mice, depending on whether the agonists are administered I.P., I.T., or I.C.V.[89] Thus, it appears that the loss of beta-endorphin has affected mu receptor pharmacology but that change appears to be at a functional level and not at the level of expression or ligand binding. Yet, we have measured an inwardly rectifying K^+ channel following DAMGO treatment of hypothalamic slices and the two genotypes produce identical dose-response curves.[90] Any change in mu receptor function in the beta-endorphin null mutant does not involve coupling to K^+ channels, at least not in the arcuate nucleus of the hypothalmus.

A large increase in mu receptor binding using [^{125}I]DAMGO was found in amygdala and pallidum of enkephalin null mutant mice, while a more modest increase in delta receptor binding using [^{125}I]DELT-II was also found in amygdala and pallidum.[92] The changes in receptor binding were found in brain regions that are normally rich in enkephalin. The limbic regions where changes in receptor binding were found are also regions that mediate the emotional outputs that were altered in the mutant mice. These alterations suggest that enkephalin may also normally bind the mu receptor as well as the delta receptor. Alternatively, it supports the data from the receptor null mutants suggesting that the delta and mu receptors are coregulated in some fashion.

Much of the classical pharmacology has been confirmed in the opioid-receptor null mutant mouse, but the data from these mice have suggested than the interdependence of opioid systems is more significant than what may have been previously appreciated. Receptor interactions, even dimerization or oligomerization, may also play a role in the pharmacology of the opioid system. In addition, the gene structure and potential splice variants of the opioid receptors suggest that the complicated pharmacology of the opioid system may, in part, be due to genetic variations of the receptors. Gene deletion studies are useful for analysis of the genetic contributions to opioid pharmacology and gene replacement or transgenic studies will be useful in confirming opioid pharmacology *in vivo*.

3.4 VIRAL VECTORS

Most behavioral techniques for rodents were developed using rats, which is a major drawback to using transgenic and targeted mutation techniques for experiments examining the opioid system. Many antinociceptive and self-administration techniques have been successfully adapted to mice but the investigator should be aware that behavioral measurement techniques developed for rat experiments may not translate to mouse behavior. If one finds it necessary to use rats, there are considerably fewer options for experiments involving genetic manipulations. Although a few transgenic rats have been produced, it has been with great effort and expense. Primarily, one is limited to the use of recombinant viral-based expression systems.

Introduced gene-expression as transgenesis can be accomplished by the use of recombinant viruses *in vivo*.[93] Engineered viruses used for transgenic expression *in vivo* have limited uses primarily due to their short duration of expression, restricted regions of expression, and possible neurotoxicity. Several viral-based expression systems have been developed to improve upon these problems but each system has unique limitations. Generally, the murine retroviral viruses offer longer-lasting expression but are limited to dividing cells and so are not useful for studies on postmitotic neurons. Some nonpathogenic retroviruses based on lentivirus are able to infect nondividing cells, including neurons, and so may be useful for neuroscience research.[94] Dividing neurons *in utero* could be a target for retroviral infection except that it appears that embryonic neurons silence their expression.[95] This problem has been overcome with an engineered retrovirus lacking a long terminal repeat (LTR) and was successfully used to deliver neuronal precursor gene expression *in utero*, which also required an ultrasound image-guided injection.[96] The duration of this expression remains to be examined.

Recombinant herpes simplex virus-based expression systems have been used for some studies of the opioid system. Enkephalin was expressed in the rat amygdala to produce antinociception, demonstrating the modulatory role of this nucleus in central nervous system-produced antinociception.[97] GluR1 was expressed in the ventral tegmentum of rats to increase sensitization to morphine,[98] and CREB was expressed in the nucleus accumbens to block behavioral reward to cocaine.[99] Although viral expression can be sustained for long periods in some tissues, transgenics have the advantage of permanent expression beginning during development. Additionally, traditional transgenic expression can be directed in particular cells depending on the promoter used whereas viral expression is often site directed and limited to that region. Tissue-specific promoters have been used to drive expression in tissue such as the liver,[30] but the brain probably represents more of a challenge because of the blood–brain barrier and unique immune system. Finally, the only way to eliminate the function of a targeted gene with a viral expression system is through antisense, which is less efficient than a genomic mutation present in all cells. Thus, although viral expression systems are a powerful tool and provide a technique for gene expression in animals other than mice there are still several limitations.

3.5 CONDITIONAL TARGETED MUTATIONS USING *IN VIVO* RECOMBINASES

Mutations can be created with temporal and/or anatomical restrictions (a conditional null mutant) by using site-specific recombinases. The Cre recombinase from bacteriophage P1 will excise DNA sequences flanked on either side by a recognition sequence (*lox*P) in the same orientation. An eukaryotic recombinase from yeast is the Flp recombinase, which excises DNA flanked by the *frt* recognition sequences. Both recombinases will invert the intervening sequence if the recognition sequences are in opposite orientation.[100]

3.5.1 TISSUE-SPECIFIC NULL MUTATIONS

The recombinases are used to produce a null mutation by generating a mouse with a gene that contains critical exons flanked by the recognition sites: "floxed" for flanked with *lox*P sites or "flrted" for flanked by *frt* sites. These mutations are introduced by means of homologous recombination in ES cells similar to producing a null mutant. This mouse is then crossed with a transgenic mouse harboring a transgene that expresses the appropriate recombinase under the control of a tissue-specific promoter. When crossed, the recombinase will remove the flanked exons in only those cells where the transgene is expressed and in this way a tissue-specific null mutant is created. This has been done by targeting Cre expression to thymocytes and eliminating DNA polymerase beta expression only in thymocytes.[101,102]

The Cre recombinase has been used quite extensively *in vivo* while there are only a few reports of Flp recombinase activity in mice. Experience seems to suggest that Flp recombination is less efficient. An early study showed that the recombination by *flp in vivo* was only slightly less that that of Cre.[103] This study used the human beta-actin promoter to drive the expression of the *flp* recombinase. Another study adopted an elegant design that used a construct containing both *lox*P and *frt* sites in different regions of the *fgf*8 gene to generate three different mouse lines from the same progenitor line, termed allelogenic.[104] Using the human beta-actin promoter to drive the expression of both *flp* and *cre* recombinases they found that embryonic Cre-mediated recombination was essentially 100% while Flp-mediated recombination was less than that. Because of incomplete recombination during embryogenesis, a chimeric generation was produced that had to be mated to generate offspring with the ubiquitous recombination. This study demonstrates a use for both recombinases to generate an allelogenic line of mice. Another study also describes an allelogenic line of mice using only the Cre-recombinase and an introduced point mutation by a replacement type vector.[105] This technique provides a method for generating two different mutations in the same gene and allows for an *in vivo* analysis of these two mutations alone and combined, all in the same line and conditionally. Thus, both Cre and Flp can be used together but if one recombinase is required, experience suggests that Cre is the better choice.

Recombinases can be used to generate an allelogenic line of mice as well as a conditional mutant. Other potential uses for recombinases are creation of knockins as will be discussed below and for generating site-specific chromosomal integration of transgenes. A consistent problem with transgenic studies is that the site of integration can have profound effects on levels of expression by the influence of surrounding enhancer or repressor elements or by chromatin structure, so-called "position effects." Currently, the only way to avoid this problem is to generate several lines of mice and analyze them individually. One way to avoid this problem would be to use recombinases to integrate a transgene at a marked location on a chromosome where position effects are favorable. This has been done in culture and it will only be a matter of time before mice harboring these marked loci will be available for transgenic production.[106]

3.5.2 INDUCIBLE TARGETED MUTATIONS

For studies not directly focused on development, the conventional null mutant can be problematic because the mouse develops in the absence of that gene. This can result in a lethal phenotype early in development, which is useless for studies involving behavior or physiology. Perhaps more complicated, compensations by other gene products may be a result of the null mutant and so an observed phenotype may be produced only indirectly by the mutation. One of the most powerful uses of the Cre or Flp recombinases is to express them from an inducible transgene so that a temporal null mutant can be created. This technique provides a way of eliminating a gene product without concerns for developmental abnormalities or compensations. In addition, the recombinase transgene can have a tissue-specific promoter so that recombination can be controlled spatially as well as temporally. The tetracycline-controlled system has been used to induce expression of Cre with temporal control.[20,107] It has also been used to control Cre expression with temporal and spatial control by making a transgenic with a tissue-specific promoter to drive the transactivator.[108] The EBD system has been adapted to express the Cre recombinase (Cre-ER).[25,109] Low background expression of Cre recombinase is the strength of the EBD system. However, Cre recombinase activity induced by the EBD was not complete in the cell type that was targeted[25] and concentrations of tamoxifen necessary for recombinase activity were in danger of interfering with pregnancy.[109] The GR^{dex} system has been adapted to express Cre (Cre-GR^{dex}), which also demonstrated a low background of recombinase activity and efficient inducible recombinase activity in culture but hasn't been tested in mice yet.[26] The mutant hPR that binds RU486 but not progesterone has not been used to express Cre yet, so it will be interesting to see how its activity compares with those of the other steroid hormone receptor systems or the tetracycline system.

A future possibility will be the integration of Cre recombinase preceded by an internal ribosomal entry site (IRES), which will allow for targeting a locus for tissue-specific expression.[110] This technique will allow for directed Cre expression by an endogenous locus without disrupting the expression of that gene. A collective effort has been put into establishing a database of these mice with information on their tissue-specific expression (http://www.mshri.on.ca/develop/nagy.Cre.htm).

3.6 OTHER TARGETED MUTATIONS

One can also target a mutation in mice to alter gene function rather than eliminate it; this is commonly known as a knockin. Several techniques have been described to perform this type of mutation. Similar to a targeted null mutation, these techniques require the modification of an allele in ES cells as the starting point. One technique that has been described is the cDNA homologous insertion protocol (CHIP) which fuses the cDNA of a gene to an exon of a genomic fragment that is from the targeted allele.[111] This uses an insertion vector and may be a simple method of introducing a mutated version of a protein into its own locus, providing the loss of intronic sequences won't disrupt its expression pattern. Other techniques rely entirely on using genomic fragments.

Mutations can be targeted by using the hit-and-run/in-and-out method.[112,113] These techniques combine the endogenous activity of repair enzymes and negative selection with selectable markers to replace stretches of DNA by using an insertion vector. Thus, they require two separate selections of the ES cells, the second one being a relatively rare event but not requiring an additional transfection. Another method is the double replacement method[114,115] or the tag-and-exchange method.[116] This technique requires two separate targeting constructs that are used for two sequential transfections. An added benefit with this method is that the intermediate clone is a marked locus that can be transfected with a number of different constructs. The disadvantage of this technique is that two sequential transfections result in ES cells maintained in culture long enough to lose their pluripotency. In fact, our experience with the double replacement technique was that it was difficult to generate final clones that could be passed through the germ line.[37] The final procedure for generating a knockin is by a replacement vector with a floxed negative and positive selection cassette (e.g., Neo and Thymidine kinase). Following homologous recombination of the construct the selection cassettes can then be removed by a transient transfection with Cre recombinase and negative selection.

This procedure is being used to introduce a point mutation into the mu receptor that causes antagonist binding to activate the receptor instead of blocking its function.[117] Thus, the construct is designed to have the serine at position 196 replaced by leucine in vivo using Cre-loxP to remove the neo cassette.[118]

The many techniques described to produce targeted mutants have their own advantages and disadvantages but the main disadvantage is common to all of the techniques and that is the second round of selection that endangers the pluripotency of the ES cells. The double-replacement type strategy allows for the generation of many knockin mice from an intermediate step but the extra transfection step is the riskiest of them all. The hit-and-run type procedures depend upon a recombination event that is rare. In our opinion, the removal of the selection cassette by the Cre recombinase is the procedure most likely to produce an ES cell clone retaining pluripotency.

3.7 CONSTRUCT DESIGN

One of the technical hurdles for building a targeting construct is the requirement of cloning and mapping genomic fragments from a suitable library. Two techniques describe methods that allow for minimal mapping of restriction fragments. In one method, the genomic fragment is cloned into a yeast shuttle vector while a plasmid containing the *neo* cassette also contains short flanking recombinagenic arms that are produced by PCR from the region of the gene to be targeted.[119] Using the powerful recombinant activity of yeast, the *neo* cassette (or any other genetic element) is inserted via homologous recombination into the genomic fragment at the site targeted by the recombinagenic arms. The recombinant DNA construct is then recovered from the yeast and used for homologous recombination into the gene's chromosomal site in ES cells. Another technique describes the use of a transposon flanked by commonly used restriction endonuclease sites. Colonies containing a cloned genomic fragment have undergone the transposition reaction with this element and will have convenient restriction sites introduced in the exon where *neo* or some other element can now be easily introduced.[120] The acquisition of a genomic fragment from a genomic library has also been made more convenient with the commercial availability of bacteria artificial chromosomes (BAC) containing genomic fragments much larger than the inserts found in the conventional lambda phage library. Considerable time is saved in not screening a lambda phage library, which generally requires at least three successive rounds of plaque purification to isolate a single pure clone. It is also unlikely that more than one single clone will ever be required for the building of a targeting construct because of the size of a BAC clone. Tools with which to manipulate unwieldy genomic fragments have made the construct assembly process simpler, although a thorough strategy is still the most important tool.

An important factor in designing most constructs is the inclusion of the *neo* selection cassette used during the cloning of correctly targeted ES cells. However, the *neo* cassette contains cryptic splice sites that can be utilized by the targeted gene.[121,122] Both a canonical splice acceptor sequence and a splice donor site are found in the noncoding strand of *neo*, which are therefore more active when the *neo* cassette is in the reverse transcriptional orientation relative to the targeted allele. This activity has been utilized in at least two studies where hypomorphic alleles were generated by inserting the *neo* cassette in the same orientation as the targeted allele.[104,105] This technique could, in fact, be used to analyze the alternatively spliced forms of the mu opioid receptor or other receptors by allowing for the generation of one splice form(s) and the exclusion of others.

3.8 GENETIC BACKGROUND OF MUTANT MICE

The complicated behaviors that are modulated by the opioid system are undoubtedly polygenic. Thus, the genetic context in which opioid pathways are studied is likely to have a significant influence on these behaviors and, consequently, experimental results. Different strains of mice diverge for their basal sensitivities in nociceptive assays, their analgesic sensitivities to opioid drugs, and their propensity to self-administer opioid drugs. Since these behaviors are influenced by multiple neurotransmitter systems, it is expected that the source of strain variation will not be found in only one gene. Experiments using targeted genetic mutations must be designed carefully so that a genetically equivalent wild-type control is used, and interpretations of experimental results must be in the context of the particular mouse strain used. We also must be aware of the potential role genetic background can have in determining phenotypes and some of the idiosyncrasies of specific mouse strains. Of utmost importance is the adoption of standards for maintaining mutant strains of mice.

A flawed, but commonly used, technique is to maintain a line of mutant mice and a parallel line of wild-type mice by inbreeding F_2 homozygous littermates. Super-ficially, this is an enticing method because it eliminates the need for genotyping and is economical. However, this method is flawed because the genetic background of such inbred mice is not known and cannot be reproduced. In fact, after 20 generations of brother–sister matings, new recombinant inbred lines will be generated whose genetic makeup, in addition to the targeted locus, can differ drastically. These differences arise because of random segregation and fixation of polymorphic alleles, including genes linked to the targeted locus, that are unique to the wild-type or the targeted line. Thus, it is preferable to avoid this scheme at all costs. If this method has been used, then the targeted mutation should be transferred to standard inbred backgrounds, which can be done in 10 generations of backcrosses.

3.8.1 Backcrossing Techniques

The method the Banbury Conference recommends is maintaining a mutant mouse line on a known genetic background. For example, chimeras are mated to both a 129 strain (origin of the ES cells and mutant locus) and a C57BL/6 strain (origin of the blastocyst).[123] Heterozygous mice are then repeatedly backcrossed in parallel on the inbred C57BL/6 and 129 lines, such that congenic heterozygotes are created on both backgrounds. Heterozygotes from the two congenic strains are then crossed to produce F_1 hybrid mice in Mendelian proportions of 1 wild-type:2 heterozygotes:2 homozygotes for experiments. F_1 hybrid mice have the advantages of reproductive vigor and the absence of deleterious recessive traits from both of the background strains. Unfortunately, this breeding procedure is often not possible because some behaviors to be studied are only present in certain inbred strains, and the procedure is prohibitively expensive for most laboratories. For example, studying a double mutant by this method would require massive numbers of mating pairs because double mutants would only be 1/16 of the F_1 progeny of double heterozygous parents. A less expensive alternative is to backcross the chimeric mice onto only one inbred strain and use littermates from heterozygote matings as the experimental subjects. However, this method requires some knowledge of the strain to be used and should not be done without a comprehensive literature review and pilot study.

Ten backcrosses are required to produce 99.9% congenic mice. However, after only 5 generations the genome is nearly 97% congenic to the background strains. These mice could be used, at the least for pilot studies, although the backcrossing procedure should continue indefinitely. It is possible to expedite this process to as little as 3 to 4 generations with a procedure known as speed congenics.[124,125] This process requires PCR genotyping with strain-specific genetic markers. Genotyping assists in selecting for mice that are represented by the maximum amount of chromosomes from the recipient strain. Approximately 150 different markers were used resulting in a map density of about 10 cM.[124] In addition, we have postulated that the typical interval between generations can be shortened. Heterozygote 25-day-old females are immediately genotyped and superovulated by standard methods.[18] They are then mated with wild-type, mature males from the recipient strain and mating plugs are checked. Parallel females of an outbred strain are mated with vasectomized males to use as host mothers. On the fourth day following coitus, blastocyst stage embryos can be harvested and transferred to the pseudopregnant females via inter-uterine transfer. We estimate that the average time between generations can be shortened by at least 2 weeks. Additionally, if C57BL/6 mice are the recipient strain, one will find that the first pregnancy is often lost or not nursed. The use of host mother from an outbred strain increases the chances of a healthy litter.

3.8.2 PHENOTYPIC DIFFERENCES OF SOME INBRED STRAINS

Many studies of null mutations have reported learning deficits, but many inbred lines have phenotypic abnormalities in learning themselves. For example, several 129 lines have spatial learning deficits,[126] and most 129 and DBA strains show poor hippocampal-dependent learning.[127,128] Mice from the C57BL/6 line become deaf to certain frequencies at an early age,[129] and they are poor avoidance learners.[130] Studies that are especially relevant to experiments on the opioid system have shown that the C57BL/6 strain is considerably more tolerant to the analgesic effects of mu-selective opioids such as morphine and DAMGO compared to other strains.[131,132] These strain differences are not limited to performances on behavioral tests. For example, kainic acid-induced seizures occur in the absence of hippocampal neuronal degeneration in C57BL/6 and Balb/C mice,[133,134] and the corpus callosum is defective in many mice from BALB and 129 substrains.[135–137] An F_1 hybrid cross of inbred lines often eliminates some of these abnormalities because homozygous recessive genes are often present in the inbred strains. For example, of all inbred lines tested, the C57BL/6 is the best performer on the Morris water maze, yet all F_1 hybrid lines performed better than the C57BL/6 mice.[128] A null mutation of the Thy-1 gene was made in ES cells from both C57BL/6 and 129/Sv/Ev and resultant outbred lines of all combinations showed that genetic background strongly influenced initial learning of the Morris water maze.[41] If the Morris water maze is an intended assay, then an F_1 hybrid might be the preferred genetic background for studying the effects of single gene deletions that are expected to decrease spatial learning.

A fundamental problem of each of the breeding strategies described is the phenomenon of "hitchhiking genes" closely linked to the targeted genetic locus.[138] These alleles are derived only from the genome of the ES cells and even after 12 generations of backcrossing to produce a congenic line they may represent as much as 16 cM of the ES genome, or 1% of the total genome containing 50 to 100 genes.[123] Virtually all reliable ES cells are derived from 129 substrain mice, so these closely linked hitchhiking genes are 129 alleles. In addition to these closely linked genes the substrains of 129 can vary significantly.[40] Standardized behavioral testing of these substrains is incomplete and many are no longer commercially available. Moreover, certain substrains are notorious for inherited neurological disorders[126] as well as low fecundity. There are at least two instances in which phenotypic differences between induced mutant mice and wild-type control mice have been attributed to closely linked gene alleles originating from the 129 background strain, rather than the targeted mutation itself.[133,134,139] Remedies for this problem are not easy. The most desirable, but still not available, solution will be the development of a new series of ES cells derived from a panel of different inbred strains that can be used for gene targeting and are easily available.

3.9   CONCLUSIONS

The first wave of genetically engineered mice that contain targeted mutations in the genes encoding opioid peptides and opioid receptors has had important implications for our understanding of the physiological function of these molecules. These new models are complementary to pharmacological and antisense strategies, not a replacement. Converging data suggest that beta-endorphin and not enkephalins are the critical endogenous opioid peptides involved in stress-induced analgesia. The mu opioid receptor has a fundamental role in the analgesic and rewarding actions of morphine and likely is also essential for supraspinal analgesia mediated by delta agonists. Kappa receptors appear to be vital in establishing the tonic nociceptive threshold for visceral pain.

These studies have also demonstrated important limitations of the current technology and the need for improved experimental standards. To make valid comparisons among the mouse models generated by different laboratories, it is essential that better documentation of genetic background be provided in publications. Initial results using F_2 hybrid mice must be followed by more comprehensive analyses in congenic or inbred strains to truly evaluate the consequences of the single gene mutation. Finally, there needs to be a continued effort to develop reliable techniques for introducing more subtle mutations in specific genes and to permit the spatial and temporal control of gene inactivation and reactivation *in vivo*.

ACKNOWLEDGMENTS

The authors thank members of the Low Lab who contributed to some of the experiments described here. This work was supported by NIH grants F32DA05841 (M.D.H.) and P01DK55819 (M.J.L.).

REFERENCES

1. Chen, Y. et al., Molecular cloning and functional expression of a μ-opioid receptor from rat brain, *Mol. Pharm.*, 44, 8, 1993.
2. Evans, C.J. et al., Cloning of a delta opioid receptor by functional expression, *Science*, 258, 1952, 1992.
3. Keiffer, B.L. et al., The δ-opioid receptor: isolation of a cDNA by expression and pharmacological characterization, *Proc. Natl. Acad. Sci. USA*, 89, 12048, 1992.
4. Meng, F. et al., Cloning and pharmacologicial characterization of a rat κ opioid receptor, *Proc. Natl. Acad. Sci. USA*, 90, 9954, 1993.

5. Thompson, R.C. et al., Cloning and pharmacological characterization of a rat μ opioid receptor, *Neuron*, 11, 903, 1993.
6. Reisine, T. and Bell, G. I., Molecular biology of opioid receptors, *Trends Neurosci.*, 16, 506, 1993.
7. Brownstein, M. J., A brief history of opiates, opioid peptides, and opioid receptors, *Proc. Natl. Acad. Sci. USA*, 90, 5391, 1993.
8. Christie, M.J. et al., Cellular mechanisms of opioid tolerance: studies in single brain neurons, *Mol. Pharmacol.*, 32, 633, 1987.
9. Mansour, A. et al., Opioid-receptor mRNA expression in the rat CNS: anatomical and functional implications, *Trends Neurosci.*, 18, 22, 1995.
10. Reisine, T. and Pasternak, G., Opioid analgesics and antagonists, in *Goodman and Gilman's The Pharmacological Basis of Therapeutics*, Hardman, J.G., Gilman, A.G., and Limbird, L.E., Eds., McGraw-Hill, New York, 1996, 521.
11. Schoffelmeer, A.N.M. et al., β-Endorphin: a highly selective endogenous opioid agonist for presynaptic *mu* opioid receptors, *J. Pharmcol. Exp. Ther.*, 258, 237, 1991.
12. Zadina, J.E. et al., A potent and selective endogenous agonist for the μ-opiate receptor, *Nature*, 386, 499, 1997.
13. Iles, P., Modulation of transmitter and hormone release by multiple neuronal opioid receptors, *Rev. Physiol. Biochem. Pharmacol.*, 112, 140, 1989.
14. Yamada, K. and Nabeshima, T., Stress-induced behavioral responses in multiple opioid systems in the brain, *Behav. Brain. Res.*, 67, 133, 1995.
15. Olson, G.A. et al., Endogenous opiates: 1995, *Peptides*, 17, 1421, 1995.
16. Low, M.J., The identification of neuropeptide gene regulatory elements in transgenic mice, in *Methods in Molecular Biology*, Longstaff, A. and Revest, P., Eds., The Humana Press, Totowa, New Jersey, 1992, 181.
17. Kozak, M., At least six nucleotides preceding the AUG initiator codon enhances translation in mammalian cells, *J. Mol. Biol.*, 196, 947, 1987.
18. Hogan, B. et al., *Manipulating the Mouse Embryo*, 2nd ed., Cold Spring Harbor Laboratory Press, Plainview, New York, 1994.
19. Gossen, M. and Bujard, H., Tight control of gene expression in mammalian cells by tetracycline-responsive promoters, *Proc. Natl. Acad. Sci. USA*, 89, 5547, 1992.
20. Furth, P.A. et al., Temporal control of gene expression in transgenic mice by a tetracycline-responsive promoter, *Proc. Natl. Acad. Sci. USA*, 91, 9302, 1994.
21. Gossen, M. et al., Transcriptional activation by tetrayclines in mammalian cells, *Science*, 268, 1766, 1995.
22. Baron, U. et al., Generation of conditional mutants in higher eukaryotes by switching between the expression of two genes, *Proc. Natl. Acad. Sci. USA*, 96, 1013, 1999.
23. No, D. et al., Ecdysone-inducible gene expression in mammalian cells and transgenic mice, *Proc. Natl. Acad. Sci. USA*, 93, 3346, 1996.
24. Littlewood, T.D. et al., A modified oestrogen receptor ligand-binding domain as an improved switch for the regulation of heterologous proteins, *Nuc. Acids Res.*, 23, 1686, 1995.
25. Schwenk, F. et al., Temporally and spatially regulated somatic mutagenisis in mice, *Nuc. Acids Res.*, 26, 1427, 1998.
26. Brocard, J. et al., A chimeric Cre recombinase inducible by synthetic, but not natural ligands of the glucocorticoid receptor, *Nuc. Acids Res.*, 26, 4086, 1998.
27. Wang, Y. et al., Ligand-inducible and liver-specific target gene expression in transgenic mice, *Nat. Biotechnol.*, 15, 239, 1997.
28. Wang, Y. et al., Antiprogestin regulable gene switch for induction of gene expression in vivo, *Methods Enzymol.*, 306, 281, 1999.

29. Wang, Y. et al., An antiprogestin regulable gene switch for induction of gene expression in vivo, *Adv. Pharmacol.*, 47, 343, 2000.
30. Burcin, M.M. et al., Adenovirus-mediated regulable target gene expression in vivo, *Proc. Natl. Acad. Sci. USA*, 96, 355, 1999.
31. Donovan, D.M. et al., Preproenkephalin "cassette" confers brain expression and synaptic regulation in transgenic mice, *Proc. Nat. Acad. Sci. USA*, 89, 2345, 1992.
32. Donovan, D.M. et al., Transgenic mice: preproenkephalin and the mu opiate receptor, *Ann. NY Acad. Sci.*, 780, 19, 1996.
33. Borsook, D. et al., Proenkephalin gene regulation in the paraventricular nucleus by GABA: interactions with opioid systems in a transgenic model, *J. Neurochem.*, 70, 604, 1998.
34. Van Koughnet, K. et al., Proenkephalin transgene regulation in the paraventricular nucleus of the hypothalamus by lipopolysaccharide and interleukin-1beta, *J. Comp. Neurol.*, 405, 199, 1999.
35. Hu, X. et al., Promoter activity of mouse kappa opioid receptor gene in transgenic mouse, *Brain Res. Mol. Brain Res.*, 69, 35, 1999.
36. Young, J.I. et al., Authentic cell-specific and developmentally regulated expression of pro-opioimelanocortin genomic fragments in hypothalamic and hindbrain neurons of transgenic mice, *J. Neurosci.*, 18, 6631, 1998.
37. Hayward, M.D. and Low, M.J., unpubished data, 1999.
38. Soriano, P., Gene targeting in ES cells, *Annu. Rev. Neurosci.*, 18, 1, 1995.
39. Thomas, K.R. and Capecchi, M.R., Site-directed mutagenesis by gene targeting in mouse embryo-derived stem cells, *Cell*, 51, 503, 1987.
40. Simpson, E. et al., Genetic variation among 129 substrains and its importance for targeted mutagenesis in mice, *Nat. Genet.*, 16, 19, 1997.
41. Mayeux-Portas, V. et al., Mice lacking the cell adhesion molecule Thy-1 fail to use socially transmitted cues to direct their choice of food, *Curr. Biol.*, 10, 68, 1999.
42. Dinkel, A. et al., Efficient generation of transgenic BALB/c mice using BALB/c embryonic stem cells, *J. Immunol. Meth.*, 223, 255, 1999.
43. Matthes, H.W.D. et al., Loss of morphine-induced analgesia, reward effect and withdrawal symptoms in mice lacking the μ-opioid-receptor gene, *Nature*, 383, 819, 1996.
44. Loh, H.H. et al., Opioid receptor knockout in mice: effects on ligand-induced analgesia and morphine lethality, *Mol. Brain Res.*, 54, 321, 1998.
45. Sora, I. et al., Opiate receptor knockout mice define μ receptor roles in endogenous nociceptive respones and morphine-induced analgesia, *Proc. Natl. Acad. Sci. USA*, 94, 1544, 1997.
46. Schuller, A.G. et al., Retention of heroin and morphine-6 beta-glucuronide analgesia in a new line of mice lacking exon 1 of MOR-1, *Nat. Neurosci.*, 2, 151, 1999.
47. Tian, M. et al., Altered hematopoiesis, behavior, and sexual function in μ opioid receptor-deficient mice, *J. Exp. Med.*, 185, 1517, 1997.
48. Simonin, F. et al., Disruption of the κ-opioid receptor gene in mice enhances sensitivity to chemical visceral pain, impairs pharmacological actions of the selective κ-agonist U-50,488H and attenuates morphine withdrawal, *EMBO J.*, 17, 886, 1998.
49. Zhu, Y. et al., Retention of supraspinal delta-like analgesia and loss of morphine tolerance in ∂ opioid receptor knockout mice, *Neuron*, 24, 243, 1999.
50. Rubinstein, M. et al., Absence of opioid stress-induced analgesia in mice lacking β-endorphin by site-directed mutagenesis, *Proc. Natl. Acad. Sci. USA*, 93, 2577, 1996.
51. König, M. et al., Pain responses, anxiety, and aggression in mice deficient in preproenkephalin, *Nature*, 383, 535, 1996.
52. Sharifi, N. et al., Generation of dynorphin knockout mice, *Mol. Brain Res.*, 86, 70, 2001.

52a. Wang, Z. et al., Pronociceptive actions of dynorphin maintain chronic neuropathic pain, *J. Neurosci.*, 21, 1779, 2001.

53. Lai, J. et al. Prodynorphin "Knock-out" mice do not show sustained neuropathic pain or spinal opioid tolerance, *Int. Narcotics Res. Conf.*, Seattle, WA, 45, 04-18, 2000.

54. Low, M.J. and Pintar, J., unpublished data, 1999.

55. Czyzyk, T.A. et al. Production of combinatorial opioid receptor knock out mice, *Soc. Neurosci. Annu. Mtg.*, 73.1, 1999.

56. Mogil, J.S. et al., Opioid and nonopioid swim stress-induced analgesia: a parametric analysis in mice, *Physiol. Behav.*, 59, 123, 1996.

57. Rubinstein, M. et al., High efficiency introduction of a point mutation into the mouse genome in embryonic stem cells by homologous recombination using a replacement type vector and selectable markers, *Nucl. Acids Res.*, 21, 2613, 1993.

58. Kitchen, I. et al., Quantitative autoradiographic mapping of μ-, ∂, and κ-opioid receptors in knock out mice lacking the μ-opioid receptor gene, *Brain Res.*, 778, 73, 1997.

59. Sutters, K.A. et al., Analgesic synergy and improved motor function produced by combinations of μ-δ and μ-κ-opioids, *Brain Res.*, 530, 290, 1990.

60. Jiang, Q. et al., Modulation of the potency and efficacy of mu-mediated anti-nociception by delta agonists in the mouse, *J. Pharmacol. Exp.*, 254, 683, 1990.

61. Maldonado, R. et al. Anticoceptive responses induced by opioid compounds in two different lines of MOR knockout mice, *Int. Narcotics Res. Conf.*, Seattle, WA, O4-18, 30, 2000.

62. Minami, M. et al., DAMGO, a μ-opioid receptor selective ligand, distinguishes between μ- and κ-opioid receptors at a different region from that for the distinction between μ- and δ-opioid receptors, *FEBS Lett.*, 364, 23, 1995.

63. Abbadie, C. et al., Differential distribution in rat brain of mu opioid receptor carboxy terminal splice variants MOR-1C-like and MOR-1-like immunoreactivity: evidence for region-specific processing, *J. Comp. Neurol.*, 419, 244, 2000.

64. Pan, Y.X. et al., Isolation and expression of a novel alternatively spliced mu opioid receptor isoform, MOR-1F, *FEBS Lett.*, 466, 337, 2000.

65. Pan, X.-X. et al., Differential expression of eight splicing variants are directed by a new promoter of the mouse mu opioid receptor gene (MOR-1), *Int. Narcotics Res. Conf.*, Seattle, WA, Mon08, 2000.

66. Stolerman, I., Drugs of abuse: behavioural principles, methods and terms, *Trends Pharmacol.*, 13, 170, 1992.

67. Koob, G.F., Drugs of abuse: anatomy, pharmacology and function of reward pathways, *Trends Pharmacol.*, 13, 177, 1992.

68. Koob, G.F. et al., Neural substrates of opiate withdrawal, *Trends Neurosci.*, 15, 186, 1992.

69. Di Chiara, G. and North, R. A., Neurobiology of opiate abuse, *Trends Pharmacol.*, 13, 185, 1992.

70. Di Chiara, G. and Imperato, A., Drugs abused by humans preferentially increase synaptic dopamine concentrations in the mesolimbic system of freely moving rats, *Proc. Natl. Acad. Sci. USA*, 85, 5274, 1988.

71. Fang, F.G. et al., Action at the mu receptor is sufficient to explain the supraspinal analgesic effect of opiates, *J. Pharmacol. Exp. Ther.*, 238, 1039, 1986.

72. Baamonde, A.I. et al., Systemic administration of [Tyr-D-Ser(O-Tert-Butyl)-Gly-Phe-Leu-Thr(O-Tert-Butyl)], a highly selective delta opioid agonist, induces mu receptor-mediated analgesia in mice, *J. Pharmacol. Exp. Ther.*, 257, 767, 1991.

73. Funada, M. et al., Blockade of morphine reward through the activation of κ-opioid receptors in mice, *Neuropharmacology*, 32, 1315, 1993.

74. Székely, J.I., μ-Agonist induced euphoria as opposed to dysphoria elicited by κ-agonists in humans and experimental animals, in *Opioid Peptides in Substance Abuse*, CRC Press, Boca Raton, Florida, 1994, 55.

75. Mucha, R.F. and Herz, A., Motivational properties of kappa and mu opioid receptor agonists studied with place and taste preference conditioning, *Psychopharmacology*, 86, 274, 1985.

76. Nabeshima, T. et al., Opioid κ receptors correlate with the development of conditioned suppression of motility in mice, *Eur. J. Pharmacol.*, 152, 129, 1988.

77. Bals-Kubik, R. et al., Evidence that the aversive effects of opioid antagonists and κ-agonists are centrally mediated, *Psychopharmacology*, 98, 203, 1989.

78. Devine, D.P. et al., Differential involvement of ventral tegmental mu, delta and kappa opioid receptors in modulation of basal mesolimbic dopamine release: in vivo microdialysis studies, *J. Pharmacol. Exp. Ther.*, 266, 1236, 1993.

79. Di Chiara, G. and Imperato, A., Opposite effects of mu and kappa opiate agonists on dopamine release in the nucleus accumbens and in the dorsal caudate of freely moving rats, *J. Pharmacol. Exp. Ther.*, 244, 1067, 1988.

80. Suzuki, T. et al., Effects of nor-binaltorphimine on the development of analgesic tolerance to and physical dependence on morphine, *Eur. J. Pharmacol.*, 213, 91, 1992.

81. Mogil, J. and Low, M.J., unpublished data, 1996.

82. Fuchs, P.N. et al., Characterization of mechanical withdrawal responses and effects of mu-, delta- and kappa-opioid agonists in normal and mu-opioid receptor knockout mice, *Brain Res.*, 821, 480, 1999.

83. Matthes, H.W.D. et al., Activity of the ∂-opioid receptor is partially reduced, whereas activity of the κ-receptor is maintained in mice lacking the μ-receptor, *J. Neurosci.*, 18, 7285, 1998.

84. Jordan, B.A. and Devi, L.A., G-protein-coupled receptor heterodimerization modulates receptor function, *Nature*, 399, 697, 1999.

85. Traynor, J.R. and Elliott, J., Delta-opioid receptor subtypes and cross-talk with mu-receptors, *Trends Pharmacol. Sci.*, 14, 84, 1993.

86. Slowe, S.J. et al., Quantitative autoradiography of μ, ∂, κ₁ opioid receptors in κ-opioid receptor knockout mice, *Brain Res.*, 818, 335, 1999.

87. Matthes, H.W.D. et al., Functional response of delta- and kappa-opioid receptors in mu-opioid receptor knock-out mice, *Soc. Neurosci. Abstr.*, 1997.

88. Sora, I. et al., The μ-opioid receptor is necessary for [D-Pen², D-Pen⁵] enkephalin-induced analgesia, *Eur. J. Pharmacol.*, 256, 281, 1997.

89. Mogil, J.S. et al., Disparate spinal and supraspinal opioid antinociceptive responses in β-endorphin-deficient mutant mice, *Neuroscience*, 101, 709, 2000.

90. Slugg, R.M. et al., Effect of the μ-opioid agonist DAMGO on medial basal hypothalamic neurons in β-endorphin knock-out mice, *Neuroendocrinology*, 72, 208, 2000.

91. Hayward, M.D. and Low, M.J., Targeted mutagenesis of the murine opioid system, in *Regulatory Peptides and Cognate Receptors*, D. Richter, Ed., Springer-Verlag, Berlin, 1999, 169.

92. Brady, L.S. et al., Region-specific up-regulation of opioid receptor binding in enkephalin knockout mice, *Mol. Brain Res.*, 68, 193, 1999.

93. Simonato, M. et al., Gene transfer into neurones for the moleculat analysis of behaviour: focus on herpes simplex vectors, *Trends Neurosci.*, 23, 183, 2000.

94. Naldini, L. et al., In vivo gene delivery and stable transduction of nondividing cells by a lentiviral vector [see comments], *Science*, 272, 263, 1996.

95. Pice, J., A *sonic* boom for gene delivery, *Nat. Neurosci.*, 2, 779, 1999.

96. Gaiano, N. et al., A method for rapid gain-of-function studies in the mouse embryonic nervous system, *Nat. Neurosci.*, 2, 812, 1999.
97. Kang, W. et al., Herpes virus-mediated preproenkephalin gene transfer to the amygdala is antinociceptive, *Brain Res.*, 792, 133, 1998.
98. Carlezon, W.A. et al., Sensitization to morphine induced by viral-mediated gene transfer, *Science*, 277, 812, 1997.
99. Carlezon, W.A. et al., Regulation of cocaine reward by CREB, *Science*, 282, 2272, 1998.
100. Kilby, N. et al., Site-specific recombinases: tools for genome engineering, *Trends Genet.*, 9, 413, 1993.
101. Orban, P.C. et al., Tissue-and site-specific DNA recombination in transgenic mice, *Proc. Natl. Acad. Sci. USA*, 89, 6861, 1992.
102. Gu, H. et al., Deletion of a DNA polymerase β gene segment in T cells using cell type-specific gene targeting, *Science*, 265, 103, 1994.
103. Dymecki, S.M., Flp recombinase promotes site-specific DNA recombination in embryonic stem cells and transgenic mice, *Proc. Natl. Acad. Sci. USA*, 93, 6191, 1996.
104. Meyers, E. et al., An Fgf8 mutant allelic series generated by Cre- and Flp-mediated recombination, *Nat. Genet.*, 18, 136, 1998.
105. Nagy, A. et al., Dissecting the role of N-*myc* in development using a single targeting vector to generate a series of alleles, *Curr. Biol.*, 8, 661, 1998.
106. Feng, T.-Q. et al., Site-specific chromosomal integration in mammalian cells: highly efficient CRE recombinase-mediated cassette exchange, *J. Mol. Biol.*, 292, 779, 1999.
107. St-Onge, L. et al., Temporal control of the Cre recombinase in transgenic mice by a tetracycline responsive promoter, *Nucl. Acids Res.*, 24, 3875, 1996.
108. Kistner, A. et al., Doxycycline-mediated quantitative and tissue-specific control of gene expression in transgenic mice, *Proc. Natl. Acad. Sci. USA*, 93, 10933, 1996.
109. Danielian, P.S. et al., Modification of gene activity in mouse embryos in utero by a tamoxifen-inducible form of Cre recombinase, *Curr. Biol.*, 8, 1323, 1998.
110. Gorski, J.A. and Jones, K.R., Efficient bicistronic expression of *cre* in mammalian cells, *Nuc. Acids Res.*, 27, 2059, 1999.
111. Tucker, K.L. et al., Germ-line passage is required for establishment of methylation and expression patterns of imprinted but not of nonimprinted genes, *Genes Dev.*, 10, 1008, 1996.
112. Valancius, V. and Smithies, O., Testing an "in-out" targeting procedure for making subtle genomic modifications in mouse embryonic stem cells, *Mol. Cell Biol.*, 11, 1402, 1991.
113. Hasty, P. et al., Introduction of a subtle mutation into the *Hox-2.6* locus in embryonic stem cells, *Nature*, 350, 243, 1991.
114. Wu, H. et al., Double replacement: strategy for efficient introduction of subtle mutations into the murine *Colla-1* gene by homologous recombination in embryonic stem cells, *Proc. Natl. Acad. Sci. USA*, 91, 2819, 1994.
115. Liu, X. et al., A targeted mutation at the known collagen cleavage site in mouse type I collagen impairs tissue remodeling, *J. Cell. Biol.*, 130, 227, 1995.
116. Askew, G. et al., Site-directed point mutations in embryonic stem cells: a gene-targeting tag-and-exchange strategy, *Mol. Cell Biol.*, 13, 4115, 1993.
117. Claude, P.A. et al., Mutation of a conserved serine in TM4 of opioid receptors confers full agonistic properties to classical antagonists, *Proc. Natl. Acad. Sci. USA*, 93, 5715, 1996.
118. Yang, W. et al. Study of S196A mutation of mu opioid receptor using knock-in strategy, *Int. Narcotics Res. Conf.*, Seattle, WA, 31, 57, 2000.

119. Storck, T. et al., Rapid construction in yeast of complex targeting vectors for gene manipulation in the mouse, *Nuc. Acids Res.*, 24, 4594, 1996.

120. Westphal, C.H. and Leder, P., Transposon-generated "knock-out" and "knock-in" gene-targeting constructs for use in mice, *Curr. Biol.*, 7, 530, 1997.

121. Jacks, T. et al., Tumour predisposition in mice heterozygous for a targeted mutation in Nf1, *Nat. Genet.*, 7, 353, 1994.

122. Carmeliet, P. et al., Abnormal blood vessel development and lethality in embryos lacking a single VEGF allele, *Nature*, 380, 435, 1996.

123. Banbury Conference, Mutant mice and neuroscience: recommendations concerning genetic background, *Neuron*, 19, 755, 1997.

124. Markel, P. et al., Theoretical and empirical issues for marker-assisted breeding of congenic mouse strains, *Nat. Genet.*, 17, 280, 1997.

125. Lander, E.S. and Schork, N.J., Genetic dissection of complex traits, *Science*, 265, 2037, 1994.

126. Gerlai, R., Gene-targeting studies of mammalian behavior: is it the mutation or the background genotype?, *Trends Neurosci.*, 19, 177, 1996.

127. Wolfer, D.P. et al., Assessing the effects of the 129Sv genetic background on swimming navigation learning in transgenic mutants: a study using mice with a modified β-amyloid precursor gene, *Brain Res.*, 771, 1, 1997.

128. Upchurch, M. and Wehner, J.M., Differences between inbred strains of mice in Morris water maze performance, *Behav. Genet.*, 18, 55, 1988.

129. Willot, J.F., Effects of aging, hearing loss, and anatomical location on thresholds of inferior colliculus neurons in C57BL/6 and CBA mice, *J. Neurophysiol.*, 56, 391, 1986.

130. Schwegler, J. and Lipp, H.-P., Hereditary covariations of neuronal circuitry and behavior: correlations between the proportions of hippocampal synaptic fields in the regio inferior and two-way avoidance in mice and rats, *Behav. Brain Res.*, 7, 1, 1983.

131. Mogil, J.S. and Wilson, S.G., Nociceptive and morphine antinociceptive sensitivity of 129 and C57BL6 inbred mouse strains: implications for transgenic knock-out studies, *Eur. J. Pain*, 1, 293, 1997.

132. Belknap, J.K. et al., Localization to chromosome 10 of a locus influencing morphine analgesia in crosses derived from C57BL/6 and DBA/2 strains, *Life Sci.*, 57, 117, 1995.

133. Schauwecker, P.E. and Steward, O., Genetic determinants of susceptibility to excitotoxic cell death: implications for gene targeting approaches, *Proc. Natl. Acad. Sci. USA*, 94, 4103, 1997.

134. Choi, D.W., Background genes: out of sight, but not out of brain, *Trends Neurosci.*, 20, 499, 1997.

135. Wahlsten, D. and Bulman-Fleming, B., Retarded growth of the medial septum: a major gene effect in acallosal mice, *Dev. Brain Res.*, 77, 203, 1994.

136. Wahlsten, D. and Schalomon, P.M., A new hybrid mouse model for agenesis of the corpus callosum, *Behav. Brain Res.*, 64, 111, 1994.

137. Livy, D.J. and Wahlsten, D., Tests of genetic allelism between four inbred mouse strains with absent corpus callosum, *J. Hered.*, 82, 459, 1991.

138. Crusio, W.E., Gene-targeting studies: new methods, old problems, *Trends Neurol. Sci.*, 19, 186, 1996.

139. Kelly, M.A. et al., Locomotor activity in D2 dopamine receptor-deficient mice is determined by gene dosage, genetic background, and developmental adaptations, *J. Neurosci.*, 18, 3470, 1998.

140. Bunzow, J.R. et al., Characterization and distribution of a cloned rat μ-opioid receptor, *J. Neurochem.*, 64, 14, 1995.

4 Animal Models of Pain

Gary J. Bennett

CONTENTS

0-8493-0035-5/01/$0.00+$1.50
© 2001 by CRC Press LLC

4.1 INTRODUCTION

Experimental models of pain in animals presuppose that we can measure pain in animals. Can we do this, and do such measures have any value in the human context? The problem is fundamentally similar to the question of whether we can measure pain in a human being.

Philosophers tell us that pain, indeed any sensory experience, is a private and subjective phenomenon, and thus cannot be measured. Strictly speaking, this is true (at least for now), but strictly speaking, it is not very important. The human subject can report his sensations to us. He does so with an act, some sort of behavior — the spoken word, a pencil mark on a ruled line, etc. This act is a public and objective phenomenon that is measurable to any degree of precision for which we wish to pay. The relationship between the private subjective experience and the public objective act is the field of study of psychophysics. Beginning with Fechner and Weber, over a century of research leaves no doubt that psychophysical relationships are lawful and of great practical usefulness. It may well be that my optometrist does not have, and can never have, certain knowledge of my subjective visual sensations. Nevertheless, he supplied me with an undeniably useful set of bifocals. He did so by using a simple series of psychophysical forced-choice paradigms (holding two lenses in succession before my eye and asking which gives the sharper image). We can measure human sensation, not with absolute certainty, but in a practical and useful way.

What then of measuring sensation in an animal? The optometrist's procedure is based on the implicit assumption that my private subjective experience (a "sharper" image) is the same as what he would experience under the same circumstances. We constantly make this assumption — that other people are experiencing the world in the same way that we do. We realize that there are exceptions (the other person might be color blind, or hallucinating), but we know that the assumption is practical; it is true enough (imagine driving on the highway if it were not). We assume that other people see like us because they look like us. Rats do not look like us. Can we make the assumption that a rat's private and subjective experience is like ours? In its broadest sense, the question is difficult to answer and depends on exactly what kind of sensory experience we are discussing. But let us consider just pain sensation.

It is not particularly difficult to do psychophysical experiments on pain sensation in a rat. For example, we can measure the rat's pain threshold by determining the minimum amount of heat applied to the bottom of its paw that evokes the hind limb withdrawal response. We find that the average rat heat-pain threshold is about 45°C. We could use the identical equipment and procedure to apply heat to our fingertip and we would observe a similar withdrawal response with a threshold that was very close, if not identical, to the rat's threshold. The threshold value is not arbitrary; it is chemistry. The threshold for denaturation of many proteins is 45°C; it is the temperature at which tissue damage begins. This highlights an important point: Under normal circumstances, the sensation of pain is tightly (although not perfectly) related to tissue damage. It is reasonable to argue that this relationship has obvious evolutionary value. It is also an obviously primitive relationship (one can pinch a worm and evoke a withdrawal response) that is likely to be highly conserved in man, rat, other mammals, and probably in all animals with a nervous system. Even bacteria will swim away from an environment that is injuriously hot or acidic.

There is pharmacological evidence that argues for the similarity between pain in man and other mammals. This evidence is especially persuasive because it is so arbitrary (why would rats and men show the same response to morphine if they did not share some deep similarity in the way their nervous systems respond to tissue injury?). One can rank order the potency of a large chemical series of opioids by their ability to raise the threshold of the tail-withdrawal response (the tail-flick) in mice and compare it to the rank order of the drugs' potencies in suppressing human pain. The two orderings will be nearly identical. One could do the same for the potency of non-steroidal antiinflammatory drugs (NSAIDS) using tests of the withdrawal response evoked by squeezing an animal's inflamed paw. Data from exactly these kinds of experiments on pain sensation in rodents form the foundation of today's multi-billion dollar pharmaceutical business in analgesics.

We can conclude that while we may never know with certainty a rat's subjective experience of pain, we can measure the animal's behavioral response to stimuli that cause pain to us, and such measurements yield practical information that is of use in the human context. In other words, we can measure acute stimulus-evoked pain in animals. Note the distinction between acute stimulus-evoked pain and acute ongoing or spontaneous pain. In addition, note that the conclusion refers to acute "normal" pain, by which I mean pain that is evoked in an animal that is free of pathology. Can we reach the same conclusion for pain that is chronic, or for the abnormal (neuropathic) pain that we see when the nervous system is damaged or dysfunctional?

Neither human beings nor animals have reflexes or other easily measured behaviors associated with chronic ongoing pain. Heart rate, blood pressure, plasma corticoid concentrations, and dozens of other physiological variables have been examined and none has been found to have any useful relationship to chronic pain. The human animal responds to chronic pain with certain general behavioral tendencies: loss of appetite, fitful sleep, and reduced activity; but these are pronounced only when the pain is quite severe, and even then can be quite variable. There is good evidence that chronic pain produces the same general behavioral tendencies in animals (excepting disturbed sleep, for which I am not aware of a single investigation), but again they appear to be pronounced only when the pain is severe and are highly variable. These general behavioral tendencies have not proven to be useful experimental measures of chronic pain in animals or man. In the human being there appears to be a nearly universal change in mood: The patient becomes irritable, depressed, and pre-occupied with his symptoms. Clinical studies of chronic pain use paper and pencil measures such as the Minnesota Multiphasic Personality Inventory (MMPI) to quantify these changes. More often, clinical studies rely on the patient's verbal report of the intensity of his chronic pain but, of course, animals cannot speak or complete the MMPI.

All of these difficulties arise from the emphasis on chronic ongoing pain. Animal models of chronic inflammatory pain (see below) are straightforward when we emphasize the incident pain that accompanies chronic pain states. For example, the patient with osteoarthritis may experience relatively little pain at rest, but suffer when he attempts to move the joint or when an experimenter squeezes the joint. We can, and do, measure this incident or stimulus-evoked pain in the arthritic rat. The rat shows the same phenomenon as the human patient — a lowered pain threshold for pressure applied across the joint, and its threshold can be normalized with an opiate or an NSAID. Thus, the argument is the same as that given above for acute stimulus-evoked pain in the normal animal. In addition, if the animal has the same stimulus-evoked pain abnormality, then it seems reasonable to assume that the animal also experiences the same ongoing pain, even if this is difficult or impossible to measure.

Neuropathic pain might seem to be a special case because the pain is by definition abnormal. In fact, patients insist that it is abnormal; they tell us that the pain is strange and qualitatively different from the pain that they have experienced previously. Neuropathic pain is chronic, and there are both abnormal ongoing or spontaneous pains and abnormal stimulus-evoked pains.[1] Ongoing neuropathic pain is as difficult to measure in the animal as any other chronic ongoing pain. But neuropathic stimulus-evoked pains present no special problems. For example, one of the most debilitating problems seen in the neuropathic patient is pain evoked by gentle touch. This is called mechano-allodynia and it is not infrequently so severe that the patient cannot tolerate the contact of his skin with clothing or bed linens. We can easily measure the rat's mechanical pain threshold by touching the glabrous skin with von Frey hairs (a series of successively stiffer nylon monofilaments). The pressure exerted by the hair that produces the withdrawal reflex defines the animal's pain threshold. Prior to peripheral nerve injury we find that the withdrawal reflex is evoked by a hair that produces a slight pricking sensation when applied to our volar wrist (where the skin is approximately the same thickness as that on the rat's plantar hind paw). Following nerve injury, the animal withdraws when stimulated with weaker hairs, hairs that it had previously ignored. These weaker hairs elicit a clearly innocuous tactile stimulus when applied to our own skin, but evoke pain in the patient with neuropathy. We can apply heat to the skin of the neuropathic rat and show that the withdrawal threshold is clearly lower than normal, and we can show exactly the same thing with many neuropathic patients (not all, because heat-hyperalgesia is present in only a subset of patients). We may thus extend the argument given above: Normal people and normal rats act the same way when stimulated with a painful stimulus, and neuropathic people and neuropathic rats act the same way when stimulated with painful stimuli and previously innocuous stimuli. Here, pharmacology supports the argument even more strongly. Drugs (e.g., gabapentin) that have no effect on normal acute or chronic pain in man or rat suppress neuropathic pain in both.

In summary, the measurement of pain in animals is both possible and useful (albeit imprecise). Our most satisfactory knowledge comes when we use psychophysical methods to examine stimulus-evoked pain. One might note here that the study of clinical pain often relies on psychophysical measurements that are not clearly defined in terms of ongoing pain vs. stimulus-evoked pain. In the typical study, the patient is given drug or placebo and then periodically asked to rate his pain on a visual analog scale marked 1 = no pain to 10 = worst pain ever. What are the patients doing? Are they rating their ongoing pain? The stimulus-evoked pain that they encounter with the activities of daily life? If their ongoing pain was reduced, might they have become more active and encountered more stimulus-evoked pain (imagine the low-back pain patient feeling well enough to try a round of golf)? If they are rating some sort of average amount of ongoing and stimulus-evoked pain experienced during the previous day or week, what kind of arithmetic do they use? I do not mean to imply that such studies give conclusions that are wrong (in fact, they work surprisingly well), but rather to point out that we often obtain better pain measurements from rats than from people!

4.2 EXPERIMENTAL METHODS FOR MEASURING PAIN IN ANIMALS

It is traditional to divide pain models into "acute" and "chronic" categories. The distinction is of course somewhat arbitrary. In the summary that follows, "acute" refers to pain that lasts for seconds to about a day; "chronic" refers to pain that lasts for at least several days. It should be noted that there is also a somewhat arbitrary distinction between acute and chronic when discussing clinical pain states and that the definition of chronic is rarely congruent with the experimental situation. For example, one's chest will hurt for a week or more after a thoracotomy, but this would not be considered a chronic clinical pain syndrome. The clinic also presents very common chronic conditions that present as repeated episodes of pain, e.g., migraine. It is noteworthy that we have no animal model for any sort of headache.

In theory, one could produce any sort of injury to any body part in the animal and declare that one had a pain model (an impression that one sometimes gets when following the scientific literature). It is easy to be critical of this trend, but it is also important to note that pain from different causes and from different tissues may be dissimilar in important ways. The case for such differences is particularly strong for visceral pain,[2] and an excellent example of the importance of the distinction is the recent discovery that abdominal pain may be uniquely modulated by drugs that block a κ opioid-like receptor.[3] No attempt will be made here to list all proposed models of pain. Instead, I focus on those that are either especially well studied, or present with unusual features.

4.3 ACUTE STIMULUS-EVOKED PAIN

The same tests may be used to measure stimulus-evoked pain in the normal animal and in animals with chronic inflammatory or neuropathic pain. Tests of acute cutaneous pain rely on either the withdrawal reflex or the animal's vocalization (squeaking or squealing) as the index of the pain threshold. Vocalization clearly requires stronger stimulus intensities and is thus probably a measure of suprathreshold pain. However, rats can vocalize at frequencies that are considerably above what human beings can hear.[4] It is not known whether pain-evoked high-frequency vocalization has the same threshold as those that we can hear.

Note that all the tests of stimulus-evoked pain allow the animal a significant amount of control over the duration of the pain. The animal's pain response terminates the stimulus. The tests thus meet one of the important criteria of the International

Association for the Study of Pain's guidelines for ethical experimentation.[5] Note also that when testing potential analgesics a cut-off latency is used to prevent tissue damage. Additional details on these tests can be found in a recent methodological review.[6]

4.3.1 THE TAIL-FLICK TEST

The classic pain test in animals is the tail-flick test of D'Amour and Smith.[7] The tip of the tail is heated by a light bulb shining through a small aperture, and the withdrawal response is detected and timed by a photocell circuit. The withdrawal is a classic nocifensive reflex that removes the body part from the source of pain. It is organized at the spinal level, but it is under supraspinal control. The intensity of the heat stimulus is adjusted at the beginning of the experiment to yield withdrawal latencies of about 7 to 10 s. Stronger intensities (shorter latencies) create a "floor" effect that makes it difficult to detect hyperalgesia. Weaker intensities (longer latencies) make it difficult to detect analgesia, and increase the probability of a spurious non-reflexive tail movement. It is traditional to blacken the tip of the tail to reduce variability due to light reflection.

There are four oddities about the tail-flick. First, the initial trial of a series always yields an abnormally long latency. To my knowledge, the reason for this has never been determined. Second, apart from the first trial anomaly, successive tests in a series, or successive series, do not show any sign of a learning curve, provided that the stimulus intensity and inter-trial interval are adjusted to prevent minor tissue injury and primary afferent sensitization. When the stimulus is applied to our fingertip the pain threshold is clearly preceded by a sensation of increasing heat ("warm, warmer, hot, hotter, pain"). Thus, cues are available for the rat or mouse to learn to avoid the approaching pain, but they never seem to learn. The test does not work in a monkey, who learns to avoid the pain after a single trial or two. Third, it has been shown to be very important to control for the temperature of the tip of the tail.[8] A warmer tail takes less time to heat to the pain threshold, simply because it starts at a higher temperature; of course, a colder tail takes more time. This confound is likely to be present in drug trials where the drug, but not the vehicle control, causes vasodilatation or vasoconstriction. The solution is to pre-warm all tails to the same temperature prior to testing (about 32°C ought to be sufficient, as this is about as warm as a tail normally gets). The fourth oddity is of unknown significance, but it should be noted that the skin of a rat or mouse's tail is quite unusual. The tail's skin is distinctly scaly (keratinized) and covered with very short, thick, conical hairs that are unlike those found anywhere else.

4.3.2 THE HOT-PLATE TEST

This test uses a metal plate, usually made of copper, which is heated by hot water circulating beneath it. Experience has shown that a plate temperature of 50 to 55°C is most useful. The animal is placed on the hot-plate and the latency until pain-related behavior occurs is measured with a stopwatch. There are several problems with the hot-plate test. Most importantly, it is difficult to standardize what constitutes a pain-related behavior: Licking the paws? Rearing and trying to climb out of the cylinder used to confine the animal on the plate? Standing with one paw in the air? Another difficulty is that the heat stimulus is not delivered in a controlled fashion; it will vary from area-to-area and from paw-to-paw depending on whether the animal is standing still, walking, running, etc.

The hot-plate test is said to evoke behaviors that are integrated supraspinally (a distinction that is also made for vocalization). This is true, but it is not correct to contrast this with the tail-flick response (or any other withdrawal reflex) and say that one is comparing spinally mediated with supraspinally mediated pain. The withdrawal reflex is spinally mediated, but modulated supraspinally. The tail-flick threshold is significantly shorter following spinalization.

4.3.3 THE FORMALIN TEST

The formalin test is a frequently used pain model that involves the injection of dilute formalin beneath the skin of the hind paw. It was first introduced by Dubuisson and Dennis in 1977[9] and has attracted enduring interest because of the unusual biphasic (actually triphasic) pattern of responses that it evokes.[10,11] Within seconds, the rat begins to flinch (or "flick") the paw rapidly in an agitated manner and frequently licks the paw as well. The frequency of flinches and licking is initially high, but after about 5 min they stop temporarily. This is called the Phase I response. It is generally believed to reflect acute nociceptive (i.e., normal) pain caused by formalin-evoked discharge in C-fiber nociceptors. It is suppressed by traditional analgesics, while drugs that work for neuropathic pain (e.g., gabapentin and N-methyl-D-aspartate receptor antagonists) have little or no effect. In the normal rat, Phase I is followed by several minutes with little or no flinching, and this is called the Q (quiescent) Phase. Subsequently, the rats again begin to flinch at high frequency for another 10 to 15 min or more. This is called Phase II and it is generally assumed to be due to mechanisms similar to those of chronic and/or neuropathic pain, in particular to changes in the responsiveness of spinal cord nociceptive neurons (central sensitization) that are evoked by the C-fiber discharge during the initial phase. Drugs that suppress Phase II response have little or no effect against normal acute pain but suppress neuropathic pain. Several methods for scoring the behaviors evoked in the formalin test have been described.[9,11,12]

The formalin test is generally administered to normal animals, but it has been shown that diabetic rats respond differently.[13,14] Rats made diabetic via an injection of streptozotocin have a greater than normal frequency of flinching in Phase I and Phase II, and they continue to respond at a fairly high rate during what is ordinarily the Q Phase.

4.3.4 THE PAW-FLICK TEST

Hargreaves et al.[15] made a simple modification to the tail-flick test such that the plantar hind paws could be tested. The rat or mouse is confined beneath a small cage made of clear plastic set atop an elevated floor made of ordinary window glass. A source of radiant heat is placed beneath the floor such that the heat is directed to the plantar surface of the paw through a small aperture. As in the tail-flick test, the heat source is usually the light bulb from a 35-mm slide projector; the heat intensity is controlled with a rheostat. A light guide (a length of hypodermic tubing or optical fiber) is positioned at the corner of the aperture. Light reflected from the paw is thus directed onto a photocell that stops the stimulus and a timer when paw withdrawal interrupts the reflection.

This minor modification of the tail-flick has proven to be immensely useful. Unlike the tail-flick test, both hind paws can be tested. If only one is inflamed or nerve-injured, the other can serve as a control (the side opposite a nerve injury may, in fact, not be normal).[6] For the tail-flick test, the animal is restrained by wrapping it in toweling or placing it within a narrow plastic cylinder, both of which may be presumed to be stressful because the animals frequently struggle against the restraint. For the paw-flick, the animal is confined beneath an inverted plastic cage that allows some room for it to walk about. Once habituated to this confinement it appears that the animal is not stressed (in fact, one often has to wake up the animal to perform a trial). Unlike the scaly tail skin, the plantar skin of a rat or mouse paw is typical glabrous skin, excepting the cornified tori (foot pads). Unlike in the hot-plate test, the heat is delivered with considerably more precision and to a defined location. There is almost no uncertainty about the behavior to be measured. "Almost" because the animal sometimes steps away from the light; but if the response is observed carefully, it is nearly always possible to differentiate stepping from reflexive withdrawal.

As with any procedure, the paw-flick has its fine points. The first trial anomaly noted above occurs, curiously, in each paw when the animals are tested in succession. The temperature of the glass must be warmed to about 30°C, not only to avoid the confound of differing skin temperature, but also to prevent paw cooling (the large expanse of glass is a considerable heat-sink). It is also essential to be certain that the skin of the paw is contacting the glass. If it is even very slightly elevated above the glass, it will heat more quickly than when it is in contact, again because of the glass's heat-sink effect.[16,17]

4.3.5 IMMERSION TEST FOR THERMAL HYPERSENSITIVITY

It is possible to determine the heat–pain threshold for the paw or tail by immersing it into hot water bath(s) and timing the latency to withdrawal, or the percentage of trials that evoke a response. The cold–pain threshold can be measured similarly with a cold water bath(s). The immersion test has the distinct advantage of requiring no special equipment other than a thermometer, but there are two disadvantages. First, it is cumbersome to test successively hotter baths to find the threshold, and thus the typical endpoint is the latency to withdraw from a suprathreshold temperature. Second, the animal is usually held suspended in air, without its paws being in contact with a substrate. This evokes a strong extensor reflex in all four limbs (the "I'm falling" posture); one is thus trying to evoke a flexor withdrawal reflex on top of an extensor reflex. This may be a particularly important problem when testing nerve-injured rats. Testing chronic constriction injury (CCI) rats with both the paw-flick and paw-dipping methods (in counterbalanced order) gives different results: clear heat-hyperalgesia onset on the second day post-injury with the paw-flick, but with the dipping method for the very same rats the onset is not apparent until 5 to 7 days post-injury.[18]

4.3.6 COLD-ALLODYNIA TESTS

The immersion method can also be used to detect cold-allodynia, i.e., a withdrawal response to a cold temperature that does not normally evoke a withdrawal. Another method places the rat on a metal floor cooled by cold water circulating beneath it. Again, normal rats simply walk about, but nerve-injured rats hold the ipsilateral paw off the floor.[19] This method has been criticized because it exposes the rat to a cold environment that is likely to activate the sympathetic nervous system.[20] A superior method has the rat standing on an elevated wire mesh floor. A drop of acetone is placed on the plantar skin and produces a distinct cooling stimulus as it evaporates.[20] Normal rats either ignore this stimulus, or respond with a flick of the paw that is very small in magnitude and very short in duration. However, nerve-injured rats almost always respond; the response is clearly exaggerated in magnitude and duration, and often presents as repetitive withdrawals. These responses can be simply counted or their durations measured.

4.3.7 THE PIN-PRICK TEST FOR MECHANO-HYPERALGESIA

With the animal confined beneath a clear plastic cage atop an elevated mesh floor, the point of a safety pin is pressed slowly against the plantar skin until a dimple is seen. This is exactly the pricking pain test of the bedside neurological exam. The pin is not jabbed and there is no penetration of the skin; a hypodermic needle may not be used because it is too sharp. The normal rat responds with a small brisk withdrawal reflex. The hyperalgesic animal responds with a withdrawal that is clearly amplified in both magnitude and duration.[21]

4.3.8 THE RANDALL-SELLITO PAW PRESSURE TEST

This is an old method[22] that uses a motor-driven levered piston that is slowly advanced against the paw or tail, which is supported on an underlying hard surface. The end of the piston that contacts the skin is usually a conical blunt-tipped stylus, but a narrow rectangular plinth with rounded edges may be set at a right angle across the tail.[23] The pressure is increased gradually at an even rate until the animal tries to withdraw from the stimulus. This device was originally designed to measure the mechanical pain threshold in inflamed and swollen hind paws. When the paw is not swollen, the placement of the stylus can cause large differences; e.g., different thresholds are obtained when the piston is squeezing the soft tissues between the metatarsals or when the pressure is applied directly above a bone or joint. For reasons that are not clear, mechanical pain thresholds determined with this device are initially relatively high and variable, but considerably lower and more stable with "training."[24]

It is not always clear whether the hyper-responsiveness shown in this test with animals with inflammatory or neuropathic pain states should be called mechano-hyperalgesia or mechano-allodynia (the same problem occurs with some other tests), and there is no agreement in the literature as to which label is best. The official definition of hyperalgesia is "pain of abnormal intensity evoked by a normally noxious stimulus," while allodynia is "pain evoked by a normally innocuous stimulus."[25] The problem in applying these definitions lies in defining what is normal. The normal pain threshold is a distribution of values (even though we customarily speak of it in terms of a point estimate such as the distribution's mean). Strictly speaking, a hyperalgesic threshold would be one with a distribution that partly overlapped the normal pain threshold distribution, while an allodynic threshold would not have any overlap. In practice, the distributions are almost never sampled extensively, and when they are, they are found to be rather broad. A reasonable solution might be to arbitrarily use "hyperalgesia" when the new mean threshold differs from the normal mean threshold by less than 3 standard deviations, and "allodynia" when the difference is greater.

4.3.9 VON FREY HAIR TEST FOR MECHANO-ALLODYNIA

von Frey used a series of actual hairs: stiff bristles from a hog's snout, wispy hairs from a squirrel's tail, etc. Today the hairs are a series of nylon monofilaments of different thicknesses that exert varying degrees of pressure when pressed against the skin to the point where they bend. The series of von Frey hairs that is nearly always employed is the one developed by Semmes et al.[26] to examine tactile sensation in patients with brain damage. Several procedures are used, and there is no clear evidence that one is superior. In the first procedure, testing begins with a hair in the middle of the series. If the animal withdraws from the stimulus, the next thinner hair is tested; if it does not, the next stiffest hair is tested, and the procedure repeated until the threshold is determined.[21] The second method employs the up-down method of Dixon, where the threshold is interpolated from the pattern of responses vs. non-responses to stimuli in the vicinity of the threshold.[27] The third method[28] tests 3 to 4 different hairs representing the range of available stimuli (e.g., weak–medium–strong) and tabulates the percentage of responses to 10 applications of each stimulus.

4.3.10 THE WRITHING TEST

The oldest test for acute visceral pain uses an intraperitoneal injection of dilute acid (usually acetic acid) in mice. The response that is evoked is generally (but not very accurately) called writhing. The mouse responds with strong contractions of the abdominal muscles, and the animal may sit in a hunched posture, or lie on its side and roll from side-to-side (the writhing). These responses are not unlike those that would be seen in a person with severe colic or other abdominal pain. The writhing test has been justifiably criticized because an injection into the peritoneum exposes both visceral and somatic tissues, and because the stimulus is not natural.

4.3.11 DISTENSION OF A HOLLOW VISCUS

More recent tests of acute visceral pain have employed more natural stimuli that are clearly confined to viscera; they have used a variety of outcome measures. For example, a latex balloon can be inserted into the rat's vagina or uterus and inflated to produce noxious distension. Berkley et al.[29] have shown that rats trained to bar press to avoid noxious heat stimulation to the tail will subsequently bar press in order to terminate strong distension of the vagina. In contrast, only a few rats will bar press in response to severe uterine distension. The difference probably reflects the different physiological states of the vagina and uterus in the non-pregnant animal. As one might expect, the response thresholds differ during different stages of the estrous cycle.[30]

A very thorough series of experiments have validated the use of colorectal distension in the rat as a model of visceral pain.[31,32] Strong distension produces behavioral signs of pain (contraction of the abdominal wall) and animals will learn a passive avoidance response to terminate it. It also produces a cardiovascular pressor response together with tachycardia (the only instance to my knowledge where a sympathetic nervous system response has been shown to be a useful pain measure). Abdominal contraction and the cardiovascular response occur at stimulus intensities that are clearly greater than those necessary to activate the anal sphincter, showing that the responses are not likely to be due simply to stimulus-evoked defecation.

4.3.12 MUSCLE PAIN

Pain and tenderness in muscle is a common human complaint, but there has been practically no experimental analysis of the problem. Kehl et al.[33] have devised a novel model of this condition that is noteworthy for the way in which pain is measured. Carrageenan, a glycoprotein from seaweed that is a commonly used inflammatory stimulus, is injected into the muscle of one of the rat's forelimbs. Muscle pain is tested by having the rat grasp two handles, one for each forepaw, that are connected to strain gauges. Grasping the tail, the rat is pulled away from the handles until one or both forepaws releases its grip. Prior to intramuscular inflammation, the rat releases both forepaws with nearly the same threshold, but after inflammation it nearly always releases first on the inflamed side. The threshold on the inflamed side is normalized by opiates and NSAIDs. The assumption in this test is that the decreased threshold for release on the inflamed side is due to the pain caused by sustained effort (muscle contraction) in the presence of inflammation, which is what would happen with a sore human muscle.

4.4 MODELS OF CHRONIC INFLAMMATORY PAIN

4.4.1 ADJUVANT-INDUCED ARTHRITIS

An intravenous injection of Complete Freund's Adjuvant (CFA, a mineral oil emulsion containing heat-killed tuberculosis bacteria) produces a multi-joint arthritis. The joint inflammation occurs with a delay of several days, it is progressive, and it leads to dramatic swelling and permanent destruction of the joint tissues. Work with this model has been reviewed thoroughly elsewhere.[34] In brief, there is little doubt that the animals have chronic pain. They limp, as would an arthritic person, and they have a clearly lowered threshold for limb withdrawal and vocalization when pressure is applied across a joint. The adjuvant-induced arthritis model is seldom used in pain research today. The objection has been that, unlike a person with rheumatoid or osteo-arthritis, the rats have an obvious systemic illness. They lose weight, walk and sit with an abnormal hunch-back posture, and have piloerection even when the ambient temperature is warm (presumably due to fever).

4.4.2 Unilateral Arthritis

The model of systemic arthritis uses an intravenous injection of CFA, a similar subcutaneous injection given to one hind paw evokes a unilateral inflammatory response that lasts for 10 days or longer.[35,36] "Arthritis" is used correctly to describe this response because there is extensive damage to the toe and ankle joints, but there is also a very pronounced edema in the paw that is not typically seen with osteo- or rheumatoid arthritis. The skin of the injected paw becomes extremely sensitive to heat and mechanical stimulation; pain sensitivity on the other hind paw is normal (or, perhaps, approximately normal). This model has proven to be very useful for the analysis of several phenomena. For example, because the pain hypersensitivity begins within just a few hours after the injection, it is possible to record from individual dorsal horn neurons before the injection and then follow their responses as hyperalgesia develops. It has been shown that many pain-responsive dorsal horn neurons develop progressively enlarged receptive fields as hyperalgesia progresses.[37]

4.4.3 Inflammation of a Hollow Viscus

Inflammatory and irritant stimuli have been instilled into several organs to produce pain.[38-40] Most studies have examined the uterus, colon, or urinary bladder. The most commonly used stimulus has been mustard oil (n-ally-isothiocyanate) or turpentine, both of which produce a clear inflammatory response with an onset of several hours. If the animals are allowed to survive, these subsequent inflammatory pain states last for several days and the models can thus be considered to reflect acute to semi-chronic pain. As in man, inflammation of the urinary bladder causes hyperexcitable reflexes to bladder distension, spontaneous bladder contractions, and problems with micturition. The animals are also hypersensitive to stimulation of the lower abdomen and rectal areas, a phenomenon that may reflect the referred pain that is commonly noted in the clinic.

4.4.4 Ureteral Calculosis

Passing a kidney stone can be one of life's least pleasant surprises, and having a stone lodged in the ureter can be torture. Giamberardino and her colleagues[41] have reproduced calculosis in the rat by injecting the ureter with a small amount of dental acrylic cement (useful in this context because it will set when wet). Continuous videotaping over periods of 4 to 14 days revealed that the animals had repeated episodes (as many as 60) of the writhing behaviors like those seen after intraperitoneal acid injections, with each episode lasting a minute or so, or up to 45 min. These episodes were most common and most severe during the initial 3 days. In addition, the animals had decreased vocalization thresholds to electrical stimulation of the abdominal wall, and there was a strong positive correlation between the severity of writhing and the amount of decrease in the pain threshold. Thus, these animals appear to have exactly the sort of "spontaneous" and referred pain that one sees in a human patient with ureteral calculosis.

4.5 NEUROPATHIC PAIN MODELS

One would expect that damage to a sensory nerve would produce sensory loss, with the degree of loss roughly proportional to the degree of damage. This is what usually happens. But when peripheral somatosensory nerves are damaged a small percentage of cases develop positive symptoms, and these are almost always various kinds of pain. Such pains are said to be neuropathic because they are believed to be due to a dysfunctional nervous system. Painful peripheral neuropathy ranges from mild and dysesthetic to excruciating torture. The conditions are usually chronic and often fail to respond to any of the therapies that we have today. Neuropathic pain also sometimes follows injury to the CNS, particularly when the injury involves the pain processing circuits. This is called central pain (formerly, and erroneously, the thalamic pain syndrome). Central pain has also been modeled in animals; the reader is referred elsewhere for a review of this topic.[1]

Several types of painful peripheral neuropathy are distinguished in the clinic. It is suspected, but not proven, that they all share at least some underlying pathogenic mechanisms. Post-traumatic painful peripheral neuropathy is a general term given to the pain that arises when nerves are damaged by penetrating wounds, crush and stretch injuries, surgical misadventure, etc. Several diseases produce nerve damage which sometimes results in neuropathic pain; chief amongst these are postherpetic neuralgia and painful diabetic neuropathy. Painful peripheral neuropathy is seen in the cancer patient when tumors compress or stretch a nerve, or when nerves are poisoned by anti-neoplastic drugs like paclitaxel and vincristine. Patients develop painful peripheral neuropathy following injury to the dorsal roots in the lumbosacral spine. The number of chronic low-back pain patients with neuropathic pain is unknown, but the number will be over a million even if the percentage is small because chronic low back-pain affects over 20 million Americans.

Work with animal models has contributed greatly to our understanding of the mechanisms that underlie neuropathic pain states. Just as importantly, these models have been used as quick and easy pre-clinical screens for new drugs to treat neuropathic pain. Our increased understanding of the basic pathophysiological processes and the use of an efficient pre-clinical screen have led to the discovery of several new classes of drugs that suppress neuropathic pain. Interestingly, these new drugs have little or no effect on normal acute pain sensation; they would thus have been impossible to discover using the animal models traditionally used to discover analgesics. Indeed, because these drugs affect only abnormal pain, they are best called anti-hyperalgesics and anti-allodynics, rather than analgesics.

4.5.1 EXPERIMENTAL ANESTHESIA DOLOROSA

The experimental analysis of neuropathic pain began with the work of Wall and his colleagues[42] with a model of anesthesia dolorosa. Anesthesia dolorosa refers to the pain that occurs in an intact but totally deafferented body part. The phenomenon is akin to phantom pain, where the nerve injury is due to amputation of the body part. Rats whose hind limbs were completely deafferented by transection of the sciatic and saphenous nerves began to self-mutilate the insensate hind paw.

The self-mutilating behavior, called autotomy, was first noted in rats following transection of multiple dorsal roots.[43] It generally begins with damage to the tips of the toes and then progresses to encompass the entire paw. Autotomy has an average onset of about 20 days after the nerve injury and severe damage to the digits and paw occurs within about 40 to 60 days. The original interpretation was that autotomy is the animal's response to dysesthetic or painful sensations felt in the paw, analogous to the sensations felt in a phantom limb. Presumably, the animal is either trying to rid itself of a persistent source of pain, or the animal is biting the affected region as a person might rub or scratch a painful lesion. Since the paw is insensate, these responses would be unchecked by the pain that scratching or biting would ordinarily evoke, and they would thus proceed to injury.

Several objections were raised to this interpretation.[44] First, it was noted that people with totally deafferented body parts do not self-mutilate. More recent work shows that there are very rare cases where this does occur.[45] In any case, the objection is easily rebutted by noting that unlike the rat, the human being refrains from such behavior because he is aware of the consequences of such actions (or a caretaker is when the patient is a child or mentally retarded person). A more serious objection was raised by the observation that a deafferented male rat would not self-mutilate if housed with a female rat, and would cease ongoing autotomy if he were subsequently housed with a female rat. It is exceedingly difficult to imagine why pain would be modified by the introduction of a female. Phantom pain does not cease in the company of the opposite sex. An alternative interpretation[44] was that the rat's attack was not directed at a source of pain, but rather at an appendage that was useless, or "strange" in the sense of being "non-self." The obvious way to test this interpretation is to anesthetize the paw for a week or two and see whether the absence of sensory input is sufficient to induce autotomy. This is actually quite difficult to do, and such experiments have not provided an unequivocal answer to the question.[46–48]

Work with the complete nerve transection model has been extremely important in showing us the changes that occur in the damaged afferent axons. For example, this model was the first to show clearly that axotomized somatosensory afferent neurons begin to discharge spontaneously and that the discharge originates from at least two ectopic foci: the site of axonal lesion and the parent cell body in the dorsal root ganglion.[49] Work with this model was also the first to show that axotomized primary afferent neurons develop an excitatory response to norepinephrine: a possible substrate for the production and aggravation of neuropathic pain produced by activity from the sympathetic nervous system.[50] The model continues to fascinate and perplex us. For example, Seltzer et al.[51] have shown that a few seconds of painful electrical stimulation of the paw preceding the nerve transection dramatically shortens the latency to autotomy onset and accelerates its progression, although the onset is still delayed for several days after the lesion.

A powerful animal model of a disease state must clearly reproduce the important features of the clinical picture. The complete nerve transection model cannot do this.

Autotomy is objective, but the sensation (or lack of sensation?) that evokes it appears to be impossible to determine with certainty. Moreover, the model's behavioral endpoint is a spontaneous behavior caused by a variable that is unidentified and thus uncontrollable experimentally. What was needed was a model of partial nerve injury, where sufficient innervation survived so that the animal's response to experimental stimuli could be measured. In other words, it was necessary to shift our attention from neuropathic spontaneous pain (as problematic and difficult to measure in humans as it is in rats) to neuropathic stimulus-evoked pains, which are readily measurable and relatively unambiguous.

The first such model[19] was introduced nearly a decade after the seminal contribution of Wall and his colleagues. Current work in the field most often employs this and two other models of pain following partial nerve injury. All three are models of neuropathic pain following partial peripheral nerve damage. All are most closely related to the clinical syndrome of post-traumatic painful peripheral neuropathy, but are believed to have a more general relevance that includes all types of painful peripheral neuropathy.

4.5.2 EXPERIMENTAL MODELS OF PAINFUL PERIPHERAL NEUROPATHY DUE TO TRAUMATIC, PARTIAL NERVE DAMAGE

4.5.2.1 Chronic Constriction Injury

The chronic constriction injury (CCI) model of Bennett and Xie[19] is produced by tieing loosely constrictive ligatures around the rat's sciatic nerve at mid-thigh level. The ligatures evoke intraneural edema; the swelling is opposed by the ligatures, and the nerve self-strangulates. The CCI produces a differential deafferentation: Almost all the large myelinated (Aβ) afferents and a very large majority of the small thinly myelinated (Aδ) afferents are interrupted at the site of nerve injury, but a large percentage of the unmyelinated (C-fiber) afferents survive.[52–54] Thus, the sciatic nerve territory is innervated by a reduced population of C-fibers and a very small remnant of the Aδ fibers; all or almost all of the Aβ low-threshold mechanoreceptors (touch fibers) are disconnected from their peripheral receptors. A neuroma-in-continuity forms at the site of constriction. A similar model evokes the constriction by placing a snugly fitted cuff (a short length of polyethylene tubing) around the nerve.[55] Others have used different types of sutures, but it seems very likely that the immunogenic properties of the chromic gut suture play an important role.[56,57] It has been noted above that the measurement of chronic ongoing pain in an animal is difficult to do and difficult to interpret. However, a rating scale of spontaneous pain behaviors for CCI rats has been shown to be useful.[58]

4.5.2.2 Partial Nerve Transection Injury

The partial nerve transection (PNT) model of Seltzer and his colleagues[59] is produced by piercing the rat's sciatic nerve at the level of the upper thigh with an ophthalmic needle and suture (8-0 silicon-treated silk; 3/8 curved reverse-cutting needle) and tightly ligating (and thus transecting) about 1/3 to 1/2 of the nerve's volume. The

PNT thus produces a partial, but not a differential, deafferentation. The sciatic nerve territory is innervated by a reduced (1/2 to 2/3) population of all fiber types. A neuroma-in-continuity forms at the site of partial transection.

4.5.2.3 Spinal Nerve Transection Injury

The spinal nerve transection (SNT) model developed by Kim and Chung[28] involves tight ligation (and hence transection) of the L5 and L6 spinal nerves close to their respective ganglia. It is not known whether a neuroma-in-continuity forms at the transection sites. This procedure produces a partial, but not a differential, deafferentation of the nerves that have axons traveling in the L5 and L6 roots, chiefly the sciatic and saphenous nerves. The hind paw is innervated by a reduced (approximately 50%) population of all fiber types. The amount of sciatic innervation that is transected varies because of the significant rat-to-rat variation in the percentage of sciatic afferents that come from the L4 ganglion.[60]

4.5.2.4 Comparison of the Models

All three models produce signs of abnormal stimulus-evoked pain that resemble those found in patients.[19–21,28,58,59] Heat-hyperalgesia, mechano-hyperalgesia, mechano-allodynia, and cold-allodynia are all present. Signs of spontaneous or ongoing pain (e.g., limping and guarding the affected hind paw) are also present. The behavioral outcomes of the three models have been compared directly.[61] The peak severity of the abnormal pain responses is remarkably similar in all three models, but times of onset and durations are different.

The onset of abnormal pain in the SNT and PNT models appears to be within hours of surgery, but abnormal pain in the CCI model is not detected until 2 or more days post-surgery. This difference may be of little significance because the actual nerve injury (i.e., the constriction) in the CCI case is not created at the time of surgery; it develops over the course of a day or so as the nerve swells. Abnormal pains are present in the CCI model for 2 to 3 months;[19,58] for unknown reasons, other laboratories report a significantly shorter duration. The abnormal pain syndrome produced by the SNT model lasts for 1 to 2 months, and that of the PNT model lasts for 6 months or more. These are all relatively long durations, considering that the rat's life span is about 2.5 years.

4.5.2.5 A Cautionary Tale

The development of the PNT model has been greatly retarded by a failure to replicate the effect outside of the laboratory where it was initiated. The reason for this has now been identified, at least in part. The story teaches a valuable lesson in the virtues of humility and perseverance.

Soon after the initial publication describing the model,[59] researchers at Johns Hopkins University tried to replicate it, but were unable to do so. One of the model's developers went to Baltimore to help. Frustratingly, he also was unable to replicate it, until he returned to Jerusalem where he could easily replicate the original results. After 4 to 5 years of obtaining consistently good results, the Jerusalem laboratory also began to see effects that were much weaker than normal. Eventually, it was realized that the animal's diet had changed.

Commercial suppliers of rat chow control the diet's composition with regard to percentage of protein, carbohydrates, fat, etc., but the source of the protein varies with availability and cost. The protein comes from various mixtures of soy, alfalfa, fish meal, casein (milk protein), etc. The consumer is generally unaware when the sources or the mix have changed; all that is reported is the percentage protein.

Working together, the Johns Hopkins and Jerusalem laboratories showed that rats fed a diet containing soy as a protein source developed little or no sign of neuropathic pain after PNT, while those fed on a soy-free diet developed the robust symptoms described in the original report.[62] The best results were obtained when the rats were fed bread and fresh cucumbers! The mechanism of the pain suppression by dietary soy is unknown, but suspicion rests with the phytoestrogens found in soy and many other plants.

4.5.3 Idiopathic Trigeminal Neuralgia

Two types of painful peripheral neuropathy are known to present in the distribution of the trigeminal nerve; they are usually, although not always, easily differentiated. The first and best known is tic douloureux: sudden paroxysms of electric shock-like pain that are either apparently spontaneous, or evoked by cutaneous stimulation. It is believed to be due to mechanical irritation within the trigeminal root due to aberrant arterial anatomy. Tic douloureux is unlike other painful peripheral neuropathies in at least two important respects. First, it is the only one with a clear post-ictal period, i.e., once a paroxysm of pain is evoked another cannot be until seconds or minutes have passed. Second, it is the only one that responds very well (in about 50% of patients) to carbamazepine. There is no animal model of tic douloureux.

The second is idiopathic trigeminal neuropathy. Its symptoms are similar, if not identical, to the post-traumatic painful peripheral neuropathies that occur when other nerves are injured. Two laboratories have worked with rat models of this condition, usually produced by a CCI-like lesion to the infraorbital nerve.[63,64] Animals with a CCI-like injury to the infraorbital nerve show abnormal stimulus-evoked pain (withdrawal from normally innocuous touching of the vibrissae and what is reasonably presumed to be pain-related spontaneous behavior, excessive grooming of the face). The latter has been quantified via analysis of videotapes of the animals in their home cages.[63]

Testing the effects of a stimulus applied to the animal's face is extremely difficult because the rat can literally see it coming. A clever way to prevent this is to have the rat in a restrainer with its muzzle poking through a hole that has blinders positioned to block its vision.[64]

4.5.4 Experimental Models of Painful Diabetic Neuropathy

The most thoroughly studied model of painful diabetic neuropathy is produced by destroying the pancreatic islet cells with an injection of streptozocin.[13,14,65] The animals develop hyperalgesia and allodynia after several weeks of hyperglycemia.

Another model uses the obese sand rat.[66] This is a desert-dwelling species of gerbil that grows obese and diabetic when it feasts on ordinary laboratory rodent chow, in a manner analogous to the very high prevalence of diabetes in desert-adapted tribes of Amerindians who have switched to a modern high-fat diet. There is a report[67] that describes abnormal pain sensitivity in the SOD mouse, a mutant that spontaneously develops diabetes without obesity. However, mice of this strain that had not developed clinically detectable diabetes were also hyperalgesic, suggesting that the strain carries a genetic factor affecting pain sensitivity that is separate from the diabetes gene.

Objections have been raised as to the clinical relevance of the commonly used streptozocin model. Pain develops quickly in the animals, while the typical human syndrome of a painful, distal symmetrical neuropathy does not appear until a decade or more after the onset of diabetes. In addition, the typical diabetic patient complains of numbness and spontaneous pain in the distal extremities, not hyperalgesia and allodynia. However, it has been shown that tricyclic antidepressants and Na^{2+} channel blockers (e.g., lidocaine and mexiletine), drugs that have known efficacy in human painful diabetic neuropathy, suppress the abnormal pain responses seen in diabetic rats.[65,68,69]

Neurophysiological studies using the streptozocin-treated rat highlight the value of animal models of painful peripheral neuropathy.[70–72] Recordings from single C-fiber nociceptors showed that many of these fibers respond to a sustained (1 min) suprathreshold mechanical stimulus with a discharge frequency that was clearly greater than normal, although there was no change in their thresholds to noxious stimulation and they did not have spontaneous discharge. The effect is thus quite different from the primary afferent sensitization that occurs after tissue injury. In injured tissue, C-fiber nociceptors are said to be sensitized in that they acquire three abnormal characteristics: spontaneous discharge, a decreased threshold, and a leftward shift of the stimulus–response curve. Only the latter change is comparable to the hyper-responsiveness to suprathreshold stimulation seen in the diabetic rat. Interestingly, a very similar, if not identical, abnormality of C-nociceptor responsiveness is seen in rats with vincristine-evoked painful peripheral neuropathy (see below).

4.5.5 EXPERIMENTAL MODELS OF POSTHERPETIC NEURALGIA

The *varicella-zoster* virus causes chicken pox. When the pox resolves, the viral genome remains behind, stably integrated within the DNA of cells in the dorsal root ganglia (the cell bodies of somatosensory neurons and/or the satellite cells that surround them). For reasons that are only partly understood (although impaired immune surveillance is clearly a factor), the virus can be reactivated decades later. The virus multiplies wildly in the ganglia and is transported down somatosensory axons. The viral load may directly injure cell bodies and axons, and the immune response to extracellular virus causes extensive injury to neurons and to the ganglionic and neural vasculature. The virus eventually escapes from the tips of the afferent axon and enters the skin, at which time the characteristic rash (shingles) is seen. It is important to note that the rash and accompanying skin damage are secondary effects; the primary injury is to the ganglion and nerve. In the majority of cases, the acute infection subsides, the virus re-enters its latent stage, the rash heals, and the patient is left with nothing more than unsightly (and anesthetic) scars. But in a significant number of patients, the rash heals and the virus goes latent, but the patient is left with chronic pain in the previously infected dermatome(s). This is called postherpetic neuralgia and it is one of the most common causes of painful peripheral neuropathy. The incidence is strongly linked to age, those over 65 years of age when they have shingles have a 50% or greater chance of developing the chronic syndrome.

It is not particularly difficult to infect an animal with the herpes zoster virus and establish an acute infection. However, no one has succeeded in establishing a latent infection that is subsequently reactivated; i.e., there are no animal models of shingles or postherpetic neuralgia.

The acute phase of the zoster infection can be exceedingly painful, often for as much as month or more, and the pain is often difficult or impossible to control even with strong opiates (not unexpected because at least some of the pain is neuropathic due to the ongoing injury to the nerve). This acute zoster-associated pain has been modeled in rats with injection of herpes virus into the nervous system,[73,74] and these animals have hyperalgesia and allodynia in the infected dermatomes.

4.5.6 CHEMOTHERAPY-EVOKED PAINFUL PERIPHERAL NEUROPATHY

Painful peripheral neuropathy is a common, although seldom acknowledged, side effect of cancer chemotherapy. For several of the most commonly used drugs the pain limits the dose that can be used to a level that is less than the maximum tumor-killing dose. In addition, the appearance of pain often prevents using the drug in subsequent chemotherapy series. The pain usually appears as part of a distal, symmetrical neuropathy that appears first in the feet, or sometimes simultaneously in the hands and feet, and it is most often described as burning. In some cases, the pain resolves within days or weeks after treatment, but the pain in some cases is very long-lasting. The mechanisms that produce the nerve injury in general, and the neuropathic pain in particular, are unknown. As with other causes of painful peripheral neuropathy, the nerve damage is accompanied by pain in only a subset of patients.

There are very many experimental studies of the neurotoxic effects of chemotherapeutic agents, especially for the commonly used drugs, paclitaxel (the generic name was formerly taxol), vincristine (and other vinca alkaloids), and cisplatin. These studies have used doses that produced severe nerve injury (axonal degeneration), anesthesia, and even paralysis; in many cases, the doses have been near the lethal dose. The relevance of such studies to the vast majority of clinical cases is unclear. An additional problem in this area is that clinical experience strongly suggests that there are a large number of variables that influence the incidence and severity of the neuropathy. Different single dose intensities, different cumulative doses, different IV infusion rates (which will yield different peak-plasma levels), and different dosing schedules (i.e., whether the drug is given daily or with days of rest between injections) are all suspected to be important.

Recent progress in the experimental analysis of chemotherapy-evoked neuropathic pain has been made in studies using vincristine and paclitaxel. The key appears to be the use of doses that are considerably lower than those used previously. In these models[23,75,76] there is relatively little or no deterioration in the animal's general health, preservation of motor and sensory function, and no neural degeneration, but the animals are hyperalgesic and allodynic.

Aley et al.[75] injected (IV) rats with 20, 100, or 200 µg/kg vincristine 5 days per week for 2 weeks (cumulative doses were thus 0.2, 1.0, or 2.0 mg/kg). They found that vincristine treatment produced both acute and chronic effects on pain sensitivity. Acute mechano-hyperalgesia and mechano-allodynia were seen following the second dose and continued for each of the subsequent 8 doses. The relevance of these acute effects to the human problem is uncertain. While the injection itself is often painful, there are no documented instances of acute hyperalgesia or allodynia in human patients. Of course, chronic vincristine-evoked pain hypersensitivity has obvious relevance.

Aley et al. found that doses of 20, 100, or 200 µg/kg produced a clear-cut chronic mechano-hyperalgesia. The 100 µg/kg dose was examined for a chronic effect on touch sensitivity and mechano-allodynia was found with an onset after the 5th injection and a duration of 2 to 3 weeks, with a normalized threshold thereafter. A small but statistically significant chronic heat-hyperalgesia was also reported, but the dose used was not reported. Two weeks after beginning dosing, rota-rod performance was unaffected by the 100 µg/kg dose. However, there was a very severe disruption of rota-rod performance with 200 µg/kg, which suggests a significant injury to large myelinated axons (motor axons and/or proprioceptive sensory axons). Authier et al.[76] have repeated these studies, but with a slightly different dosing schedule (10 consecutive daily doses, rather than 5 consecutive daily doses with 2 days off and then another 5 consecutive doses). With the 100 µg/kg dose used by Aley et al.[75] they found a significant deterioration of general health and significant mortality during the dosing protocol. A dose of 75 µg/kg produced a significant reduction in the rate of weight gain and prevented deaths during the protocol, but a quarter of the animals died after dosing ceased. A dose of 50 µg/kg did not produce any mortality at any time, and although the rate of weight gain was significantly reduced (–9% vs. controls), the animals otherwise appeared to be in good health. Doses of 50 and 75 µg/kg produced significant mechano-hyperalgesia beginning around the time of the last injection on day 10 and continuing for at least 12 days after dosing ceased. Significant mechano-allodynia with a similar time course was seen only with 75 µg/kg. Interestingly, both doses produced a significantly increased threshold to heat-evoked pain (i.e., hypoalgesia).

The vincristine model (100 µg/kg) is not accompanied by any light- or electron-microscopic evidence of axonal loss.[77,78] However, quantitative high-magnification electron-microscopy has revealed very subtle effects. Both myelinated and unmyelinated fibers had a decreased density of microtubules that appeared to be due almost entirely to the axons being slightly swollen. In addition, there was a striking increase in the number of tangentially sectioned microtubules in nerve cross sections, suggesting that there is disruption of the normal linear organization of the microtubules that runs parallel to the long axis of the axon.

Polomano et al.[23] recently described a paclitaxel-evoked painful peripheral neuropathy in the rat that is not associated with any evidence of injury to sensory or motor axons and that is not accompanied by significant effects on the animals' general health. Rats were treated with paclitaxel via 4 IP injections given on alternate days with doses of 0.5, 1.0, or 2.0 mg/kg (cumulative doses were thus 2.0, 4.0, or 8.0 mg/kg). All three doses produced heat-hyperalgesia, mechano-hyperalgesia, mechano-allodynia, and cold-allodynia. The abnormal pain sensations began within several days of the initiation of treatment and lasted for at least several weeks afterward. There was no effect on rota-rod performance. Light-microscopic analyses revealed no sign of structural damage in the sciatic nerve or lumbar roots or ganglia. Rats in all dose groups continued to gain weight at a normal rate (relative to controls) during and after the dosing regimen. There was only one death, and this was believed to be unrelated to paclitaxel. Otherwise, there was no indication of significant generalized debilitation of the animals' health. It is probable that this model is relevant to the early symptoms of paclitaxel-evoked neuropathy, especially those cases that develop neuropathic pain.

4.5.7 NEUROPATHIC PAIN FROM NERVE INFLAMMATION

Injury to a nerve evokes a response from the immune system. This is obviously true when the injury is secondary to infection (e.g., herpes zoster), but even a completely sterile injury will evoke a response because cellular debris is an immune stimulus. In order to study the role of the immune response separate from that of structural injury to axons, Eliav and his colleagues[57] have developed an experimental model of a neuritis. The rat sciatic nerve is exposed and loosely wrapped with oxidized cellulose that is saturated with CFA. Within 24 to 48 h the animals develop heat-hyperalgesia, mechano-hyperalgesia, mechano-allodynia, and (to a lesser degree) cold-allodynia for stimuli applied to the ipsilateral plantar hind paw. These abnormal stimulus-evoked pains last until 5 to 6 days after treatment, after which responses all return to normal. Microscopic analysis of the nerve taken at the time of peak symptom severity (3 to 4 days post-treatment) finds either no sign of axonal injury or at most injury to a few tens of axons. The extra-neural immune stimulus evokes an intraneural infiltration of immune cells (neutrophils, macrophages, and T-lymphocytes). These cells appear to arrive via the nerve's blood supply, rather than via migration across the epineurium. The cause(s) of the abnormal pains in the neuritis model are unknown, but there is a strong suspicion that pro-inflammatory cytokines like tumor necrosis factor α are involved.

REFERENCES

1. Bennett, G.J., Neuropathic pain, in *Textbook of Pain*, 3rd ed., Wall, P.D. and Melzack, R., Eds., Churchill Livingstone, Edinburgh, 1994, chap. 10.
2. Cervero, F. and Laird, J.M., Visceral pain, *Lancet*, 353, 2145, 1999.
3. Su, X., Wachtel, R.E., and Gebhart, G.F., Inhibition of calcium currents in rat colon sensory neurons by K- but not mu- or delta-opioids, *J. Neurophysiol.*, 80, 3112, 1998.
4. Jourdan, D. et al., Audible and ultrasonic vocalization elicited by single electrical nociceptive stimuli to the tail in the rat, *Pain*, 63, 237, 1995.

5. Zimmermann, M., Ethical guidelines for investigations of experimental pain in conscious animals, *Pain*, 16, 109, 1983.

6. Bennett, G.J., Chung, J.-M., and Seltzer, Z., Models of neuropathic pain in the rat, *Curr. Protocols, Neurosci.*, in press.

7. D'Amour, F.E. and Smith, D.L., A method for determining the loss of pain sensation, *J. Pharmacol. Exp. Ther.*, 72, 74, 1941.

8. Hole, K. and Tjolsen, A., The tail-flick and formalin tests in rodents: changes in skin temperature as a confounding factor, *Pain*, 53, 247, 1993.

9. Dubuisson, D. and Dennis, S.G., The formalin test: a quantitative study of the analgesic effects of morphine, meperidine, and brain stem stimulation in rats and cats, *Pain*, 4, 161, 1977.

10. Abbott, F.V., Franklin, K.B., and Westbrook, R.F., The formalin test: scoring properties of the first and second phases of the pain response in rats, *Pain*, 60, 91, 1995.

11. Tjolsen, A. et al., The formalin test: an evaluation of the method, *Pain*, 51, 5, 1992.

12. Abbott, F.V. et al., Improving the efficiency of the formalin test, *Pain*, 83, 561, 1999.

13. Calcutt, N.A. et al., Tactile allodynia and formalin hyperalgesia in streptozotocin-diabetic rats: effects of insulin, aldose reductase inhibition and lidocaine, *Pain*, 68, 293, 1996.

14. Courteix, C., Eschalier, A., and Lavarenne, J., Streptozocin-induced diabetic rats: behavioral evidence for a model of chronic pain, *Pain*, 53, 81, 1993.

15. Hargreaves, K. et al., A new and sensitive method for measuring thermal nociception in cutaneous hyperalgesia, *Pain*, 32, 77, 1988.

16. Hirata, H. et al., A model of peripheral mononeuropathy in the rat, *Pain*, 42, 253, 1990.

17. Bennett, G.J. and Hargreaves, K.M., Reply to Hirata and his colleagues, *Pain*, 42, 255, 1990.

18. Xiao, W.-H. and Bennett, G.J., Unpublished data.

19. Bennett, G.J. and Xie, Y.-K., A peripheral mononeuropathy in rat that produces disorders of pain sensation like those seen in man, *Pain*, 33, 87, 1988.

20. Choi, Y. et al., Behavioral signs of ongoing pain and cold allodynia in a rat model of neuropathic pain, *Pain*, 59, 369, 1994.

21. Tal, M. and Bennett, G.J., Extra-territorial pain in rats with a peripheral mononeuro-pathy: mechano-hyperalgesia and mechano-allodynia in the territory of an uninjured nerve, *Pain*, 57, 375, 1994.

22. Randall, L.O. and Selitto, J.J., A method for measurement of analgesic activity on inflamed tissue, *Arch. Int. Pharmacodyn.*, 4, 409, 1957.

23. Polomano, R.C. et al., A painful peripheral neuropathy in the rat produced by the chemotherapeutic drug, paclitaxel. Submitted for publication, 2001.

24. Taiwo, Y.O., Coderre, T.J., and Levine, J.D., The contribution of training to sensitivity in the nociceptive paw-withdrawal test, *Brain Res.*, 487, 148, 1989.

25. Merskey, H. and Bogduk, N., *Classification of Chronic Pain: Descriptions of Chronic Pain Syndromes and Definitions of Pain Terms*, IASP Press, Seattle, 1994, 209.

26. Semmes, J. et al., *Somatosensory Changes after Penetrating Brain Wounds in Man*, Harvard University Press, Cambridge, 1960.

27. Chaplan, S.R. et al., Quantitative assessment of tactile allodynia in the rat paw, *J. Neurosci. Methods*, 53, 55, 1994.

28. Kim, S.H. and Chung, J.M., An experimental model for peripheral neuropathy produced by segmental spinal nerve ligation in the rat, *Pain*, 50, 355, 1992.

29. Berkley, K.J. et al., Behavioral responses to uterine or vaginal distension in the rat, *Pain*, 61, 121, 1995.

30. Bradshaw, H.B. et al., Estrous variations in behavioral responses to vaginal and uterine distension in the rat, *Pain*, 82, 187, 1999.
31. Ness, T.J. and Gebhart, G.F., Colorectal distension as a noxious visceral stimulus: physiological and pharmacological characterization of pseudaffective reflexes in the rat, *Brain Res.*, 450, 153, 1988.
32. Ness, T.J. et al., Further behavioral evidence that colorectal distension is a "noxious" stimulus in the rat, *Neurosci. Lett.*, 131, 113, 1991.
33. Kehl, L.J., Trempe, T.M., and Hargreaves, K.L., A new animal model for assessing mechanisms and management of muscle hyperalgesia, *Pain*, 85, 333, 2000.
34. Colpaert, F.C., Evidence that adjuvant arthritis in the rat is associated with chronic pain, *Pain*, 28, 201, 1987.
35. Stein, C., Millan, M.J., and Herz, A., Unilateral inflammation of the hind paw in rats as a model of prolonged noxious stimulation: alterations in behavior and nociceptive thresholds, *Pharmacol. Biochem. Behav.*, 31, 455, 1988.
36. Iadarola, M.J. et al., Enhancement of dynorphin gene expression in spinal cord following experimental inflammation: stimulus specificity, behavioral parameters and opioid receptor binding, *Pain*, 35, 313, 1988.
37. Hylden, J.L. et al., Expansion of receptive fields of spinal lamina I projection neurons in rats with unilateral adjuvant-induced inflammation: the contribution of dorsal horn mechanisms, *Pain*, 37, 229, 1989.
38. Coutinho, S.V., Meller, S.T., and Gebhart, G.F., Intracolonic zymosan produces visceral hyperalgesia in the rat that is mediated by spinal; NMDA and non-NMDA receptors, *Brain Res.*, 736, 7, 1996.
39. McMahon, S.B. and Abel, C., A model for the study of visceral pain states: chronic inflammation in chronic decerebrate rat urinary bladder by irritant chemicals, *Pain*, 28, 109, 1987.
40. Wesselmann, U. et al., Uterine inflammation as a noxious visceral stimulus: behavioral characterization in the rat, *Neurosci. Lett.*, 246, 73, 1998.
41. Giamberardino, M.A. et al., Artificial ureteral calculosis in rats: behavioral characterization of pain episodes and their relationship with referred lumbar muscle hyperalgesia, *Pain*, 61, 459, 1995.
42. Wall, P.D. et al., Autotomy following peripheral nerve lesions: experimental anesthesia dolorosa, *Pain*, 7, 103, 1979.
43. Basbaum, A.I., Effects of central lesions on disorders produced by multiple dorsal rhizotomy in rats, *Exp. Neurol.*, 42, 490, 1974.
44. Rodin, B.E. and Kruger, L., Deafferentation in animals as a model for the study of pain: an alternative hypothesis, *Brain Res.*, 319, 213, 1984.
45. Mailis, A., Compulsive targeted self-injurious behaviour in humans with neuropathic pain: a counterpart of animal autotomy? Four case reports and literature review, *Pain*, 64, 569, 1996.
46. Blumenkopf, B. and Lipman, J.J., Studies in autotomy: its pathophysiology and usefulness as a model of chronic pain, *Pain*, 13, 307, 1991.
47. Kruger, L., The non-sensory basis of autotomy in rats: a reply to the editorial by Devor and the article by Blumenkopf and Lipman, *Pain*, 49, 153, 1992.
48. Devor, M., Sense and nonsense, Reply to L. Kruger, *Pain*, 49, 156, 1992.
49. Devor, M., The pathophysiology of damaged peripheral nerves, in *Textbook of Pain*, 3rd ed., Wall, P.D. and Melzack, R., Eds., Churchill Livingstone, Edinburgh, 1994, chap. 4.
50. Wall, P.D. and Gutnick, M., Ongoing activity in peripheral nerves: the physiology and pharmacology of impulses originating from a neuroma, *Exptl. Neurol.*, 43, 580, 1974.

51. Seltzer, Z. et al., The role of injury discharge in the induction of neuropathic pain in rats, *Pain*, 46, 327, 1991.
52. Munger, B.L., Bennett, G.J., and Kajander, K.C., The chronic constriction model of peripheral neuropathy. I. Axonal pathology in the sciatic nerve, *Exptl. Neurol.*, 118, 204, 1992.
53. Gautron, M. et al., Alterations in myelinated fibres in the sciatic nerve of rats after constriction: possible relationships between the presence of abnormal small myelinated fibres and pain-related behaviour, *Neurosci. Lett.*, 111, 28, 1990.
54. Basbaum, A.I. et al., The spectrum of fiber loss in a model of neuropathic pain in the rat: an electron microscopic study, *Pain*, 47, 359, 1991.
55. Mosconi, T. and Kruger, L., Fixed-diameter polyethylene cuffs applied to the rat sciatic nerve induce a painful neuropathy: ultrastructural morphometric analysis of axonal alterations, *Pain*, 64, 37, 1996.
56. Kajander, K.C., Pollock, C.H., and Berg, H., Evaluation of hind paw position in rats during chronic constriction injury (CCI) produced with different suture materials, *Somatosens. Motor Res.*, 13, 95, 1996.
57. Eliav, E. et al., Neuropathic pain from an experimental neuritis of the rat sciatic nerve, *Pain*, 83, 169, 1999.
58. Attal, N. et al., Further evidence for "pain-related" behaviours in a model of unilateral peripheral mononeuropathy, *Pain*, 41, 235, 1990.
59. Seltzer, Z., Dubner, R., and Shir, Y., A novel behavioral model of neuropathic pain disorders produced in rats by partial sciatic nerve injury, *Pain*, 43, 205, 1990.
60. Devor, M., Proliferation of primary sensory neurons in adult rat dorsal root: the kinetics of retrograde cell loss after sciatic nerve section, *Somatosens. Res.*, 3, 139, 1985.
61. Kim, K.J., Yoon, Y.W., and Chung, J.M., Comparison of three rodent neuropathic pain models, *Exptl. Brain Res.*, 113, 200, 1997.
62. Shir, Y. et al., Neuropathic pain following partial nerve injury in rats is suppressed by dietary soy, *Neurosci. Lett.*, 240, 73, 1998.
63. Vos, B.P., Strassman, A.M., and Maciewicz, R.J., Behavioral evidence of trigeminal neuropathic pain following chronic constriction injury to the rat's infraorbital nerve, *J. Neurosci.*, 14, 2708, 1994.
64. Imamura, Y., Kawamoto, H., and Nakanishi, O., Characterization of heat hyperalgesia in an experimental trigeminal neuropathy in rats, *Exp. Brain Res.*, 116, 97, 1997.
65. Courteix, C. et al., Study of the sensitivity of the diabetes-induced pain model in rats to a range of analgesics, *Pain*, 57, 153, 1994.
66. Wuarin-Bierman, L. et al., Hyperalgesia in spontaneous and experimental models of diabetic neuropathy, *Diabetologia*, 30, 653, 1987.
67. Davar, G. et al., Behavioral evidence of thermal hyperalgesia in non-obese diabetic mice with and without insulin dependent diabetes, *Neurosci. Lett.*, 190, 171, 1995.
68. Calcutt, N.A. and Chaplan, S.R., Spinal pharmacology of tactile allodynia in diabetic rats, *Brit. J. Pharmacol.*, 122, 1478, 1997.
69. Kamei, J. et al., Antinociceptive effect of mexiletine in diabetic mice, *Res. Comm. Chem. Pathol. Pharmacol.*, 77, 245, 1992.
70. Ahlgren, S.C. and Levine, J.D., Mechanical hyperalgesia in streptozotocin-diabetic rats, *Neurosci.*, 52, 1049, 1993.
71. Ahlgren, S.C., Wang, J.-F., and Levine, J.D., C-fiber mechanical stimulus-response functions are different in inflammatory versus neuropathic hyperalgesia in the rat, *Neurosci.*, 76, 285, 1997.

72. Ahlgren, S.C., White, D.M., and Levine, J.D., Increased responsiveness of sensory neurons in the saphenous nerve of the streptozotocin-diabetic rat, *J. Neurophysiol.*, 68, 2077, 1992.

73. Fleetwood-Walker, S.M. et al., Behavioural changes in the rat following infection with varicella-zoster virus, *J. Gen. Virol.*, 80, 2433, 1999.

74. Takasaki, I. et al., Allodynia and hyperalgesia induced by herpes simplex type-1 infection in mice, *Pain*, 86, 95, 2000.

75. Aley, K.O., Reichling, D.B., and Levine, J.D., Vincristine hyperalgesia in the rat: a model of painful vincristine neuropathy in humans, *Neurosci.*, 73, 259, 1996.

76. Authier, N. et al., Pain related behaviour during vincristine-induced neuropathy in rats, *Neuroreport*, 10, 965, 1999.

77. Tanner, K.D., Levine, J.D., and Topp, K.S., Microtuble disorientation and axonal swelling in unmyelinated sensory axons during vincristine-induced painful neuropathy in rat, *J. Comp. Neurol.*, 395, 481, 1998.

78. Topp, K.S., Tanner, K.D., and Levine, J.D., Damage to the cytoskeleton of large diameter sensory neurons and myelinated axons in vincristine-induced painful peripheral neuropathy in the rat, *J. Comp. Neurol.*, 424, 563, 2000.

5 Methods in Visceral Pain Research

Timothy J. Ness and Gerald F. Gebhart

CONTENTS

5.1 INTRODUCTION

The recent past has seen an increase in the study of mechanisms of visceral pain, principally due to the development of models that reasonably replicate human visceral pain. Prior to the characterization of such models, the chemically induced writhing model in mice or rats was considered to represent visceral pain. The chemical stimulus, injected into the peritoneal cavity, did not selectively activate the viscera, yielded false positives when used to screen potential analgesic drugs, and is, moreover, associated with a persistent stimulus from which the animal cannot escape. Current models of visceral pain utilize mechanical (e.g., distending) stimuli of controllable duration or chemical stimuli applied directly to relevant targets, thus permitting selectivity with respect to site and intensity of stimulation.

A key consideration in the development of models for the study of visceral pain relates to understanding what constitutes an adequate stimulus. Sherrington[1] provided the definition of an adequate noxious stimulus as applied to somatic, principally cutaneous structures. However, those same mechanical and thermal stimuli that damage or threaten to damage skin, and are thus adequate for activation of cutaneous nociceptors, are not adequate when applied to the viscera because they are not associated with a conscious appreciation of pain in humans. Rather, mechanical distending or obstructive stimuli and irritating/inflammatory stimuli applied to

0-8493-0035-5/01/$0.00+$1.50

hollow viscera are associated with pain in humans and behaviors in non-human animals consistent with visceral pain.

Balloon distension of hollow organs, principally along the gastrointestinal tract, is the most widely used experimental stimulus of the viscera. As reviewed previously,[2] experimental balloon distension of the gastrointestinal tract in humans has been established to reproduce pathologically experienced pain in terms of intensity, quality, and area to which the sensation is referred. Whereas distending stimuli have been established as adequate for hollow organs, occlusive, ischemic, and irritant stimuli have been tested as adequate stimuli in other organs. Because inflammation of the urinary bladder is commonly associated with reports of pain and urgency in humans, experimental models of bladder irritation, including a model of cystitis, have been developed in rats and mice. Kidney stones are undeniably painful in humans and a model of artificial ureteral calculosis has been developed in rats. Occlusion of blood supply to most viscera is associated with pain and ischemia/anoxia is thus considered an adequate stimulus in the viscera. Accordingly, models of coronary artery occlusion and ischemia of abdominal visceral organs have been reported.

In assessing models of visceral pain as to whether they are appropriate and useful in the study of visceral pain mechanisms, important criteria should be considered. To the extent possible, the stimulus should reproduce a natural one for the tissue or organ. Balloon distension of the gastrointestinal tract, artificial kidney stones, chemical irritation of the urinary bladder, and ischemic stimuli all meet this criterion. Second, the stimulus should be reproducible in terms of onset, intensity, and duration. Electrical stimulation of nerves or organs meets this criterion, but is neither natural nor selective in effect produced. The responses produced in the animal should be quantifiable and reliable, and ideally, reproducible. Last, and perhaps most important, the stimulus/model should produce in the unanesthetized animal behavior consistent with organ-appropriate nociception. Below are described several models of visceral pain, highlighting their strengths and weaknesses as models and indicating, where appropriate, where additional characterization of the model should be considered. Characteristics of these models are summarized in Table 5.1.

5.2 VISCERAL PAIN MODELS

5.2.1 MODEL #1: COLONIC-RECTAL DISTENSION

Balloon distension of the distal gastrointestinal tract in the rat is widely used as a generic model of visceral pain; the same stimulus has also been used in horses, dogs, cats, and rabbits. Distension of the colon/rectum (colorectal distension, CRD) produces in rats several readily quantifiable pseudaffective responses (e.g., changes in blood pressure, heart rate, and in contractility of the abdominal musculature). The model has been characterized behaviorally and has been replicated in a number of laboratories.

Before starting such studies, balloon length and means of distension require consideration. Lewis[3] commented that distension of hollow organs was most painful when long continuous segments of the gut were simultaneously distended. Accordingly, when the authors[4] developed CRD as a model of visceral pain in the rat, latex

TABLE 5.1
Summary of Visceral Pain Models

Model	Stimulus Type — Duration	Response Measure
Colorectal distension	Mechanical — Brief	Simple behavioral
		Motor and autonomic threshold and vigor
		Neuronal
Small bowel distension	Mechanical — Brief	Graded behaviors
		Autonomic vigor
Artificial kidney stones	Mechanical — Prolonged	Graded behaviors
		Frequency of behaviors
Urinary bladder distension	Mechanical — Brief	Motor and autonomic threshold and vigor
		Neuronal
Urinary bladder irritants	Chemical — Intermediate	Motor and autonomic threshold and vigor
		Graded/freq. behaviors
		Secondary hyperalgesia
		Neuronal
Vaginal/uterine distension	Mechanical — Brief	Motor and autonomic threshold and vigor
		Simple and complex behavior
		Neuronal
Uterine mustard oil	Chemical — Prolonged	Graded/freq. behaviors
		Secondary hyperalgesia
Arterial occlusion	Ischemic — Brief	Behavioral
		Autonomic
		Neuronal

balloons of 7 to 8 cm length were used. Others have subsequently used balloons 3 to 4 cm long and obtained qualitatively similar results, but generally require higher intensities of stimulation to evoke quantitatively similar responses. Balloon distension of hollow organs can be applied as either a constant volume stimulus or a constant pressure stimulus. As reviewed previously,[2] numerous studies have established that the relation between constant pressure distension and sensations in humans is reproducible and reliable. Constant volume distension of hollow organs is associated with an inconstant intensity of stimulation as organ musculature relaxes in response to the stimulus.

In rats, a flexible latex balloon fixed to a pliable catheter is placed into the descending colon and/or rectum transanally, securing the catheter to the tail with tape. Methods of construction of the balloon, fenestration of the catheter tubing, and distending apparatus have been described previously.[5] Briefly, either a latex condom or a finger from a latex glove may be used as the balloon. The catheter we use in rats is Tygon® flexible tubing (3/16 inch OD and 1/8 inch ID). For a 7 to 8-cm long balloon, 6 cm of one end of the flexible tubing is repeatedly perforated with a #35 hole punch (20 to 25 holes), inserted into the balloon, and tied tightly with silk suture. Before use, it is necessary to overinflate the balloons overnight to make them flaccid and reduce their resistance to distension. When ready for use, the diameter of the balloon is greater than the diameter of the rat colon and does not contribute resistance to distension.

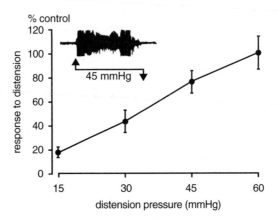

FIGURE 5.1 Mean stimulus-response function to graded intensities of colonic distension in the mouse. The inset gives an example of the electromyographic (EMG) record taken from the external oblique musculature to 45-mmHg colonic distension. The EMG record is truncated at high amplitude in this example. (From C. Jones, E. Kamp, and G.F. Gebhart, unpublished. With permission.)

In rabbits[6] and dogs,[7,8] similar assemblies and placement have been described. In mice, we use a 2-cm long polyethylene bag and PE 10 tubing and the same apparatus to distend the colon. In horses,[9,10] cecal distension has been produced by passage of a similar rubber balloon through a cecal fistula placed a month before the experiment. In these studies as well as in human studies,[11] cardiovascular responses (blood pressure, heart rate) and visceromotor responses (contraction of the abdominal musculature) to distension have been measured. When a constant pressure distending stimulus is employed, pressor, tachycardic, and visceromotor responses to CRD in the unanesthetized animal are graded with the intensity of the stimulus. The stimulus-response functions are incrementing, monotonic functions from which a response threshold can be extrapolated, and both a slope and maximum response can be determined. Figure 5.1 illustrates the intensity-dependent visceromotor response to colonic distension recorded in mice. In studies on responses to CRD, one can thus determine drug effects on response threshold, slope of the stimulus-response function, and response magnitude.[2,12–14]

In the rat, a variety of irritants/inflamogens instilled into the colon have been shown to enhance responses to CRD and thus represent a visceral hyperalgesia. Trinitrobenzene sulfonic acid (TNBS),[15,16] turpentine,[17,18] acetic acid,[19] and zymosan[20] have been applied to the colorectal mucosa, usually several hours to days (TNBS) before an experiment. Typically, the stimulus-response function to CRD in the presence of colonic inflammation is shifted leftward; response threshold is decreased and response magnitude is increased, often at all distension intensities studied.[20] Accordingly, the CRD model has been widely used to study both peripheral and central mechanisms of sensitization and hyperalgesia.

Because CRD produces pain in humans, aversive responses in non-human animals, and reliable, reproducible, and readily quantifiable responses attenuated by analgesic (but not non-analgesic) drugs, it fulfills the necessary criteria for a valid model of visceral pain. The model of CRD has been better characterized than other models of visceral pain and has been replicated in many laboratories. However, CRD is a model of mechanical stimulation of an organ with complex intrinsic and extrinsic innervation and motor activity. Other models described below may be more appropriate for other experimental objectives.

5.2.2 MODEL #2: SMALL BOWEL DISTENSION

An alternative visceral site for stimulation is the upper gastrointestinal tract. Distension of the esophagus and stomach have both been utilized as stimuli, but the most complete characterization to date is that which used small bowel distension. Colburn et al.[21] characterized behavioral responses that could be evoked in unanesthetized rats chronically implanted with a duodenal balloon. The responses were reliable and inhibited by morphine,[21] nonsteroidal anti-inflammatory drugs (NSAIDs),[22] and spinal corticosteroids.[23] Small bowel distension also evoked fos-protein expression in the spinal cord.[24] The apparatus used for these studies consists of a 7.5 to 11-cm long distensible latex rubber balloon placed 4 to 5 days prior to testing. The balloon/cannula assembly is placed in the first portion of the duodenum and a flexible connecting catheter exits the gastrointestinal tract via a surgical gastrostomy. It is then tunneled subcutaneously and is externalized at the base of the skull using a silicone sleeve sutured to the dermis. Testing consists of inflating and deflating the duodenal balloon 5 times with 0.5 to 0.7 cc of saline during a 30-s period of time and then distending the balloon for 1 min. Behavioral responses are scored on a scale of 0 to 4: 0 = normal behavior defined as exploration, escape attempts, and resting; 1 = slightly modified behavior defined as cessation of exploration, focusing, wet-dog shake, excessive facial grooming, teeth chattering, and deep breathing; 2 = mildly to moderately modified behavior defined as wrenching-like activity, hunching, abdominal grooming or nipping, and immobility of hindlimbs; 3 = severely modified behavior defined as stretching of the hindlimbs, arching, and dorsiflexion of the hind paws; and 4 = intensive visceromotor activity defined as repetitive stretching of the body, extension of the hindlimbs and pelvis, and frequent rotation sideways in a fashion similar to the writhing response which is evoked by intraperitoneal irritants.

Other investigators have used distension of the duodenum,[25,26] jejunum,[27] and proximal ileum[28,29] to evoke decreases in blood pressure in anesthetized rat preparations. In those preparations, rats are deeply anesthetized (pentobarbital, urethane, and/or α-chloralose), vascular and tracheal cannulae are placed, and the intestines are exposed by laparotomy. Segments of bowel (3 to 10 cm) are isolated by ligation and intraluminally cannulated, leaving neurovascular connections intact. Fluid is used to distend the bowel to intraluminal pressures of 10 to 100 mmHg. Vigorous and reliable depressor responses of 15 to 40 mmHg have been noted which are inhibited by morphine, codeine, kappa-opioids, 5-HT$_3$ receptor antagonists, bilateral vagotomy, spinal transection, neonatal capsaicin treatment, and topical local anesthetics.

5.2.3 MODEL #3: ARTIFICIAL KIDNEY STONES

A clinically relevant model of pain due to kidney stones is that of Giamberardino et al.[30] who examined responses to artificial ureteral calculosis in rats. A key strength of this model is that it correlates well with the human experience of pain due to kidney stones in that muscular hyperalgesia has been noted to occur and the pain itself appears to be episodic in nature. Following surgical exposure, an artificial stone is placed into the upper third of the ureter by injecting 0.02 cc of dental resin cement (while still liquid) through a fine needle. Rats are allowed to recover from surgical anesthesia and then continuously observed (video monitoring) for 4 or more days for visceral episodes. These episodes consist of 3 or more behaviors which must occur in succession and which must last at least 2 min. The duration of each episode is measured and separate episodes require an inter-episode period of at least 1 min in which rats behave normally. Brief (less than 2 min) occurrences or those with less than three behaviors do not count as episodes. A total of 6 behaviors are formally observed. As described by Giamberardino et al.,[30] the behaviors are the following: (1) a humped back position; (2) licking of the lower abdomen and/or the left flank; (3) repeated waves of contraction of the flank muscles associated with inward movements of the hindlimb, mostly ipsilaterally to the implanted ureter; (4) stretching of the body; (5) squashing of the lower abdomen against the floor; and (6) supine position with the left hindlimb adducted and compressed against the abdomen. The number of visceral episodes varies from rat to rat (0 to ≥ 60 per rat), each episode lasting from a few to 45 min. Complexity of each visceral episode is graded using an arbitrary 4-point scale (3 types of behavior = 1, 4 types of behavior = 2, and so forth.)

Ipsilateral lumbar muscular hyperalgesia can also be determined as decreased electrical thresholds for the evocation of vocalization: every 2 s, a 200-Hz, 250-msec train of 1 msec square wave pulses is administered, increasing in intensity in 0.3 mA steps until vocalization occurs. Then, the current is decreased/increased by 0.1 mA until a reliable threshold value is obtained. Neurophysiological studies have also observed alterations in the excitability of spinal neurons receiving ureteric input after exposure to an artificial kidney stone.[31] Daily morphine injections produce a dose-dependent reduction of the number of visceral episodes in this model.

Methodologically, this model of visceral pain is reliable, but quantitatively variable from animal to animal and requires significant skill in the execution of the experiments. It carries the ethical concern that the noxious stimulus is inescapable, is of long duration (multiple days), and is intended as a model of a highly painful human disease.

Using a stimulus related to symptomatic kidney stones, Brasch and Zetler[32] have characterized hemodynamic (depressor) responses to renal pelvis distension in the pentobarbital-anesthetized rat. They demonstrated inhibitory effects of subcutaneous injections of morphine and the peptide caerulein on these depressor responses. Roza and Laird[33] demonstrated vigorous pressor responses to ureteric distension in pentobarbital-anesthetized rats. In their preparation, the ureter is cannulated near the bladder and may be ligated at the ureteric–pelvic junction. The pressor responses to ureteral distension are reliable, reproducible, and dose-dependently inhibited by morphine and NMDA receptor antagonists.

5.2.4 MODEL #4: URINARY BLADDER DISTENSION

Few studies have utilized the stimulus of urinary bladder distension in models of pain despite the demonstration of vigorous cardiovascular and visceromotor responses in numerous species. Rather, studies have focused upon the evocation of micturition reflexes and their modulation by pharmacological manipulations.[34] Neurophysiological studies have examined the substrates of sensation related to the urinary bladder and various studies have investigated the effects of inflammation of the urinary bladder (see 5.2.5).

We have recently utilized the stimulus of urinary bladder distension (UBD) as a correlate to our previous studies of visceral sensation that employed colorectal distension as a visceral stimulus. Vigorous neuronal and pseudaffective reflex responses can be evoked by UBD[35,36] (examples given in Fig. 5.2) which are inhibited by morphine, kappa opioid receptor agonists, lidocaine, and NMDA receptor antagonists. In halothane-anesthetized female rats, we cannulate the urinary bladder via the urethra using a lubricated 22-gauge angiocatheter which is anchored to the animal by a urethral ligature. Rats are mechanically ventilated and the concentration of anesthetic is slowly decreased until vigorous flexion-withdrawal responses are present, but no spontaneous escape behaviors occur (typically 0.6 to 0.7% halothane). Two forms of UBD are utilized. For slow volume-controlled UBD, we utilize normal saline and a constant rate infusion pump, infusing at a rate of 50 or 100 µl min. Intravesical pressure is measured using either an in-line pressure transducer or via a separate 22-g angiocatheter placed transurethrally. For phasic UBD (rapid onset/rapid termination), we utilize constant pressure-controlled air at intensities of 10 to 80 mmHg. Reliable and reproducible pressor responses and contractions of the abdominal musculature (a visceromotor response measured by electromyogram) can be evoked by 20-s, 60-mmHg UBDs. Gender differences and hormonal influences are apparent; the most vigorous responses are in female rats in their proestrus stage. Inflammation, produced by intravesical treatment with inflammatory compounds such as zymosan or nerve growth factor lead to more vigorous responses to UBD, particularly at low intensities of stimulation.[36]

5.2.5 MODEL #5: URINARY BLADDER IRRITANTS

Independent of an evocative stimulus such as distension, inflammation of the bladder commonly produces reports of pain and urgency in patients suffering from a urinary tract infection. Experiments in non-human animals have artificially inflamed the bladder with the intravesical administration of irritants including turpentine, mustard oil, croton oil, acetic acid, acetone, xylene, or capsaicin and its analogues. Such irritation leads to fos-protein induction in the spinal cord,[37] alterations in the spontaneous activity of primary afferent neurons,[38] and spontaneous behavioral responses. Most studies examining the effects of bladder inflammation have also used bladder filling as an evocative stimulus. A majority of studies have been performed in rats, although behavioral and neurophysiological studies have been performed in primates[39] and cats.[40]

McMahon and Abel[41] performed an extensive characterization of visceromotor and micturition reflexes in chronically decerebrated rats following inflammation of the bladder with 25% turpentine, 2.5% mustard oil, or 2% croton oil administered directly into the bladder via a urethral cannula (angiocatheter similar to that used in Model #4). Subsequently, the bladder was slowly filled with saline through the urethral cannula and intravesical pressures measured, thereby generating a

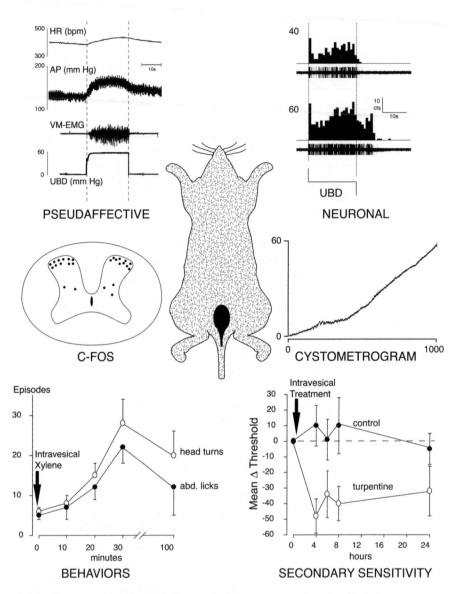

FIGURE 5.2 Examples of responses to urinary bladder stimulation using mechanical or chemical stimuli. Urinary bladder distension (UBD) can evoke pseudaffective reflexes[35] (top left) and spinal neuron responses[36] (top right) in addition to cystometrograms[35] (middle right). Pseudaffective reflexes are typically visceromotor responses that can be measured by direct visualization or electromyographically (VM-EMG) or as autonomic responses such as alterations in heart rate (HR in beats per minute, bpm) and alterations in arterial pressure (AP measured in mmHg). Neuronal responses have been demonstrated to be graded in response to graded UBD (40 or 60 mmHg in this example) and can be quantified as activity per unit time (peristimulus time histograms with 1-s bins are indicated above oscillographic tracings demonstrating neuronal action potentials). Bladder irritants have been demonstrated to induce

cystometrogram. Following administration of an irritant, increased responses to urinary bladder filling correlated with measures of inflammation such as tissue edema, plasma extravasation, and leukocyte infiltration. Rats became hypersensitive to noxious stimuli applied to the lower abdomen, perineum, and tail as measured by the number of kicks evoked by a given stimulus.

This model has been modified[42–44] for use in both anesthetized and unanesthetized rat preparations. In those models, the focus of study was novel mechanisms related to visceral hyper-reflexia (altered cystometrograms) or secondary hyper-algesia (decreased thresholds to heat stimuli in hindlimbs; see Fig. 5.2 lower right). Intact female rats are anesthetized and a 25% solution of turpentine is instilled into the bladder for 1 h, using a transurethral angiocatheter. Subsequently, repeat cysto-metrograms or thermal testing of the hind paws are performed 1 to 24 hours after the intravesical treatment in awake or anesthetized preparations. Modulatory effects of glutamate-receptor antagonists, nitric oxide synthase inhibitors, and bradykinin receptor antagonists have all been noted, but the effects of traditional analgesics such as morphine have not been investigated.

A second model using bladder irritant administration is that described by Abelli and co-workers,[45,46] and modified by Craft and colleagues.[47–49] This model examines immediate effects of the irritants on animal behaviors. In this preparation a midline laparotomy is performed 24 h prior to testing and a 1-mm diameter polyethylene cannula is inserted through the dome of the bladder. The cannula is held in place by a purse string suture and the opposite end is tunneled subcutaneously to the midline upper back where it is externalized and anchored by a skin button. On the day of testing, rats receive an injection through the cannula of 0.3 cc of xylene (30% in silicone oil vehicle), capsaicin, or its related compound resiniferatoxin (0.1 or 3.0 nmol). Immediate behavioral responses are evoked which consist of abdominal/perineal licking, headturns, hindlimb hyperextension, head grooming, biting, vocalization, defecation, scratching, and salivation (see Fig. 5.2 lower left). The time to onset, incidence, and number of individual behavioral responses are recorded for 15 min or more following intravesical drug administration. Baclofen, mu-, kappa-, and delta-opioid receptor agonists and intravesical tetracaine all inhibit these behavioral responses. Pretreatment with systemic capsaicin as adults or neonates abolishes the licking of the lower abdomen/perineum, but does not block hindlimb hyper-extension. Bladder denervation abolishes all behavioral responses. This model of acute urinary bladder pain is methodologically simple and only requires a minimally invasive surgical preparation a day prior to testing. The noxious stimulus is normally intermediate in length (15 min), but is not escapable. There is no parallel human literature related to intravesical xylene application, but there is an increasing anecdotal literature related to the painful nature of intravesical capsaicin instillation in humans.

FIGURE 5.2 (continued) c-fos in central nervous system structures[37] (stylized spinal cord slice of immunohistochemical localization — middle left), to evoke behaviors such as abdominal licking and head turns (data adapted from Abelli et al.[45]) or secondary hyperalgesia of the hindlimbs (indicated as a decrease in the thermal threshold/latency for hindlimb withdrawal [Mean Δ Threshold]) (data adapted from Jaggar et al.[64]). In all of these models, pharmacological manipulation with analgesic drugs such as morphine leads to reduced responses.

Numerous analgesic manipulations have been demonstrated to be inhibitory in this model and no nonspecific effects of non-analgesics have been noted.

A third, distinctly different, model of subacute, irritant-induced urinary bladder pain mimics clinically noted pain that occurs due to cystitis that is secondary to antineoplastic treatments.[50] In this model, the cancer chemotherapeutic agent cyclo-phosphamide (CP; 100 mg/kg) is administered intraperitoneally and subsequently metabolized and excreted as urinary acrolein, a potent irritant of the bladder. Beginning approximately 1 h after systemic administration of CP and continuing for approximately 4 h, unanesthetized rats demonstrate alterations in normal behavior. To quantify these alterations, behaviors are scored on a 6-point scale for 300 s every 20 min and a cumulative sum for each period thereby assigned. The activity scale was defined as follows: 1 = normal behavior; 2 = lacrimation; 3 = piloerection; 4 = rounded-back posture with alertness; 5 = rounded-back posture with immobility; and 6 = transient crises. Presumably the highest applicable score was utilized. Hence, 300 s of normal behavior result in a score of 300; 300 s of piloerection with no other alterations result in a score of 900, and so forth. Measures of bladder inflammation correlate with the induction of c-fos and Krox-24 proteins within the spinal cord. Data related to the reliability of the behavioral responses are not available, although Bon et al.[51] make mention of a subset of animals that failed to develop the behaviors characteristic of CP-cystitis. Statistical analyses are somewhat difficult in this model because the behavioral data are nonlinear. They are hopefully ordinal and so nonparametic analyses may be appropriate. There is the ethical concern that the stimulus is inescapable and of 4 h duration.

5.2.6 MODEL #6: FEMALE REPRODUCTIVE ORGAN STIMULATION

Numerous electrophysiological studies of primary afferents, spinal neurons, and higher order neurons[52] have characterized the substrates of sensation due to stimulation of female reproductive organs. Such stimulation produces variable reflex responses that are dependent upon the sites stimulated and the hormonal state of the animal. Vaginal probing with subsequent pressure applied to the uterine cervix can produce analgesia or evoke reproductive behaviors such as lordosis in young, hormonally cycling female animals, but can also produce aversive (escape) behaviors if administered at high intensities[53] or if administered to animals which are anestrous. Strong cardiovascular and motor responses can be elicited from non-human animals with vaginal and/or uterine stimulation. Chemical stimulation of the uterus has also been employed as a stimulus in rats by Wesselmann et al.[54] Following mustard oil instillation into the lumen of one uterine horn, rats demonstrate altered activity for 4 to 7 days. Video monitoring of the animals demonstrates behaviors that are virtually identical to those described for artificial kidney stones (hunching, licking of abdomen, squashing of abdomen on floor, etc.). Also similar to the kidney stone model, flank muscle hyperalgesia is noted following uterine inflammation. This model is mechanistically easy and so may prove valuable in the evaluation of gynecological visceral pain. Like some of the other models discussed, this model has the ethical problem that it employs a long-lasting, inescapable stimulus.

5.2.7 MODEL #7: ISCHEMIC STIMULI
(CORONARY ARTERY OCCLUSION)

Cessation of blood supply to most viscera is a life-threatening event which frequently leads to pathological pain. Coronary artery insufficiency leads to pathological pain (angina) in humans and mechanical interruption of the blood supply to the myocardium has been utilized as a visceral stimulus to evoke primary afferent, spinal, and medullary neuron responses.[55] These preparations require surgical exposure of the heart and large epicardial vessels with the subsequent placement of "snares" or occlusion cuffs for the temporary interruption of blood flow. Validation of bloodflow cessation is given by doppler flow probes placed distally to the occlusion site, by characteristic electrocardiographic changes, or by altered ventricular contractility. Numerous other methods of producing cardiac ischemia have been employed, but only rarely in studies of sensory phenomena.[56] In these models, partial or full ligation of a smaller coronary vessel is performed and then the metabolic demand on the heart increased by use of a rapid rhythm pacemaker or by other stimuli such as gall bladder stimulation, which reflexly increases cardiac activity and, therefore, oxygen and nutrient demand. At rest, the myocardium may have adequate bloodflow, but in the stimulated state, ischemia occurs.

Although a natural stimulus, experimental ischemia produces responses and evokes sensations in an unreliable fashion. This is reflected in the clinical phenomenon known as silent ischemia where electrocardiographic evidence of severe cardiac ischemia can occur in the absence of reports of pain. Cardiovascular responses to coronary artery occlusion have been demonstrated and appear to consist of alterations (normally increases) in sympathetic tone. Neurophysiological experiments have demonstrated similar variability even in repeated measures of the same neurons.[57] Studies of behavioral responses in unanesthetized dogs have demonstrated responses to coronary artery occlusion,[58] but subsequent studies using the same preparation suggested that the noxious stimulus in those studies was mechanical distortion of perivascular structures rather than ischemia.[59,60]

Longhurst and colleagues have performed extensive studies of abdominal visceral afferents excited by ischemia, and characterized the cardiovascular responses to the same stimuli.[61,62] Factors related to the activation of primary afferents and the evocation of cardiovascular reflexes include hypoxia, tissue acidosis, oxygen free radicals, prostaglandins, leukotrienes, kinins, and numerous other metabolic products. Recent studies have presented evidence that ischemia-related activation of primary afferents is not secondary to mechanical changes.[63] The variability of responses to ischemic visceral stimuli have been proposed to be related to intrinsic neural circuitry within visceral structures and to interactions of vagal, brainstem, and spinal sensory components. The delineation of the mechanisms of this variability may serve to define the overall variability of responses to all visceral stimuli. Given the clinical importance of ischemia-related pain, there is a need for further development of models of visceral ischemia with precise definition of its neurophysiological substrates, evoked responses, and identification of the mechanisms of response variability.

5.3 SUMMARY AND
RECOMMENDATIONS

Because the hollow organs of the gastrointestinal tract are readily accessible through natural orifices, classical clinical studies of visceral sensation used balloon distension of esophagus, stomach, small bowel, large bowel, and rectum.[2] The advantages of balloon distension of hollow organs are many, foremost being that balloon distension reproduces pathologically experienced pain in humans in terms of intensity, quality, and area to which the sensation is referred. Hollow organ distension at constant pressure produces sensations and responses that are reliably reproducible and easily controlled by the experimenter. When applied to non-human animals, distending stimuli of graded intensity produce easily quantifiable, graded responses although the vigor and direction of these responses may be dependent upon the preparation. For example, pressor responses to colorectal distension in unanesthetized rats are depressor responses in barbiturate-, α-chloralose-, or urethane-anesthetized rats.[4] Importantly, because stimulus duration can be easily controlled, ethical considerations associated with models in which animals cannot escape or terminate the stimulus are not a concern in models of distension. Notwithstanding the many advantages of these models, they are limited in that they employ a mechanical stimulus to organs in which inflammation and chemical mediators also contribute to sensation. Moreover, the intrinsic and extrinsic innervation of the gastrointestinal tract is complex and distension in one area of the gastrointestinal tract is typically associated with motor reflexes in another organ.

The other, non-distension models described here offer alternative strategies and stimuli for the study of visceral pain. Animal models of cystitis and artificial kidney stones reasonably replicate those conditions in humans and have contributed significantly to our understanding of visceral pain mechanisms. Perhaps the most important point to be made here is that visceral pain is not a unitary entity. It has long been appreciated that visceral pain differs from somatic pain, and it needs to be understood also that mechanisms of pain generation differ between visceral organs. As an example, the normally sterile urinary bladder becomes exquisitely painful with the invasion of bacteria into its lumen, but lying within the adjacent lumen of painless colons are a true sewer of multiple organisms. Some differences are clearly developmental in that organs which derive from midline structures (i.e., the gut) are associated with bilateral sensations, very generalized responses, and bilateral processing within the spinal dorsal horn. In contrast, those organs which derive from unilateral structures (i.e., kidneys, ureters) generally have lateralized sensations, more regionalized responses, and lateralized spinal processing. We no longer have the luxury of referring to visceral pain as a single entity, and may need to use a more mechanistic approach in our study of different types of pain so that future treatments may be more efficacious. The use of multiple models and multiple types of noxious stimuli allows us to distinguish the generalities of visceral pain from its mechanistic specifics and so development of clinically relevant, reliable models is to be encouraged.

REFERENCES

1. Sherrington, C.S., *The Integrative Action of the Nervous System*, Yale University Press, New Haven, 1906.
2. Ness, T.J. and Gebhart, G.F., Visceral pain: a review of experimental studies, *Pain*, 41, 167, 1990.
3. Lewis, T., *Pain*, MacMillan, London, 1942.
4. Ness, T.J. and Gebhart, G.F., Colorectal distension as a noxious visceral stimulus: physiologic and pharmacologic characterization of pseudaffective reflexes in the rat, *Brain Res.*, 450, 153, 1988.
5. Gebhart, G.F. and Sengupta, J.N., Evaluation of visceral pain, in *Methods in Gastrointestinal Pharmacology*, Gaginella, T.S., Ed., CRC Press, Boca Raton, 1996, 359.
6. Jensen, R.M., Madsen, J.B., Ringsted, C.V., and Christensen, A., Intestinal distension test, a method for evaluating intermittent visceral pain in the rabbit, *Life Sci.*, 43, 747, 1988.
7. Houghton, K.J., Rech, R.H., Sawyer, D.C., Durham, R.A., Adams, T., Langham, M.A., and Striler, E.L., Dose-response of intravenous butorphanol to increase visceral nociceptive threshold in dogs, *Proc. Soc. Exp. Biol. Med.*, 197, 290, 1991.
8. Sawyer, D.C., Rech, R.H., Durham, R.A., Adams, T., Richter M.A., and Striler, E.L., Dose-response to butorphanol administered subcutaneously to increase visceral nociceptive threshold in dogs, *Am. J. Vet. Res.*, 52, 1826, 1991.
9. Kohn, C.W. and Muir, W.W., Selected aspects of the clinical pharmacology of visceral analgesics and gut motility modifying drugs in the horse, *J. Vet. Int. Med.*, 2, 85, 1988.
10. Muir, W.W. and Robertson, J.T., Visceral analgesia: effects of sylazine, butorphanol, meperidine, and pentazocine in horses, *Am. J. Vet. Res.*, 46, 2081, 1985.
11. Ness, T.J., Metcalf, A.M., and Gebhart, G.F., A psychophysiological study in humans using phasic colonic distension as a noxious visceral stimulus, *Pain*, 43, 377, 1990.
12. Banner, S.E., Carter, M., and Sanger, G.J., 5-Hydroxytryptamine 3 receptor antagonism modulates a noxious visceral pseudoaffective reflex, *Neuropharmacology*, 34, 263, 1995.
13. Harada, Y., Nishioka, K., Kitahata, L.M., Kishikawa, K., and Collins, J.G., Visceral antinociceptive effects of spinal clonidine combined with morphine, [D-Pen2, D-Pen5] enkephalin, or U50,488H, *Anesthesiology*, 83, 344, 1995.
14. Burton, M.B. and Gebhart, G.F., Effects of κ-opioid receptor agonists on responses to colorectal distension in rats with and without acute colonic inflammation, *J. Pharmacol. Exp. Ther.*, 285, 707, 1998.
15. Julia, V., Mezzasalma, T., and Bueno, L., Influence of bradykinin in gastrointestinal disorders and visceral pain induced by acute or chronic inflammation in rats, *Dig. Dis. Sci.*, 40, 1913, 1995.
16. Friedrich, A. and Gebhart, G.F., Effects of spinal CCK receptor antagonists on morphine antinociception in a model of visceral pain in the rat, *J. Pharmacol. Exp. Therap.*, 292, 538, 2000.
17. Ide, Y., Maehara,Y., Tsukahara, S., Kitahata, L.M., and Collins, J.G., The effects of an intrathecal NMDA antagonist (AP5) on the behavioral changes induced by colorectal inflammation with turpentine in rats, *Life Sci.*, 60, 1359, 1997.
18. Ness, T.J., Randich, A., and Gebhart, G.F., Further behavioral evidence that colorectal distension is a noxious visceral stimulus in rats, *Neurosci. Lett.*, 131, 113, 1991.
19. Burton, M.B. and Gebhart, G.F., Effects of intracolonic acetic acid on responses to colorectal distension in the rat, *Brain Res.*, 672, 77, 1995.

20. Coutinho, S.V., Meller, S.T., and Gebhart, G.F., Intracolonic zymosan produces visceral hyperalgesia in the rat that is mediated by spinal NMDA and non-NMDA receptors, *Brain Res.*, 736, 7, 1996.
21. Colburn, R.W., Coombs, D.W., Degnen, C.C., and Rogers, L.L., Mechanical visceral pain model: chronic intermittent intestinal distension in the rat, *Physiol. Behav.*, 45, 191, 1989.
22. DeLeo, J.A., Colburn, R.W., Coombs, D.W., and Ellis, M.A., The differentiation of NSAIDS and prostaglandin action using a mechanical visceral pain model in the rat, *Pharmacol. Biochem. Behav.*, 33, 253, 1989.
23. Winfree, C.J., Coombs, D.W., DeLeo, J.A., and Colburn, R.W., Analgesic effects of intrathecally-administered 3 alpha-hydroxy-5 alpha-pregnan-20-one in a rat mechanical visceral pain model, *Life Sci.*, 50, 1007, 1992.
24. DeLeo, J.A., Coombs, D.W., and McCarthy, L.E., Differential c-fos-like protein expression in mechanically versus chemically induced visceral nociception, *Brain Res.*, 11, 167, 1991.
25. Diop, L., Riviere, P.J.M., Pascaud, X., and Junien, J.-L., Peripheral k-opioid receptors mediate the antinociceptive effect of fetodozine on the duodenal pain reflex in rat, *Eur. J. Pharmacol.*, 271, 65, 1994.
26. Moss, H.E. and Sanger, G.J., The effects of granisetron, ICS 205-930 and ondansetron on the visceral pain reflex induced by duodenal distension, *Br. J. Pharmacol.*, 100, 497, 1990.
27. Lembeck, F. and Skofitsch, G., Visceral pain reflex after pretreatment with capsaicin and morphine, *Naunyn-Schmiedeberg's Arch. Pharmacol.*, 321, 116, 1982.
28. Clark, S.J. and Smith, T.W., Opiate-induced inhibition of the visceral distension reflex by peripheral and central mechanisms, *Naunyn-Schmiedeberg's Arch. Pharmacol.*, 330, 179, 1985.
29. Clark, S.J., Follenfant, R.L., and Smith, T.W., Evaluation of opioid-induced antinociceptive effects in anaesthetized and conscious animals, *Br. J. Pharmacol.*, 95, 275, 1988.
30. Giamberardino, M.A., Valente, R., de Bigontina, P., and Vecchiet, L., Artificial ureteral calculosis in rats: behavioural characterization of visceral pain episodes and their relationship with referred lumbar muscle hyperalgesia, *Pain*, 61, 459, 1995.
31. Roza, C., Laird, J.M.A., and Cervero, F., Spinal mechanisms underlying persistent pain and referred hyperalgesia in rats with an experimental ureteric stone, *J. Neurophysiol.*, 79, 1603, 1998.
32. Brasch, H. and Zetler, G., Caerulein and morphine in a model of visceral pain: effects on the hypotensive response to renal pelvis distension in the rat, *Naunyn-Schmiedeberg's Arch. Pharmacol.*, 319, 161, 1982.
33. Roza, C. and Laird, J.M., Pressor responses to distension of the ureter in anaesthetised rats: characterisation of a model of acute visceral pain, *Neurosci. Lett.*, 198, 9, 1995.
34. Dray, A. and Nunan, L., Opioid inhibition of reflex urinary bladder contractions: dissociation of supraspinal and spinal mechanisms, *Brain Res.*, 337, 142, 1985.
35. Ness, T.J., Lewis-Sides, A., and Castroman, P.J., Characterization of pseudaffective responses to urinary bladder distension in the rat: sources of variability and effect of analgesics, *J. Urol.*, 165, 968, 2001.
36. Ness, T.J., Lewis-Sides, A., and Castroman, P.J., Effects of intravesical treatments on rat lumbosacral dorsal horn neurons excited by urinary bladder distension, *Soc. Neurosci. Abstr.*, 96, 932, 2000.
37. Birder, L.A. and de Groat, W.C., Increased c-fos expression in spinal neurons after irritation of the lower urinary tract in the rat, *J. Neurosci.*, 12, 4878, 1992.

38. Dmitrieva, N. and McMahon, S.B., Sensitisation of visceral afferents by nerve growth factor in the adult rat, *Pain*, 66, 87, 1996.

39. Ghoneim, G.M., Shaaban, A.M., and Clarke, M.R., Irritable bladder syndrome in an animal model: a continuous monitoring study, *Neurourol. Urodynam.*, 14, 657, 1995.

40. Habler, H.J., Janig, W., and Koltzenberg, M., Activation of unmyelinated afferent fibres by mechanical stimuli and inflammation of the urinary bladder in the cat, *J. Physiol. (Lond.)*, 425, 545, 1990.

41. McMahon, S.B. and Abel, C., A model for the study of visceral pain states: chronic inflammation of the chronic decerebrate rat urinary bladder by irritant chemicals, *Pain*, 28, 109, 1987.

42. Rice, A.S. and McMahon, S.B., Pre-emptive intrathecal administration of an NMDA receptor antagonist (AP-5) prevents hyper-reflexia in a model of persistent visceral pain, *Pain*, 57, 335, 1994.

43. Rice, A.S., Topical spinal administration of a nitric oxide synthase inhibitor prevents the hyper-reflexia associated with a rat model of persistent visceral pain, *Neurosci. Lett.*, 187, 111, 1995.

44. Jaggar, S.I., Habib, S., and Rice, A.S.C., The modulatory effects of bradykinin B1 and B2 receptor antagonists upon viscero-visceral hyper-reflexia in a rat model of visceral hyperalgesia, *Pain*, 75, 169, 1998.

45. Abelli, L., Conte, B., Somma, V., Maggi, C.A., Giuliani, S., Geppetti, P., Alessandri, M., Theodorsson, E., and Meli, A., The contribution of capsaicin-sensitive sensory nerves to xylene-induced visceral pain in conscious, freely moving rats, *Naunyn-Schmiede-berg's Arch. Pharmacol.*, 337, 545, 1988.

46. Abelli, L., Conte, B., Somme, V., Maggi, C.A., Girliani, S., and Meli, A., A method for studying pain arising from the urinary bladder in conscious, freely-moving rats, *J. Urol.*, 141, 148, 1989.

47. Craft, R.M., Carlisi, V.J., Mattia, A., Herman, R.M., and Porreca, F., Behavioral characterization of the excitatory and desensitizing effects of intravesical capsaicin and resiniferatoxin in the rat, *Pain*, 55, 205, 1993.

48. Craft, R.M., Henley, S.R., Haaseth, R.C., Hruby, V.J., and Porreca, F., Opioid anti-nociception in a rat model of visceral pain: systemic versus local drug administration, *J. Pharmacol. Exp. Ther.*, 275, 1535, 1995.

49. Craft, R.M. and Porreca, F., Tetracaine attenuates irritancy without attenuating desen-sitization produced by intravesical resiniferatoxin in the rat, *Pain*, 57, 351, 1994

50. Lanteri-Minet, M., Bon, K., de Pommery, J., Michiels, J.F., and Menetrey, D., Cyclo-phosphamide cystitis as a model of visceral pain in rats: model elaboration and spinal structures involved as revealed by the expression of c-fos and Krox-24 proteins, *Exp. Brain Res.*, 105, 220, 1995.

51. Bon, K., Lanteri-Minet, M., Menetrey, D., and Berkley, K.J., Sex, time of day and estrous variations in behavioral and bladder histological consequences of cyclophos-phamide-induced cystitis in rats, *Pain*, 73, 423, 1997.

52. Berkley, K.J., Benoist, J.-M., Gautron, M., and Guilbaud, G., Responses of neurons in the caudal intralaminar thalamic complex of the rat to stimulation of the uterus, vagina, cervix, colon and skin, *Brain Res.*, 695, 92, 1995.

53. Berkley, K.J., Wood, E., Scofield, S.L., and Little, M., Behavioral responses to uterine or vaginal distension in the rat, *Pain*, 61,121, 1995.

54. Wesselmann, U., Czakanski, P.P., Affaitati, G., and Giamberardino, M.A., Uterine inflammation as a noxious visceral stimulus: behavioral characterization in the rat, *Neurosci. Letts.*, 246, 73, 1998.

55. Foreman, R.D. and Ohata, C.A., Effects of coronary artery occlusion on thoracic spinal neurons receiving viscerosomatic inputs, *Am. J. Physiol.*, 238, H667, 1980.

56. Li, P., Pitsillides, K.F., Rendig, S.V., Pan, H.L., and Longhurst, J.C., Reversal of reflex-induced myocardial ischemia by median nerve stimulation: a feline model of electroacupuncture, *Circulation*, 97, 1186, 1998.

57. Gutterman, D.D., Pardubsky, P.D., Petersen, M., Marcus, M.L., and Gebhart G.F., Thoracic spinal neuron responses to repeated myocardial ischemia and epicardial bradykinin, *Brain Res.*, 790, 293, 1998.

58. Sutton, D.C. and Lueth, H.C., Pain: experimental production of pain on excitation of the heart and great vessels, *Arch. Int. Med.*, 45, 827, 1930.

59. Katz, L.N., Mayne, W., and Weinstein, W., Cardiac pain — presence of pain fibers in the nerve plexus surrounding the coronary vessels, *Arch. Int. Med.*, 55, 760, 1935.

60. Martin, S.J. and Gordham, L.W., Cardiac pain: an experimental study with reference to the tension factor, *Arch. Intern. Med.*, 62, 840, 1938.

61. Longhurst, J.C., Chemosensitive abdominal visceral afferents, in *Visceral Pain*, Gebhart, G.F., Ed., IASP Press, Seattle, 1995, 99.

62. Rendig, S.V., Chahal, P.S., and Longhurst, J.C., Cardiovascular reflex responses to ischemia during occlusion of celiac and/or superior mesenteric arteries, *Am. J. Physiol.*, 272, H791, 1997.

63. Pan, H.L., Zeisse, Z.B., and Longhurst, J.C., Mechanical stimulation is not responsible for activation of gastrointestinal afferents during ischemia, *Am. J. Physiol.*, 272, H99, 1997.

64. Jaggar, S.I., Scott, H.C.F., and Rice, A.S.C., Inflammation of the rat urinary bladder is associated with a referred hyperalgesia which is NGF dependent, *Br. J. Anaesth.*, 83, 442, 1999.

6 The Cytokine Challenge: Methods for the Detection of Central Cytokines in Rodent Models of Persistent Pain

Sarah M. Sweitzer, Janice L. Arruda, and Joyce A. DeLeo

CONTENTS

6.1 INTRODUCTION

Our laboratory has extensively characterized and analyzed the role of central proin-flammatory cytokines in the development and maintenance of persistent pain states in several rodent models. This work has required the adaptation of methodology for the analysis of cytokines within spinal cord tissue and for the quantification of small changes in cytokine expression within this tissue. This chapter will introduce the pain researcher to the importance of central proinflammatory cytokines in persistent pain as well as discuss important methodological issues relevant to the study of central cytokines. For simplicity, this chapter has been divided into three sections starting with a brief history of the study of cytokines in pain and the pain models our laboratory has utilized for the study of central cytokines. The second section examines methods utilized in the study of spinal proinflammatory cytokine at both the mRNA and protein level. The final section synthesizes the current understanding of central cytokines into the global implications of neuroinflammation in the etiology of persistent neuropathic pain states.

6.2 CYTOKINES IN THE CENTRAL NERVOUS SYSTEM

In general, cytokines can be grouped into three families: interleukins, interferons, and tumor necrosis factors. All three families of cytokines are extremely potent, small soluble proteins that regulate the amplitude and duration of immune and inflammatory states through autocrine or paracrine actions in an effort to maintain tissue homeostasis. Originally classified as lymphokines and monokines to indicate their cellular source, it is now understood that all nucleated cells are capable of synthesizing and responding to these inflammatory mediators, including cells within the central nervous system (CNS). For the scope of this chapter, cytokines will be classified in terms of their regulation of the inflammatory cascade and are thus termed either proinflammatory or antiinflammatory. This chapter will focus on tumor necrosis factor α (TNF), Interleukin (IL)-1β, and IL-6 as proinflammatory cytokines, and IL-10 in the context of being an antiinflammatory cytokine.

In general, peripheral cytokines are undetectable in the absence of tissue insult or injury. In contrast, within the CNS, cytokines are constituitively expressed at low levels with beneficial physiological functions.[1,2] This difference in basal expression suggests that cytokines expressed in the CNS may have very different actions from those expressed peripherally. Within the CNS, many cytokines are known to function as neuromodulators and as such can be beneficial and necessary or detrimental, depending on their relative concentrations. For example, in the hippocampus, IL-1β at low levels (fM) augments long-term potentiation, while at higher levels (nM-pM) IL-1β inhibits long-term potentiation.[3] Additional support for the neuromodulatory role of cytokines in the CNS come from studies which have found that IL-1β and TNF are important mediators of sleep and sickness responses.[3,4] Conversely, proinflammatory cytokines such as TNF and IL-1β have been postulated to exert detrimental effects in central nervous system disease states such as multiple sclerosis, ischemia, Alzheimer's disease, and HIV-associated dementia.[5–9]

6.2.1 CYTOKINES IN PAIN

Early work with lipopolysaccharide (LPS)-induced illness reported marked hyper-algesia (increased response to a noxious stimuli) as a hallmark of illness behaviors. LPS-induced hyperalgesia was found to correlate with increases in peripheral and central proinflammatory cytokines such as TNF, IL-1β, and IL-6.[10–12] This early work in LPS-induced hyperalgesia led to the initial hypothesis that TNF, IL-1β, and IL-6 possess nociceptive actions in both the peripheral and central nervous systems in the absence of injury. In the periphery, both intraperitoneal and intraplantar IL-1β or TNF have been reported to produce short-lived dose-dependent hyperalge-sia.[10,13–19] Furthermore, direct application of TNF on the sural nerve has been shown to produce C-fiber nociceptor sensitization,[20,21] while administration to the sciatic nerve resulted in mechanical allodynia (response to a normally non-noxious stimuli) and thermal hyperalgesia.[22] In an analogous manner, intracerebroventricular admin-istration of TNF, IL-1β, or IL-6 has been found to elicit hyperalgesia[23–25] while intrathecal IL-6 had been reported to mediate allodynia in the absence of injury.[26] Taken together, these data implicate three nociceptive processing regions at which TNF, IL-1β, and IL-6 may modulate pain responses following injury: in the periphery at the site of injury, in the spinal cord at the initial site of signal integration and processing, or in the brain at the site of supraspinal processing.

A potential role for peripheral actions of IL-1β and TNF in the genesis or maintenance of pain states following injury has been supported by observations of their upregulation in various peripheral inflammatory models that result in hyperal-gesia and allodynia.[13,15,16] Pharmacological blockade of proinflammatory cytokines with IL-1 receptor antagonist (IL-1ra), anti-TNF serum, thalidomide, or antiinflam-matory IL-10 or IL-4 has been found to delay the onset and attenuate hyperalgesia in models of peripheral inflammatory pain.[13,16,27–30] TNF has also been implicated in neuropathic pain models, such as chronic constriction injury (CCI), through the use of TNF receptor antibodies, thalidomide, and antiinflammatory IL-10.[31–33] These pharmacological interventions have been shown to decrease TNF expression endo-neurially and this decrease paralleled the attenuation of mechanical allodynia or thermal hyperalgesia. These data highlight the vast literature that supports a potential role for peripheral cytokines in the modulation of pain behaviors in both inflamma-tory and neuropathic pain models.

In addition to peripheral expression of proinflammatory cytokines following injury, our laboratory has shown that several neuropathic models induce increased TNF, IL-1β, and IL-6 mRNA and protein in the spinal cord in a manner that parallels the development of allodynia.[26,34–38] A similar upregulation of IL-1β expression in the spinal cord has been found in both formalin and zymosan peripheral inflammatory models as well as in the L5 spinal nerve transection mononeuropathy model.[39] Further support of central proinflammatory cytokines in nociception have come from the observations that intrathecal application of IL-6 evokes hyperalgesia in a sciatic cryoneurolysis model that normally does not produce hyperalgesia.[26] From studies utilizing pharmacological antagonism as well as studies in both genetically modified transgenic and knockout mice, numerous other laboratories have also implicated an important role of central cytokines in nociception. Pharmacological antagonism with intrathecal administration of IL-1ra or antibodies to the IL-1 receptor has produced attenuation in pain behaviors in both the formalin[40] and CCI[41] models, lending further support for an important role of central cytokines in the modulation of neuropathic pain. Similarly, transgenic mice overexpressing TNF on the glial fibrillary acidic protein (GFAP) promoter exhibited increased allodynia following L5 spinal nerve transection as compared to wild-type counterparts.[42] Conversely, in response to nerve injury, IL-6 knockout mice have been shown to exhibit delayed mechanical allodynia[44] without developing heat or pressure hypersensitivity.[43] These studies highlight the importance of central TNF, IL-1β, and IL-6 in nociception following nerve injury.

Unfortunately, trying to decipher the role of cytokines in the genesis of pain states is very complex. This complexity can be further highlighted by the observations that exogenous application of IL-1β can be both pro- and antinociceptive.[45–47] A recent study demonstrated that high-dose intrathecal application of IL-1β was antinociceptive in a carrageenan model of inflammation, but application of IL-1β neutralizing antibody, without the addition of exogenous IL-1β, did not alter carrageenan hyperalgesia.[46] These findings coupled with previous research in the formalin model[40] suggest that the increase in endogenous IL-1β following inflammatory injury is not sufficient to activate antinociceptive pathways and may even be pronociceptive. These observations of dose-dependent effects of IL-1β in the CNS are not unique to pain models. In ischemia models, TNF and IL-1β have been found to be both neurotoxic and neuroprotective depending on their concentration, the concentration of their endogenous antagonists, as well as the time at which they are expressed following injury.[3,5,7] Extrapolation from the ischemia literature would suggest that the study of central cytokine contributions to nociception must not only include the study of TNF, IL-1β, and IL-6 expression but also the temporal pattern of this expression as well as the balance of this expression with the endogenous antagonists.

6.2.2 ROLE OF CYTOKINES IN CENTRAL SENSITIZATION

Proinflammatory TNF, IL-1β, and IL-6 are postulated to contribute to central sensitization.[1,2] Central sensitization produces lower thresholds and spontaneous ectopic neuronal firing[48] which are manifested as hyperalgesia and allodynia at the behavioral level. The expression of cytokine receptors on neurons provides a mechanism by which TNF, IL-1β, and IL-6 may directly sensitize neurons; for example, IL-1β has been found to act directly on neurons to increase axonal transport and release of substance P, a potent nociceptive substance.[49–51] Similarly, exogenous TNF or IL-6 has been shown to induce substance P synthesis and release in sympathetic ganglia.[52,53] In addition to a direct action on neurons, proinflammatory cytokines may indirectly affect neurons via interactions with the surrounding glial population. IL-1β and substance P have been found to synergistically induce the release of IL-6 and prostaglandins from human spinal cord astrocytes.[54] Furthermore, TNF and IL-1β are known to be inducers of both microglial and astrocytic activation *in vitro* as well as *in vivo*.[7,55]

Activation of spinal glial (microglia and astrocytes) cells is important to the development of both allodynia and hyperalgesia. Administration of glial metabolic inhibitors in zymosan and formalin models has been shown to attenuate hyperalgesia.[40,59] In addition, our laboratory has repeatedly shown spinal glial activation in a number of rodent neuropathy models.[35,36,56,57] Activated glial cells synthesize and secrete proinflammatory mediators such as cytokines (TNF, IL-1β, IL-6), nitric oxide (NO, through the induction of iNOS), and prostaglandins (through the induction of COX-2),[7,58,59] all of which are potent nociceptive substances. Further neuronal sensitization is possible when TNF interacts with astrocytes, thereby increasing intracellular Ca^{2+} and inducing depolarization. Depolarization results in decreased glutamate uptake, leaving excessive glutamate in the synaptic cleft which produces aberrant or ongoing neuronal firing.[60,61] Together this suggests that glial-derived cytokines or cytokine-induced glial activation are involved in the genesis and maintenance of persistent neuropathic pain states.

6.2.3 RODENT MODELS OF PERSISTENT PAIN

There are numerous, well-established animal models of neuropathic pain.[56,62–67] Our laboratory has developed two peripheral neuropathy models in the rat using freeze lesions.[56,57,65] Although the freeze lesion is unique because it produces a regenerative permissive injury, the extent of the nerve lesion precludes early onset testing of neuropathic behaviors. In addition, it is technically challenging to selectively freeze the L5 spinal nerve with available cryoprobes. For these reasons, we adopted a modified Chung model and exclusively use this model for our neuropathic pain studies. From multitudinous studies, we have determined that a simplified L5 spinal nerve transection vs. L5 and L6 spinal nerve ligation (Chung model) produces robust mechanical allodynia and thermal hyperalgesia with quick onset and long duration in the nerve injury group as well as eliciting minimal inflammatory sham surgery behaviors. The results in this chapter utilize this L5 spinal nerve transection model for the neuropathic pain studies. As seen in Fig. 6.1, mechanical allodynia increases with time following L5 spinal nerve transection.

Our laboratory also has experience with models of low back pain associated with lumbar radiculopathy (i.e., lumbar dorsal and ventral root injury proximal to the dorsal root ganglion).[38] The mechanisms that give rise to low back pain associated with or without lumbar radiculopathy remain obscure. During the past decade, there has been increasing interest in the mechanisms of low back pain/radiculopathy. Several useful animal models of lumbar nerve root injury[68–73] established two specific mechanisms at the injury site level, specifically mechanical deformation of the nerve roots, and biologic or biochemical activity of the disc tissue. However, the exact contribution of each component and its full elucidation have thwarted the scientific community. To this end, our laboratory further characterized the original Kawakami/Weinstein rat model of radiculopathy in which we investigated the type of suture material and extent of the lumbar root injury.[38] In order to better understand radicular pain mechanisms, we examined spinal glial activation and IL-1β expression following lumbar nerve root injuries (loose ligation of only L5 dorsal and ventral roots with chromic gut or silk, or tight ligation with silk). We concluded that root injury via a chemically induced and/or mechanically induced factor evokes pain behaviors and activates glial cells and IL-1β expression.

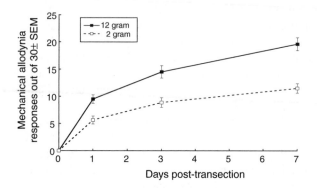

FIGURE 6.1 L5 spinal nerve transection in male Holtzman rats produces a reproducible and reliable 2-g and 12-g von Frey filament (Stoelting, Wood Dale, IL)-induced mechanical allodynia (n = 8/time point). Animals were acclimated to the testing procedure and three baseline measurements were collected before the day of surgery. Rats were subjected to 3 sets of 10 stimulations with each filament with at least 10 min between each set of stimulations. Allodynia was characterized as an intense withdrawal of the paw to this normally non-noxious stimuli. Results are displayed as the average number of paw withdrawals ± SEM.

6.3 TECHNIQUES FOR THE STUDY OF CENTRAL CYTOKINES

6.3.1 CYTOKINE mRNA DETECTION

The effects of neuropathic pain on the level of messenger ribonucleic acid (mRNA) expression can be studied through a variety of assays including *in situ* hybridization, ribonuclease protection assay (RPA), and reverse transcriptase polymerase chain reaction (RT-PCR). By utilizing these methods, information about not only cytokine modulation at the mRNA level can be acquired, but in some assays the cell source(s) of these alterations can be identified. Therefore, these methods could prove crucial in elucidating the complexities of neuroimmune interactions and their effects on the complex cytokine cascade resulting from injury-induced neuropathic pain.

6.3.1.1 *In Situ* Hybridization

Both radioactive and non-radioactive protocols exist for *in situ* hybridization (ISH). While autoradiography provides the ability to quantify mRNA it is often cumbersome and expensive with increased complexity to identify the cellular source of message. Non-radioactive ISH combined with immunohistochemistry is a technique that allows for simultaneous cell source identification/localization and semiquantitative analysis of mRNA expression.[74] ISH allows for the elucidation of the roles that the various cell types in the CNS have in the etiology of neuropathic pain states. Our laboratory has had success in utilizing nonradioactive *in situ* hybridization to clearly characterize the modulation of IL-6 mRNA in various models of neuropathic pain.[34]

In our experience, we have found *in situ* hybridization to pose several problems. First of all, our probe source was limited and did not have reproducible quality

TABLE 6.1
Comparison of Final Yield and A260/A280 Ratio of Two Commercially Available Total RNA Isolation Kits: TRI Reagent® and Pharmingen Total RNA Isolation Kit

Isolation Kit	Treatment	Yield [RNA]/Starting Tissue Weight (µg/mg)	A260/A280 Ratio
TRI Reagent	normal	6.6	1.59
Pharmingen	normal	9.0	1.64
TRI Reagent	3-h LPS treated	7.0	1.60
Pharmingen	3-h LPS treated	10.9	1.63

control. We found that the difference in detection varies dramatically for each lot of probe, from high detection to no detection on the same tissue samples run simultaneously with different probe lots. This made it very difficult to maintain detection reproducibility in our studies. Second, our laboratory utilized the nonradioactive detection method 5-Bromo-4-Chloro-3-indolyl Phosphate/Nitroblue (BCIP/NBT). Even when concentrations and incubation times were consistent, the development concentration of probe varied between runs. This made it impossible to analyze our slides by densitometry and, therefore, we could not detect changes in mRNA concentration within the cells, but only the number of cells with mRNA present. We also found that significant time and effort were needed to collect a relatively small amount of semiquantitative data. However, the value of the knowledge gained (in this case both production of mRNA and the cellular source of production) needs to be weighed against the effort and time required to attain that data.

6.3.1.2   Ribonuclease Protection Assay and Reverse Transcriptase Polymerase Chain Reaction

There are several issues to consider when using techniques that require isolation of RNA. One of the first considerations is choosing an RNA isolation procedure from the various available methods. A very exhaustive description of total, cytoplasmic, or messenger RNA isolation methods can be found in *Molecular Cloning: A Laboratory Manual*.[75] However, since these methods can be very time consuming and laborious, there are many commercially available kits that are good alternatives to these methods and will reduce the number of steps and time required for RNA isolation. Our laboratory did a comparative study of two commercially available kits, TRI Reagent® (Molecular Research Center, Inc.) and Total RNA Isolation Kit (Pharmingen), to determine if there was a difference in the purity or volume of RNA isolated by the kits. In our hands, for the same relative amount of work invested, both a higher final yield and purity of RNA were obtained from normal and LPS-treated rat spinal cord using the Pharmingen kit compared to TRI Reagent® (Table 6.1). Therefore, we utilize the Pharmingen kit for RNA isolation in our studies.

Another critical item to note when working with RNA is the sensitivity of the spectrophotometric analysis of the final product. The isolated RNA spectrometric

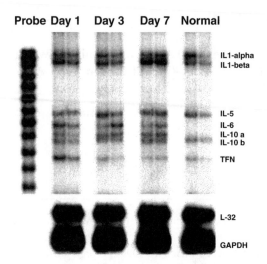

FIGURE 6.2 RPA analysis of cytokine mRNA in the L5 spinal cord on days 1, 3, and 7 post L5 spinal nerve transection as compared to normal naïve animals (n = 2/time point). TNF mRNA was transiently upregulated at day 1 post-transection while mRNA for IL-1β and IL-6 were markedly upregulated at each time point examined. In rats, IL-10 has three isoforms of which one isoform appeared constituitively expresssed (IL-10b) while a second isoform was upregulated following transection (IL-10a). Housekeeping genes, L-32 and GADPH, provided a control for gel loading.

absorbance at 260 and 280 is utilized to determine the concentration of RNA as well as the A260/A280 ratio which indicates the level of DNA and protein contamination. It is imperative that the pH of the measured sample is above 7.5 to obtain accurate readings. For example, the acceptable A260/A280 ratio is between 1.5 and 2.0, with 2.0 having the least contamination. Using an identical sample, our laboratory obtained an A260/A280 ratio of 1.32 vs. 1.91 with solutions having a pH of less than or greater than 7, respectively. As is obvious from these results, the wrong sample pH will make a significant difference in measurement accuracy.[76]

Having successfully isolated RNA, it is now possible to analyze the samples. One method for this analysis is with the Ribonuclease Protection Assay (RPA). In this assay, it is possible to simultaneously screen a panel of biologically relevant mRNAs to compare cytokine mRNA expression levels within each sample. In addition, housekeeping genes are included in each probe panel. Housekeeping genes are genes which are essential to cell function, and as such are always expressed in cells at consistent levels, for example, β-actin and glyceraldehyde-3 phosphate dehydrogenase (GADPH). Due to their presence in each probe panel, concentration comparisons can be made between samples by normalizing the data to the concentration of the housekeeping genes.

We have found RPA particularly useful for obtaining an understanding of the mRNA modulation in models of persistent pain. For example, our laboratory has recently completed time-course studies of mRNA expression in an L5 spinal nerve transection mononeuropathy model. Lumbar spinal cord was collected at 1, 3, and 7 days post-transection as well as from normal naïve animals. RPA analysis uncovered constitutive expression of IL-1α, IL-1β, and one isoform of IL-10 mRNA as well as increased mRNA expression of TNF, IL-1β, IL-6, and a second isoform of IL-10 following L5 nerve transection (Fig. 6.2). These data further support the contribution of central TNF, IL-1β, and IL-6 in the development and maintenance of persistent neuropathic pain. Of interest was the transient upregulation of TNF, such that TNF mRNA was upregulated at day 1 post-transection but returned to basal naïve levels by day 3 post-transection. Similar temporal patterns of cytokine and endogenous cytokine antagonist expression have been observed in the brain following intrastriatal injection of LPS.[77] This highlights the importance of temporal characterization in the analysis of central cytokines. For example, if the mRNA of interest was upregulated and degraded before the time point measured, it could falsely appear that no change had taken place and/or that the cytokine of interest was not involved in the pathological sequelae of the model system.

Like any assay, it is important to realize the limitations of the RNAse protection assay. First of all, the commercially available probe sets are limited. Currently, there is a wide range of human and mouse panels available, but limited rat panels. This affects which mRNAs can be analyzed and, in some cases, which species to utilize (i.e., rat vs. mouse models). Second, RPA requires a large volume of total RNA per assay (ranging from 10 to 20 μg). We have found that at least 15 μg was needed to effectively detect the presence of mRNAs with ^{32}P-labeled probe in our spinal cord tissue samples. This limits the number of RPA probe panels that can be assayed per sample. A third limitation is that it is difficult to quantify small changes in mRNA between treatments. This problem can be further compounded by the sample collection technique. For example, if the modulation of mRNA is localized, as we have seen in our models, then a tissue sample incorporating large non-modulated portions could dilute the detectable effect below assay sensitivity.

When only small volumes of RNA are available or when it is necessary to detect small changes in mRNA, a powerful tool to use is reverse transcriptase-polymerase chain reaction (RT-PCR). This technique generally uses only 1 to 2 μg of total RNA for detection and due to the amplification process of the technique is able to detect very small differences in concentrations of the mRNA of interest. This is extremely useful considering that cytokines can elicit their effects at such low concentration (fM) and, so, only a small volume of mRNA may be present to elicit an effect. As with RPAs, by utilizing housekeeping genes, it is possible to compare the concentrations of mRNA between samples and semi-quantitatively analyze the data. A detailed explanation of the protocol and methodology of RT-PCR can be found in *Molecular Cloning: A Laboratory Manual*.[75] In addition, although RT-PCR is limited to single mRNA detection in each amplification procedure, virtually any cDNA probe can be manufactured for detection when the desired DNA sequence is known.

Regardless of the means used to detect mRNA, it is important to understand how to interpret the data and what the data's limitations are. First, the data collected is a snapshot of the currently available mRNA. It does not take into account changes in either the production (transcription from DNA to RNA) or degradation (half-life/stability) rate of mRNA within the system being studied. It is known that the half-life of mRNA can be both increased and decreased depending on several influencing factors including cytokines. For example, glucocorticoids, which have been used for the treatment of chronic pain, have been reported to decrease the half-life of certain cytokine mRNAs[78,79] while LPS treatment, which has been shown to induce hyperalgesia, has been reported to increase the half-life of certain cytokine mRNAs.[80] This alteration in the half-life will affect the amount of time the mRNA is available for the cell to produce protein. Therefore, if the mRNA transcription and/or translation rate have increased to produce elevated protein levels, but the degradation of mRNA has also increased, you may, in theory, still have the same amount of mRNA in your snapshot even though there has been an increase in protein production.

A second thing to remember when interpreting mRNA data is that the presence of mRNA does not necessarily equal translation (protein production from mRNA). This has been clearly shown in experiments where various stimuli were able to increase IL-1β mRNA production, but not its translation into protein.[80–84] Several known regulators of the translation of mRNA into protein exist, including regulation of the percentage of mRNA that is rendered inactive by sequestering into messenger ribonucleoprotein particles,[85] or the binding of the mRNA active site such that translation cannot be initiated.[86] The methods described in this chapter do not distinguish between translating and non-translating mRNA, so the presence of mRNA does not always equal an alteration in protein expression levels. Related to this, the data also do not give information about the rate of mRNA translation into protein. If the mRNA is producing protein at an elevated rate, it is conceivable that no change would be detectable at the mRNA level when, in fact, there has been a change in the system's protein output.[87]

6.3.2 CYTOKINE PROTEIN DETECTION AND QUANTIFICATION

In the study of central mechanisms that contribute to persistent pain, it is not enough to understand the regulation of spinal cytokine genes, but it is also imperative to understand if those genes are translated into bioactive proteins. Thus, it is important to be able to analyze and quantify protein concentrations of cytokines in the spinal cord. This can be achieved by immunohistochemistry and enzyme-linked immunosorbent assays (ELISA). Immunohistochemistry allows visualization of cytokines in terms of the global anatomy of the spinal cord as well as in terms of cellular localization. The anatomical expression within the tissue of interest is important to determine the relevance of the observed changes to pain processing pathways. Cytokine immunohistochemistry, while able to reliably detect large changes in cytokine expression, has proven to be very difficult to utilize in the quantification and characterization of small changes between injury models or treatment protocols. Our laboratory has found it necessary to reliably measure small changes in cytokine levels and, thus, we have turned to ELISA techniques to supplement and expand on previous immunohistochemical findings.[26,37,39]

6.3.2.1 Immunohistochemistry

A large part of our assessment of neuropathic and radicular pain in rodent models relies on the information obtained from immunohistochemical staining of spinal cord tissue. Unfortunately, specificity and reproducibility of immunohistochemical staining for any single cytokine can vary greatly between commercial sources and across fixation methods. Since immunohistochemical assessment of cytokines plays a large role in our studies of persistent pain in rodent models, we determined that it was necessary to compare and evaluate available antibodies for the antigens in which we were interested, specifically TNF, IL-1β, and IL-6.

Antibodies for TNF, IL-1β, and IL-6 were obtained from four to six commercial sources and analyzed in our laboratory (Table 6.2). The antibodies were titrated to determine the best dilution for optimum staining. Optimum staining was defined as the least background for the clearest specific staining in lumbar spinal cord tissue. Reproducibility of staining was defined as the ability to achieve similar staining across numerous immunohistochemical runs. Staining quality was defined as the ease of visualization of antigen-specific staining over the inherent non-specific background staining of the antibody. Several of the antibodies produced high quality and reproducible staining in lumbar spinal cord tissue, while several did not (Table 6.2). Next, the specificity of each antibody was tested by preabsorbing the antibody with 1:10 or 1:20 molar ratio of target antigen and incubating this mixture for 1 h at room temperature before proceeding with the immunohistochemistry protocol. Surprisingly, much of the immunohistochemical staining with these commercial antibodies did not appear to be specific as it was not blocked by preabsorbing antigen to the antibody (Table 6.2). The monoclonal anti-IL-1β antibody from Serotec and the polyclonal anti-TNF antibody from Genzyme resulted in the best quality and reproducible specific staining.

Different commercial methods of antibody purification are likely one reason for immunohistochemical staining inconsistencies and lack of antibody specificity. In addition, the quality and specificity of staining may be affected by the fixation method utilized. It is important to note that in the above findings we only studied these antibodies with one fixative method entailing a 4-h post-fix in 4% paraformaldehyde.[37] We previously found that the length of post-fixation can dramatically alter immunohistochemical staining. For example, a 4-h post-fixation utilizing antibodies for Major Histocompatibility Complex II (MHC II) and Intracellular Adhesion Molecule (ICAM) results in no staining while excellent staining can be acquired with tissue post-fixed for less than 1 h. This variation appears antibody specific since we can achieve similar GFAP and OX-42 staining with either a 1- or 4-h post-fixation. It is thus necessary when utilizing immunohistochemical techniques to determine on an antibody-to-antibody basis the reproducibility, quality, and specificity of individual antibodies for the specific immunohistochemical protocol being utilized.

6.3.2.2 Enzyme-Linked Immunosorbent Assay

Recently, our laboratory has begun to utilize ELISAs for the quantification of both pro- and antiinflammatory cytokines in neuropathic and radicular models of persistent pain. The detection of spinal cord cytokines is not limited by the available commercial ELISA kits since a laboratory in theory can make its own plates using existing available specific antibodies. However, we have found it more efficient to utilize commercial kits. As with the techniques to study mRNA, the first hurdle was to isolate protein from spinal cord tissue. For the present work, spinal cord homogenization

TABLE 6.2

Analysis of Commercial Sources of TNF, IL-1β, and IL-6 Antibodies for Reproducibility, Quality, and Specificity of Immunohistochemical Staining

Source (Catalogue Number)	Antibody Description	Dilution	Reproducible Staining	Staining Quality	(1:10) Pre-Absorbed Antigen	(1:20) Pre-Absorbed Antigen
TNF						
Endogen (PR-370)	PC[a] rabbit, anti-rat	1:1500	No	Average	0[d]	Not done
Genzyme Diagnostics (IP-400)	PC rabbit, anti-mouse	1:10,000	Yes	Above average	–[f]	–
PeproTech Inc. (500-P72)	PC rabbit, anti-rat	1:1000	Yes	Below average	+[e]	Not done
R&D Systems Inc. (AF-510-NA)	PC goat, anti-rat	1:1500	No	Below average	–	–
Santa Cruz Bio. Inc. (Sc-1349)	PC goat, anti-rat	1:2500	Yes	Average	0	–
Serotec Ltd. (AAM12)	PC rabbit, anti-mouse/rat	1:3000	Yes	Above average	–	–
IL-1β						
Endogen (PR-427B)	PC rabbit, anti-rat	1:1500	No	Average	NC/SNR[c]	Not done
Genzyme Diagnostics (LP-712)	PC rabbit, anti-human	1:1500	Yes	Average	0	0
PeproTech Inc. (500-P80)	PC rabbit, anti-rat	1:1000	No	Average	+	Not done
R&D Systems Inc. (AF-501-NA)	PC goat, anti-rat	1:1000	NC	Average	–	+
Santa Cruz Bio. Inc. (Sc-1252)	PC goat, anti-rat/mouse	1:2000	Yes	Below average	NC/SNR	+
Serotec Ltd. (MCA1397)	MC[b] mouse, anti-rat	1:1000	Yes	Above average	–	Not done
IL-6						
Endogen (PR-627)	PC rabbit, anti-rat	1:1500	Yes	Below average	+	Not done
Genzyme Diagnostics (LP-716)	PC rabbit, anti-human	1:1000	Yes	Above average	–	Not done
PeproTech Inc. (400-06)	PC rabbit, anti-rat	1:1000	No	Average	NC/SNR	Not done
Santa Cruz Bio. Inc. (Sc-1266P)	PC goat, anti-rat/mouse	1:1500	Yes	Below average (perivas)	+	+

[a] PC = polyclonal antibody

[b] MC = monoclonal antibody

[c] NC/SNR = not conclusive/staining not reproducible

[d] 0 is no change in staining quality or quantity following pre-absorption of antigen

[e] + is increased staining following pre-absorption of antigen

[f] – is decreased staining following pre-absorption of antigen

A

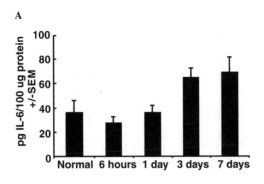

B

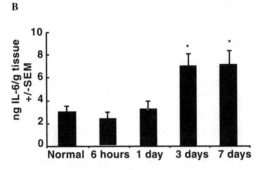

FIGURE 6.3 Analysis of IL-6 protein by ELISA produced similar increases in protein at days 3 and 7 post-transection whether IL-6 protein was quantified in terms of µg of total protein (A) or in terms of tissue weight examined (B). IL-6 protein quantified per tissue weight revealed a statistically significant ($^*p < 0.05$) increase in IL-6 in the L5 spinal cord on days 3 and 7 post-transection as compared to normal naïve animals (B). L5 spinal cord was isolated at 6 h (n = 4), 1 (n = 4), 3 (n = 8), or 7 (n = 7) d post L5 spinal nerve transection as well as from normal naïve animals (n = 6).

was adapted from the method of De La Monte et al.[88] Once protein was isolated, we stored aliquots of protein at –80°C until the ELISA was run. Storage of samples in aliquots is ideal since repeated freeze–thaw cycles can degrade protein and thus degrade the detection of the target protein by ELISA.

A review of the ELISA literature revealed two ways to report data. Data have been reported as the weight of the protein of interest per either total tissue weight analyzed or per total protein in the sample. We completed a time course following L5 spinal nerve transection and examined IL-1β and IL-6 levels by both methods. IL1-β protein concentrations were determined utilizing the quantitative sandwich enzyme immunoassay Quantikine® M rat IL-1β immunoassay (R&D Systems, Minneapolis, MN) according to the manufacturer's directions. IL-6 protein concentrations were determined utilizing Biosource IL-6 immunoassay (Biosource International, Camarillo, CA) according to the manufacturer's directions. For the determination of total protein the BCA (bicinchoninic acid, Pierce Chemical Company, Rockford, IL) assay was utilized. We found that similar trends could be obtained by both methods of data presentation (Figs. 6.3 and 6.4) but significance was only achieved when IL-6 protein was calculated in terms of total tissue weight analyzed. Therefore, to detect

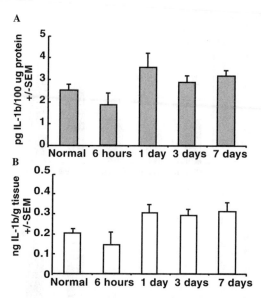

FIGURE 6.4 Analysis of IL-1β protein by ELISA produced similar trends whether IL-1β protein was quantified in terms of μg of total protein (A) or in terms of tissue weight (B) in L5 spinal cord tissue which was isolated at 6 h (n = 4), 1 (n = 4), 3 (n = 4), or 7 (n = 7) d post L5 spinal nerve transection as well as from normal naïve animals (n = 6).

and characterize small changes in cytokines, we have chosen to report data in terms of tissue weight analyzed, which may preclude the detection of small changes in a specific cytokine due to the simultaneous upregulation of a large number of proteins at a rate disproportionate with the upregulation of the protein of interest.

The time-course data clearly exhibited increases in IL-6 protein levels at 3 and 7 d post L5 spinal nerve transection (Fig. 6.3), while IL-1β appeared upregulated on days 1 and 7 post-transection (Fig. 6.4). The temporal expression pattern of IL-1β and IL-6 protein following injury is reminiscent of the temporal expression patterns of mRNA as detected by RPA and thus, warrants the same implications of time points in determining whether a difference in the protein of interest will be observed or not. Unfortunately, analysis of central cytokine contributions to nociception is not as simple as just looking at the level of a specific cytokine in the lumbar spinal cord. It must be remembered that there is endogenous regulation of cytokine expression and activity by concomitant expression of antiinflammatory cytokines, like IL-10 and IL-4, as well as endogenous cytokine antagonists. As already discussed in detail, regulation of cytokines at the message level also determines how much protein is synthesized. In addition, there are multiple points of regulation at the protein level, from the proteolytic cleavage of pro-proteins to the regulation of receptor and accessory protein expression. The complex regulation of the inflammatory cascade and the multiple proteins and signaling cascades involved highlights the challenge of studying the contribution of central cytokines to the etiology of neuropathic pain.

6.4 LOOKING BEYOND THE CONVENTIONAL BOX: IMMUNOCOMPETENCE IN THE CNS

The CNS was long considered an immunologically privileged site, a place wherein the immune system performed few functions. Over the past decade, this view has altered dramatically. The CNS is actively involved in immunological phenomena that are physiological and pathological. Work in our laboratory has focused on the mechanisms by which inflammation develops in the CNS in response to peripheral nerve and root injuries. Toward this end, we have utilized rodent models of neuro-pathic and radicular pain previously discussed in this review. In our dissection of these immunologically distinct systems, we have investigated topics relative to cell migration and trafficking, adhesion molecule expression, antigen presentation, and cytokine production. The recent discovery that the CNS broadly responds similarly, albeit not exactly, in an immunologic fashion to autoimmunity, trauma, brain abscess, and peripheral nerve or root injury that leads to chronic pain is highly innovative. For the first time, the mechanisms of CNS cellular glial and neuronal activation and leukocyte trafficking in CNS reactions following peripheral nerve and nerve root injury are being directly compared in the same laboratory.

6.4.1 POTENTIAL ROLE OF IMMUNOCOMPETENT GLIA

As alluded to earlier in this review, cytokines may play an important role in the development of central sensitization through their interaction with glia. Glial cells (microglia, astrocytes, and oligodendrocytes) constitute over 70% of the total cell population in the brain and spinal cord. Once thought of as merely a physical support system for neurons, glial cells have recently come under intense scrutiny as key neuromodulatory, neurotrophic, and neuroimmune elements in the CNS. Microglia, cells of monocytic origin, are the macrophages of the brain and as such perform a vast number of immune-related duties.[89] Pathological stimuli provoke a graded transformation of microglia from a highly ramified resting surveillance state ulti-mately to a phagocytic macrophage. Microglial activation involves a stereotypic pattern of cellular responses, such as proliferation, increased expression of immuno-molecules, recruitment to the site of injury, and functional changes that include the release of cytotoxic and/or inflammatory mediators. The initial signal for microglial activation is not well understood; however, following injury, neuronal depolarization combined with extracellular ion changes may be major stimuli.[90] Alternatively, neuronal signals such as nitric oxide (NO) or proinflammatory cytokines may provide the stimuli for this activation.[6,91] During autoimmune inflammation of the nervous system, microglia both release and respond to several cytokines including TNF, IL-1β, and IL-6, all of which are instrumental in astrocyte activation, induction of cell adhesion molecule expression, and recruitment of T-cells into the lesion. In addition to the synthesis of inflammatory cytokines, microglia act as cytotoxic effector cells by releasing harmful substances including proteases, reactive oxygen intermediates, and NO.

Glial changes in response to injury include proliferation and hypertrophy of astrocytic and microglia cells and overexpression of GFAP. GFAP increases in the spinal cord following different nerve injuries such as chronic constriction injury, nerve crush, and axotomy.[92-94] Following a peripheral nerve freeze lesion, immunoreactive GFAP expression increases at 14 d with a second major increase at 42 d, consistent with peaks in autotomy behavior and mechanical allodynia, respectively.[36] Glia are, indeed, intimately involved in the neuroimmune network following peripheral nerve injury. However, the concept that all glia are producing the same deleterious proteins and that ameliorating their function will have a beneficial outcome for persistent pain appears naïve. Using immunocytochemistry, we have demonstrated that only a subpopulation of astrocytes co-localize with specific cytokines and that a subpopulation of microglia express major histocompatibility complex class II (MHC II) and CD_{4+} expression.

6.4.2 Major Histocompatibility Complex

The MHC is a region of highly polymorphic genes whose products are expressed on the surfaces of a variety of cells. MHC genes play a central role in immune responses to protein antigens. The principal function of MHC molecules is to bind fragments of foreign proteins, thereby forming complexes that can be recognized by T-lymphocytes. T-lymphocytes do not recognize antigens in free or soluble form but instead only recognize portions of protein antigens (i.e., peptides) complexed to MHC. In the periphery, CD8 expression by T-cells is strongly correlated with MHC I, and CD4 T-helper cells are strongly correlated with MHC II. Having defined the role of MHC in the periphery, its role in the CNS is difficult to discern. In normal CNS, both MHC I and MHC II expression is minimal compared to other tissues.[89,95,96] This low expression of MHC and reduced immune surveillance are believed to contribute to the immune-privileged status of non-renewable CNS neurons.

However, microglia and astrocytes clearly express MHC II after neuronal and axonal degeneration.[97-99] The significance of this expression is still unclear and may involve functions other than antigen presentation. It has been recognized that a subpopulation of microglia become immunocompetent in response to infection or injury. One way in which this occurs is through the expression of MHC II. Cytokines are key modulators of MHC I and II genes in a wide variety of cells. Of relevance to the discussion of cytokines and pain processing, glia do not normally express MHC II but its expression on glial cells can be induced by cytokines such as TNF.[100] The mediators that act to alter the expression of the MHC II antigens are tissue specific. This has implications for possible selective immunomodulation of MHC II in tissues in which it is overexpressed without affecting MHC II expression in other tissues. We have preliminary data showing MHC II expression on cells with glial morphology after peripheral nerve injury (Fig. 6.5). These data further support the notion that central neuroinflammatory processes involving immune recognition play a role in peripheral nerve injury.

Of particular clinical interest, specific alleles for MHC II have been associated with a variety of autoimmune diseases like multiple sclerosis, rheumatoid arthritis, and systemic lupus erythematosus.[100] A genetic component of chronic pain has

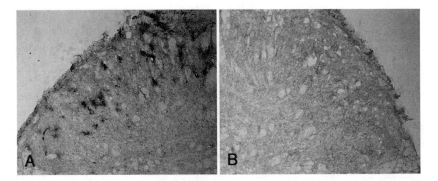

FIGURE 6.5 MHC II expression in L5 spinal cord on day 7 post L5 spinal nerve transection. MHC II expression was observed in the ipsilateral (A) but not contralateral dorsal horn (B). The slightly ramified appearance of the MHC II immunoreactivity supports a glial source of expression.

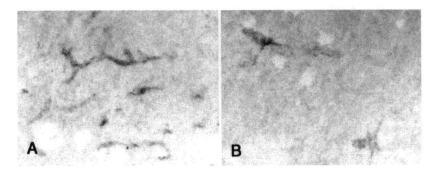

FIGURE 6.6 At day 3 post L5 spinal nerve transection an upregulation of both PECAM (A) and ICAM (B) immunoreactivity can be observed in the dorsal horn of the L5 spinal cord. Adhesion molecule staining appeared associated more with the vasculature than with glial cells.

recently emerged to help explain why all individuals with a similar injury or disease may not experience chronic pain.[101] To further support a genetic predisposition to chronic pain, a statistically significant increase of the MHC II antigen DQ1 was observed in patients with Complex Regional Pain Syndrome I as compared to control frequencies.[102] Similarly, we have put forth a provocative theory that specific polymorphisms in membrane glycoproteins such as MHC II may render an individual more susceptible to persistent pain after a root or peripheral nerve injury.[103]

6.4.3 CELLULAR ADHESION MOLECULES

In addition to the expression of MHC II on glial cells, upregulation of cellular adhesion molecules (CAM) has been observed in ischemia and multiple sclerosis.[104–106] We have preliminary data demonstrating increased intracellular adhesion molecule (ICAM) and platelet endothelial cellular adhesion molecule (PECAM) expression in the spinal cord following peripheral nerve injury (Fig. 6.6). CAMs are pivotal in the capture, rolling, adhesion, and migration of leukocytes into an area of infection. CAMs are cell-surface macromolecules that also control cell–cell interactions during the development of the nervous system by regulating such processes as neuronal adhesion and migration, neurite outgrowth, synaptogenesis, and intracellular signaling.[107] The cellular adhesion molecules are classified into four major families: the integrins, the immunoglobulin superfamily, cadherins, and selectins. ICAM and PECAM belong to the immunoglobulin superfamily.[108] TNF has been found to regulate ICAM in an autocrine manner when the CNS is exposed to an immunologic challenge.[109] TNF and IL-1β have been found to induce ICAM-1 expression on astrocytes.[55] In addition, ICAM expression is an important facilitator for astrocytes to function as antigen-presenting cells in intracerebral immune responses.[99] This means that proinflammatory cytokine expression is imperative for conversion of latent glial cells into reactive/activated immunocompetent cells.

6.5 SUMMARY

In relation to the upstream role of proinflammatory cytokines in producing spinal sensitization, there is mounting evidence of cytokines inducing release or expression of cyclooxygenase 2, inducible nitric oxide synthase, and substance P as well as enhancing capsaicin sensitivity.[50,110–112] Similarly, activated glial cells synthesize proinflammatory cytokines, proteases, NO, excess glutamate, superoxide anions, hydrogen peroxide, eicosanoids, and other toxins that act by way of the N-methyl-D-aspartate (NMDA) receptors.[8,9,113–115] Therefore, cytokines have the capacity to create downstream modulation of the CNS milieu that may indirectly enhance spinal sensitization.

In this chapter we have reviewed the evidence for the contribution of central cytokines in the etiology of persistent neuropathic pain states. We have discussed the experimental complexities inherent in the use of ISH and immunohistochemistry for the detection and quantification of cellular sources of cytokine mRNA and protein within spinal cord tissue. We have illustrated the temporal expression patterns of TNF, IL-1β, and IL-6 mRNA and protein when utilizing RPA or ELISA to detect changes in cytokine mRNA and protein expression in neuropathic and radicular pain models. Finally, we have set forth the novel hypothesis that central cytokines are important mediators in the activity of immunocompetent spinal glial cells and a neuroinflammatory cascade.

ACKNOWLEDGMENTS

The authors would like to thank Amy Rickman and Maria Rutkowski for technical assistance, Dr. William F. Hickey for monoclonal antibodies and glial expertise, and the following for grant support: National Institute of Drug Abuse grant DA11276 (J.A.D.) and DA05969 (S.M.S.); the National Institute of Arthritis and Musculoskeletal and Skin Diseases grant AR44757 (J.A.D.), and Bristol-Myers Squibb/Zimmer Orthopaedic Foundation (J.A.D.).

REFERENCES

1. Rothwell, N. and Hopkins, S., Cytokines and the nervous system II: actions and mechanisms of action, *TINS*, 18, 130, 1995.
2. Hopkins, S. and Rothwell, N., Cytokines and the nervous system I: expression and recognition, *TINS*, 18, 83, 1995.
3. Vitkovic, L., Bockaert, J., and Jacque, C., "Inflammatory" Cytokines: neuromodulators in Normal Brain?, *J. Neurochem.*, 74, 457, 2000.
4. Bluthe, R. et al., Synergy between tumor necrosis factor alpha and interleukin-1 in the induction of sickness behavior in mice, *Psychoneuroendocrinology*, 19, 197, 1994.
5. Barone, F.C. and Feurestein, G.Z., Inflammatory mediators and stroke: new opportunities for novel therapeutics, *J. Cereb. Blood Flow Metab.*, 19, 819, 1999.
6. Merrill, J. and Benveniste, E., Cytokines in inflammatory brain lesions: helpful and harmful, *TINS*, 19, 331, 1996.
7. Schubert, P. et al., Cascading glia reactions: a common pathomechanism and its differentiated control by cyclic nucleotide signaling, *Ann. NY Acad. Sci.*, 903, 24, 2000.
8. Kalaria, R., Cohen, D., and Premkumar, D., Cellular aspects of the inflammatory response in Alzheimers disease, *Neurodegeneration*, 5, 497, 1996.
9. Nottet, H. et al., A regulatory role for astrocytes in HIV-1 encephalitis: an overexpression of eicanoids, platelet-activating factor, and tumor necrosis factor-alpha by activated HIV-1 infected monocytes is attenuated by primary human astrocytes, *J. Immunol.*, 154, 3567, 1995.
10. Watkins, L., Maier, S., and Goehler, L., Immune activation: the role of pro-inflammatory cytokines in inflammation, illness responses and pathological pain states, *Pain*, 63, 289, 1995.
11. Nadeau, S. and Rivest, S., Role of microglial-derived tumor necrosis factor in mediating CD14 transcription and nuclear factor κ B activity in the brain during endotoxemia, *J. Neurosci.*, 20, 3456, 2000.
12. Kanaan, S.A. et al., Hyperalgesia and upregulation of cytokines and nerve growth factor by cutaneous leishmaniasis in mice, *Pain*, 85, 477, 2000.
13. Safieh-Garabedian, B. et al., Contribution of interleukin-1β to the inflammation-induced increase in nerve growth factor levels and inflammatory hyperalgesia, *Br. J. Pharmacol.*, 115, 1265, 1995.
14. Perkins, M., Kelly, D., and Davis, A., Bradykinin B_1 and B_2 receptor mechanisms and cytokine-induced hyperalgesia in the rat, *Can. J. Physiol. Pharmacol.*, 73, 832, 1995.
15. Poole, S. et al., Bradykinin B_1 and B_2 receptors, tumor necrosis factor α and inflammatory hyperalgesia, *Br. J. Pharmacol.*, 126, 649, 1998.
16. Woolf, C.J. et al., Cytokines, nerve growth factor and inflammatory hyperalgesia: the contribution of tumour necrosis factor α, *Br. J. Pharmacol.*, 121, 417, 1997.
17. Fukuoka, H. et al., Cutaneous hyperalgesia induced by peripheral injection of interleukin-1 beta in the rat, *Brain Res.*, 657, 133, 1994.
18. Watkins, L. et al., Mechanisms of tumor necrosis factor-α (TNF-α) hyperalgesia, *Brain Res.*, 692, 244, 1995.
19. Watkins, L. et al., Characterization of cytokine-induced hyperalgesia, *Brain Res.*, 654, 15, 1994.
20. Junger, H. and Sorkin, L., Nociceptive and inflammatory effects of subcutaneous TNFα, *Pain*, 85, 145, 2000.
21. Sorkin, L. et al., Tumour necrosis factor-alpha induces ectopic activity in nociceptive primary afferent fibres, *Neuroscience*, 81, 255, 1997.

22. Wagner, R. and Myers, R., Endoneurial injection of TNF-α produces neuropathic pain behaviors, *Neuroreport*, 7, 2897, 1996.
23. Oka, T., Aou, S., and Hori, T., Intracerebroventricular injection of interleukin-1β induces hyperalgesia in rats, *Brain Res.*, 624, 61, 1993.
24. Oka, T. et al., Intracerebroventricular injection of interleukin-6 induces thermal hyperalgesia in rats, *Brain Res.*, 692, 123, 1995.
25. Oka, T. et al., Intracerebroventricular injection of tumor necrosis factor-alpha induces thermal hyperalgesia in rats, *Neuroimmunomodulation*, 3, 135, 1996.
26. DeLeo, J.A. et al., Interleukin-6-mediated hyperalgesia/allodynia and increased spinal IL-6 expression in a rat mononeuropathy model, *J. Interferon Cytokine Res.*, 16, 695, 1996.
27. Poole, S. et al., Cytokine-mediated inflammatory hyperalgesia limited by interleukin-10, *Br. J. Pharmacol.*, 115, 684, 1995.
28. Ribeiro, R.A. et al., Analgesic effect of thalidomide on inflammatory pain, *Eur. J. Pharmacol.*, 391, 97, 2000.
29. Cunha, F.Q. et al., Cytokine-mediated inflammatory hyperalgesia limited by interleukin-4, *Br. J. Pharmacol.*, 126, 45, 1999.
30. Kitamura, Y. et al., Interleukin-4-inhibited mRNA expression in mixed rat glial and in isolated microglial cultures, *J. Neuroimmunol.*, 106, 95, 2000.
31. Sommer, C., Marziniak, M., and Myers, R., The effect of thalidomide treatment on vascular pathology and hyperalgesia caused by chronic constriction injury of rat nerve, *Pain*, 74, 83, 1998.
32. Wagner, R., Janjigian, M., and Myers, R., Anti-inflammatory interleukin-10 therapy in CCI neuropathy decreased thermal hyperalgesia, macrophage recruitment, and endoneurial TNFα expression, *Pain*, 74, 35, 1998.
33. Sommer, C., Schmidt, C., and George, A., Hyperalgesia in experimental neuropathy is dependent on the TNF Receptor 1, *Exp. Neurol.*, 151, 138, 1998.
34. Arruda, J.L. et al., Increase of interleukin-6 mRNA in the spinal cord following peripheral nerve injury in the rat: potential role of IL-6 in neuropathic pain, *Mol. Brain Res.*, 62, 228, 1998.
35. DeLeo, J.A. and Colburn, R.W., Progress in inflammation research, in *Cytokines and Pain*, Watkins, L. and Maier, S., Eds., Birkhauser Verlag, Basel, Switzerland, 1999, 159.
36. DeLeo, J.A. and Colburn, R.W., The role of cytokines in nociception and chronic pain, in *Low Back Pain: A Scientific and Clinical Overview*, Weinstein, J. and Gordon, S., Eds., American Academy of Orthopaedic Surgeons, Rosemont, IL, 1996.
37. DeLeo, J.A., Colburn, R.W., and Rickman, A.J., Cytokine and growth factor immunohistochemical spinal profiles in two animals models of mononeuropathy, *Brain Res.*, 759, 50, 1997.
38. Hashizume, H. et al., Spinal glial activation and cytokine expression after lumbar root injury in the rat, *Spine*, 25, 1206, 2000.
39. Sweitzer, S. et al., Acute peripheral inflammation induces moderate glial activation and spinal IL-1β expression that correlates with pain behavior in the rat, *Brain Res.*, 829, 209, 1999.
40. Watkins, L. et al., Evidence for the involvement of spinal cord glia in subcutaneous formalin induced hyperalgesia in the rat, *Pain*, 71, 225, 1997.
41. Sommer, C. et al., Neutralizing antibodies to interleukin 1-receptor reduce pain associated behavior in mice with experimental neuropathy, *Neurosci. Lett.*, 270, 25, 1999.
42. DeLeo, J.A. et al., Transgenic expression of TNF by astrocytes increases mechanical allodynia in a mouse neuropathy model, *Neuroreport*, 11, 599, 2000.

43. Murphy, P. et al., Endogenous interleukin-6 contributes to hypersensitivitiy to cutaneous stimuli and changes in neuropeptides associated with chronic nerve constriction in mice, *Eur. J. Neurosci.*, 11, 2243, 1999.

44. Ramer, M. et al., Spinal nerve lesion-induced mechanoallodynia and adrenergic sprouting in sensory ganglia are attenduated in interleukin-6 knockout mice, *Pain*, 78, 115, 1998.

45. Yabuchi, K. et al., Biphasic effects of intracerebroventricular interleukin-1β on mechanical nociception in the rat, *Eur. J. Pharmacol.*, 300, 59, 1996.

46. Souter, A.J. and Garry, M.G., Spinal interleukin-1β reduces inflammatory pain, *Pain*, 86, 63, 2000.

47. Kita, A., Imano, K., and Nakamura, H., Involvement of corticotropin-releasing factor in the antinociception produced by interleukin-1 in mice, *Eur. J. Pharmacol.*, 237, 317, 1993.

48. Coderre, T.J. et al., Contribution of central neuroplasticity to pathological pain: review of clinical and experimental evidence, *Pain*, 52, 259, 1993.

49. Jeanjean, A. et al., Interleukin-1β induces long term increase of axonally transported opioate receptors and substance P, *Neuroscience*, 68, 151, 1995.

50. Inoue, A. et al., Interleukin-1β induces substance P release from primary afferent neurons through the cyclooygenase-2 system, *J. Neurochem.*, 73, 2206, 1999.

51. Malcangio, M. et al., Effect of interleukin-1 beta on the release of substance P from rat isolated spinal cord, *Eur. J. Pharm.*, 299, 113, 1996.

52. Ding, M., Hart, R., and Jonakait, F., Tumor necrosis factor-α induces substance P in sympathetic ganglia through sequential induction of interleukin-1 and leukemia inhibitory factor, *J. Neurobiol.*, 28, 445, 1995.

53. Gadient, R.A. and Otten, U.H., Interleukin-6 (IL-6) — a molecule with both beneficial and destructive potentials, *Prog. Neurobiol.*, 52, 379, 1997.

54. Palma, C. et al., Functional characterization of substance P receptors on cultured human spinal cord astrocytes: synergism of substance P with cytokines in inducing interleukin-6 and prostaglandin E2 production, *Glial*, 21, 183, 1997.

55. Aschner, M., Astrocytes as mediators of immune and inflammatory responses in the CNS, *Neurotoxicology* 19, 269, 1998.

56. DeLeo, J.A. et al., Characterization of a neuropathic pain model: sciatic cryoneurolysis in the rat, *Pain*, 56, 9, 1994.

57. Colburn, R.W. et al., Dissociation of microglial activation and neuropathic pain behaviors following peripheral nerve injury in the rat, *J. Neuroimmunol.*, 79, 163, 1997.

58. Tonai, T. et al., Possible involvement of interleukin-1 in cyclooxygenase-2 induction after spinal cord injury in rats, *J. Neurochem.*, 72, 302, 1999.

59. Meller, S. et al., The possible role of glia in nociceptive processing and hyperalgesia in the spinal cord of the rat, *Neuropharmacology*, 33, 1471, 1994.

60. Fine, S. et al., Tumor necrosis factor alpha inhibits glutamate uptake by primary human astroctyes, *J. Biol. Chem.*, 271, 15303, 1996.

61. Koller, H., Thiem, K., and Siebler, M., Tumour necrosis factor-alpha increases intracellular Ca^{2+} and induces a depolarization in cultured astroglial cells, *Brain*, 119, 2021, 1996.

62. Wall, P., *Advances in Pain Research and Therapy*, Bonica, J.J., Ed., Raven Press, New York, 1983.

63. Bennett, G.J. and Xie, Y., A peripheral mononeuropathy in rat that produces disorders of pain sensation like those seen in man, *Pain*, 33, 87, 1988.

64. Seltzer, Z., Dubner, R., and Shir, Y., A novel behavioral model of neuropathic pain disorders produced in rats by partial sciatic nerve injury, *Pain*, 43, 205, 1990.

65. DeLeo, J.A. and Coombs, D., Autotomy and decreased substance P following peripheral cryogenic nerve lesion, *Cryobiology*, 28, 460, 1991.

66. Kim, S. and Chung, J., An experimental model for peripheral neuropathy produced by segmental spinal nerve ligation in the rat, *Pain*, 50, 355, 1992.

67. Vos, B., Strassman, A., and Maciewicz, R., Behavioral evidence of trigeminal neuropathic pain following chronic constriction injury to the rat's infraorbital nerve, *J. Neurosci.*, 14, 2708, 1994.

68. McCarron, R. et al., The inflammatory effect of nucleus pulposus: a possible element in the pathogenesis of low-back pain, *Spine*, 12, 760, 1987.

69. Olmarker, K. et al., Effects of experimental graded compression on blood flow in spinal nerve roots: a vital microscopic study on the porcine cauda equina, *J. Orthop. Res.*, 7, 817, 1989.

70. Olmarker, K., Rydevik, B., and Nordborg, C., Autologous nucleus pulposus induces neurophysiologic and histologic changes in porcine cauda equina nerve roots, *Spine*, 18, 1425, 1993.

71. Kawakami, M. et al., Pathomechanism of pain-related behavior produced by allografts of intervertebral disc in the rat, *Spine*, 21, 2101, 1996.

72. Kawakami, M., Weinstein, J., and Spratt, K., Experimental lumbar radiculopathy: immunohistochemical and quantitative demonstrations of pain induced by lumbar nerve root irritation of the rat, *Spine*, 19, 1780, 1994.

73. Kayama, S. et al., Incision of the anulus fibrosus induces nerve root morphologic, vascular, and functional changes: an experimental study, *Spine*, 21, 2539, 1996.

74. Emson, P., In-situ hybridization as a methodological tool for the neuroscientist, *Trends Neurosci.*, 16, 9, 1993.

75. Sambrook, J., Fritsch, E.F., and Maniatis, T., Extraction, purification, and analysis of messenger RNA from eukaryotic cells, in *Molecular Cloning: A Laboratory Manual*, 2nd ed., Nolan, C., Ed., Cold Spring Harbor Laboratory Press, New York, 1989, chaps. 7 and 14.

76. Wilfinger, W., Mackey, K., and Chomczynski, P., Effect of pH and ionic strength on the spectrophotometric assessment of nucleic acid purity, *BioTechniques*, 22, 474, 1997.

77. Stern, E.L. et al., Spatiotemporal induction patterns of cytokine and related immune signal molecule mRNAs in response to intrastriatal injection of lipopolysaccharide, *J. Neuroimmunol.*, 106, 114, 2000.

78. Busso, N. et al., Antagonist effect of RU 486 on transcription of glucocorticoid-regulated genes, *Exp. Cell Res.*, 173, 425, 1987.

79. Lee, S. et al., Glucocorticoids selectively inhibit the transcription of the interleukin 1 beta gene and decrease the stability of interleukin 1 beta mRNA, *Proc. Natl. Acad. Sci. USA*, 85, 1204, 1988.

80. Fenton, M., Review: transcriptional and post-transcriptional regulation of interleukin 1 gene expression, *Int. J. Immunopharmacol.*, 14, 401, 1992.

81. Chantry, D. et al., Modulation of cytokine production by transforming growth factor-beta, *J. Immunol.*, 142, 4295, 1989.

82. Sung, S. and Walters, J., Increased cyclic AMP levels enhance IL-1 alpha and IL-1 beta mRNA expression and protein production in human myelomonocytic cell lines and monocytes, *J. Clin. Invest.*, 88, 1915, 1991.

83. Knudsen, P., Dinarello, C., and Strom, T., Prostaglandins posttranscriptionally inhibit monocyte expression of interleukin 1 activity by increasing intracellular cyclic adenosine monophosphate, *J. Immunol.*, 137, 3189, 1986.

84. Dinarello, C., Dissociation of transcription from translation of human IL-1 beta: induction of steady state mRNA by adherence or recombinant C5a in the absence of translation, *Proc. Soc. Exptl. Biol. Med.*, 200, 228, 1992.

85. Hershey, J., Translational control in mammalian cells, *Annu. Rev. Biochem.*, 60, 717, 1991.

86. Jansen, M. et al., Translational control of gene expression, *Pediatr. Res.*, 37, 681, 1995.

87. Watkins, L. et al., Dynamic regulation of the proinflammatory cytokine interleukein-1 molecular biology for non-molecular biologists, *Life Sci.*, 65, 449, 1999.

88. De La Monte, S., Ganju, N., and Wands, J., Microtiter immunocytochemical ELISA assay, *Biotechniques*, 26, 1073, 1999.

89. Hickey, W. and Kimura, H., Perivascular microglia are bone marrow derived and present antigen *in vivo*, *Science*, 239, 290, 1988.

90. Caggiano, A. and Kraig, R., Eicosanoids and nitric oxide influence induction of reactive gliosis from spreading depression in microglia but not astrocytes, *J. Compar. Neurol.*, 369, 93, 1996.

91. Hauser, C., Regional macrophage activation after injury and compartmentalization of inflammation in trauma, *New Horiz.*, 4, 235, 1996.

92. Garrison, C. et al., Staining of glial fibrillary acidic protein (GFAP) in lumbar spinal cord increases following a sciatic nerve constriction injury, *Brain Res.*, 565, 1, 1991.

93. Hajos, F., Csillik, B., and Knyihar-Csillik, E., Alternations in glial fibrillary acidic protein immunoreactivity in the upper dorsal horn of the rat spinal cord in the course of transganglionic atropy and regenerative proliferation, *Neurosci. Lett.*, 117, 8, 1990.

94. Tetzlaff, W. et al., Increased glial fibrillary protein synthesis in astrocytes during retrograde reaction of the rat facial nucleus, *Glial*, 1, 90, 1988.

95. Raivich, G. et al., Neuroglial activation repertoire in the injured brain: graded response, molecular mechanisms and cues to physiological function, *Brain Res. Rev.*, 30, 77, 1999.

96. Kreutzberg, G., Microglia: a sensor of pathological events in the CNS, *Trends Neurosci.*, 19, 312, 1996.

97. Shrikant, P. and Benveniste, E., The central nervous system as an immunocompetent organ. Role of glial cells in antigen presentation, *J. Immunol.*, 157, 1819, 1996.

98. Vass, K. and Lassmann, H., Intrathecal application of interferon gamma. Progressive appearance of MHC antigens within the rat nervous system, *Am. J. Pathol.*, 137, 789, 1990.

99. Cornet, A. et al., Role of astrocytes in antigen presentation and naive T-cell activation, *J. Neuroimmunol.*, 106, 69, 2000.

100. Roitt, I., Brostoff, J., and Male, D., *Immunology*, 5th ed., Mosby, London, 1998.

101. Mogil, J. et al., Heritability of nociception I: responses of 11 inbred mouse strains on 12 measures of nociception, *Pain*, 80, 67, 1999.

102. Kemler, M., Vusse, A., and Berg-Loonen, E., The association between lymphocyte antigen and Complex Regional Pain Syndrome, Type I, 9th World Congress of Pain, IASP, Vienna, Austria, 1999.

103. DeLeo, J.A., Correlation of animal models to the clinical syndrome of low back pain, 9th World Congress of Pain, IASP, Vienna, Austria, 1999.

104. Yang, G. et al., Inhibition of TNFalpha attenuates infarct volume and ICAM-1 expression in ischemic mouse brain, *Neuroreport*, 9, 2131, 1998.

105. Yang, G. et al., Expression of intercellular adhesion molecule 1 (ICAM-1) is reduced in permament focal cerebral ischemic mouse brain using an adenoviral vector to induce overexpression of interleukin-1 receptor antagonist, *Brain Res.*, 65, 143, 1999.

106. Khoury, S. et al., Changes in serum levels of ICAM and TNF-R correlate with disease activity in multiple sclerosis, *Neurology*, 53, 758, 1999.

107. Lee, S. and Benveniste, E., Adhesion molecule expression and regulation on cells of the central nervous system, *J. Neuroimmunol.*, 98, 77, 1999.

108. Polverini, P., Cellular adhesion molecules. Newly identified mediators of angiogenesis, *Am. J. Pathol.*, 148, 1023, 1996.

109. Freyer, D. et al., Cerebral endothelial cells release TNF-α after stimulation with cell walls of *Streptococcus pneumoniae* and regulate inducible nitric oxide synthase and ICAM-1 expression via autocrine loops, *J. Immunol.*, 163, 4308, 1999.

110. Serou, M., DeCoster, M., and Bazan, N., Interleukin-1 beta activated expression of cyclooxygesase-2 and inducible nitric oxide synthase in primary hippocampal neuronal culture: platelet activating factor as a preferential mediator of cyclooxygease-2 expression, *J. Neurosci. Res.*, 58, 593, 1999.

111. Huang, Z.-F., Massey, J., and Via, D., Differential regulation of cyclooxygenase (COX-2) mRNA stability by interleukin-1b and tumor necrosis factor-a in human in vitro differentiated macrophages, *Biochem. Pharmacol.*, 59, 187, 2000.

112. Nicol, G., Lopshire, J., and Pafford, C., Tumor necrosis factor enhances the capsaicin sensitivity of rat sensory neurons, *J. Neurol.*, 17, 975, 1997.

113. Chao, C. and Hu, S., Tumor necrosis factor alpha potentiates glutamate neurotoxicity in human fetal brain cell cultures, *Dev. Neurosci.*, 16, 172, 1994.

114. Chao, C. et al., Interleukin-1 and tumor necrosis factor-alpha synergistically mediate neurotoxicity: involvement of nitric oxide and of N-methyl-D-aspartate receptors, *Brain Behav. Immun.*, 9, 355, 1995.

115. Mallat, M. and Chamak, B., Brain macrophages: neurotoxic or neurotrophic effector cells?, *J. Leukoc. Biol.*, 56, 416, 1994.

7 Extracellular Sampling Techniques

Igor Spigelman, Yoshizo Matsuka,
John K. Neubert, and Nigel T. Maidment

CONTENTS

7.1 INTRODUCTION

Extracellular sampling is of considerable importance in most biomedical disciplines. Many of the techniques have been devised and implemented for more than 30 years with improvements coming primarily in the form of continued trends toward miniaturization of the probes used to sample the extracellular environment. Major advances have been made particularly in the speed and accuracy of resolving substances of interest in the extracellular samples. This chapter provides an overview of the various *in vivo* extracellular sampling techniques that may be of use to pain researchers. These are followed by several recently developed methods that allow extracellular sampling with very fast, in some cases sub-millisecond time resolution, but which are currently limited to *in vitro* preparations.

133

7.2 *IN VIVO* SAMPLING TECHNIQUES

7.2.1 BLISTER SUCTION TECHNIQUE

This method was first described by Norwegian researchers in 1964,[1] and involves blister formation by using suction to raise the epidermal layer of the skin. Used predominantly in human studies, this technique allows one to sample an environment where the vast majority of injuries are encountered as a result of penetrating the protective epidermis. Prior to the introduction of the blister suction technique, pain researchers used chemically induced blisters.[2] These cantharidin-induced blisters were used primarily to study the responses of humans to putative algogens and puritorigenic agents rather than to sample the fluid in the blister base.[3] Part of this technique's success can be attributed to the ability of human subjects to quantitatively report on sensations elicited by perfusion of the blister with algogenic substances. However, chemical- or suction-induced blisters produce inflammatory injury, thereby making the necessary comparison with non-injured tissue difficult. More recently, the blister method was also shown to be useful for studies on disease induced-changes in the morphology of epidermal nerves.[4]

7.2.2 THE WICK METHOD

Developed in 1973 to measure protein concentration in interstitial fluid,[5] this method involves surgical placement of nylon wicks in the tissue of interest (e.g., subcutaneously or intramuscularly). The method is based on the assumption that the wick will attain a protein concentration that is in osmotic equilibrium with free interstitial fluid. After the wick is removed, the wick fluid is recovered and analyzed for total protein, colloid osmotic pressure, and protein composition. Depending on the treatment of wicks (e.g., dry or saline-soaked), interstitial fluid contents may equilibrate with the wick at different rates.[6] One of the problems with the technique is that wick implantation causes a transitory inflammatory reaction resulting in plasma extravasation, such that the protein content of the interstitial fluid becomes elevated.[7] The technique has not found favor with pain researchers, probably because of the availability of the blister suction technique, as well as other more efficient methods of extracellular sampling, such as the push–pull cannula and microdialysis.

7.2.3 CORTICAL CUP

One of the earliest techniques devised for *in vivo* sampling of the extracellular environment,[8] the cortical cup is conceptually similar to the blister technique in that an exposed surface of a tissue (in this case a small, anatomically defined area of cerebral cortex) is superfused and the withdrawn superfusate analyzed for substances of interest. The cortical cup technique has been successfully used to study release of neurotransmitters and other substances from the cerebral cortex of living animals; however, it is limited to studies of small areas of exposed cortex. For detailed descriptions of methodology, the reviews in References 9 and 10 are particularly useful.

7.2.4 PUSH–PULL CANNULA

This technique was first used in 1958 to sample the microenvironment of subcutaneous tissue,[11] but was quickly adapted for use in various other tissues, including the CNS.[12] It involves implanting two side-by-side or concentric cannulae and infusing fluid into one cannula while withdrawing fluid from the other cannula. In this manner, the tissue at the tip of the probe is perfused and the perfusion fluid is removed via the pull cannula and subjected to chemical analysis. For methodological details, the reviews by Myers[9,13–15] and Philippu[16] provide extensive information on the technique. Several inherent drawbacks can limit the usefulness of the technique, especially in the hands of inexperienced investigators. These drawbacks include mismatch of push–pull flow rates, blockage of cannula tips, and damage to the tissue of interest.[17] Despite such drawbacks, the push–pull cannula technique has been used successfully to sample the extracellular environment in various areas of the brain,[18–20] brainstem,[21] spinal cord,[22–24] and temporomandibular joint.[25]

7.2.5 SPINAL SUPERFUSION

This method for perfusion of the spinal subdural space was initially developed by Yaksh and Rudy in 1976[26] for administration of drugs, and subsequently, was improved upon to allow collection of perfusion samples in the anesthetized rat and cat.[27] The method is unique in that the samples are obtained from perfusion of a very large area of the spinal cord intrathecal space, yet it offers all of the advantages of an *in vivo* preparation, including in theory, activation of primary afferent neurons by natural stimuli. Both push–pull and peristaltic pump systems have been used with the spinal superfusion technique. However, it must be recognized that the analysis of spinal superfusate samples may at best provide only a relative indication of the concentrations of any extracellular substance of interest, including that of neurotransmitters released in the synaptic cleft. The book chapter by Yaksh[27] provides the methodological details, as well as various considerations for interpreting data obtained using the spinal superfusion technique.

7.2.6 MICRODIALYSIS

Microdialysis, developed by Ungerstedt and co-workers[28] in the late 1970s to early 1980s can be considered an evolution of the push–pull cannula technique. The most popular design of the microdialysis probe incorporates a concentric or side-by-side double cannula (steel and/or fused silica) enclosed by a hollow tubular dialysis membrane sealed at its tip with resin. Artificial cerebrospinal fluid or modified Ringer is perfused into the membrane through one cannula and flows out of the probe via the second cannula for collection and subsequent analysis. Substances in the extracellular milieu that are of sufficiently small size (the molecular weight cut-off can be varied by changing the membrane, but is most often about 20,000 Da) pass across the membrane by the process of simple diffusion down their concentration gradient, which is maintained by the continual flow of perfusion medium. Flow rates ranging from 0.1 to 5 µl/min have been used.

There are several advantages of this technique over push–pull. Primarily because microdialysis uses a closed system and slower flow rates, tissue damage is reduced, although disruption of the extracellular environment remains significant and undoubtedly influences the neurotransmitter processes under study.[29] Also, the incorporation of the membrane results in very clean, protein-free samples that can be injected directly onto analytical high-performance liquid chromatography (HPLC) columns. This combination of microdialysis and HPLC with electrochemical or fluorescence detection is used widely to continuously monitor fluctuations in extracellular concentrations of several neuromodulators such as dopamine,[30] norepinephrine,[31] serotonin,[32] acetylcholine,[33] glutamate,[34] GABA,[35] and adenosine[36] in brain and spinal cord. An associated disadvantage is that large peptide neuromodulators diffuse across the membrane with poor efficiency, making them difficult to detect. However, using sensitive radioimmunoassays, small peptides, including the enkephalins, endorphins, cholecystokinin, and substance P, have been detected in brain[37–39] and spinal cord[40] and, more recently, in trigeminal ganglia.[41] The text edited by Robinson and Justice remains an excellent source of information for the aspiring microdialysis user.[42]

7.2.7 ELECTROCHEMISTRY

The techniques described thus far require off-line analysis of substances removed from the extracellular environment. One such method of analysis used for quantification of biogenic amines, such as dopamine and serotonin, is HPLC coupled to an electrochemical detector incorporating a carbon-based electrode. *In vivo* electrochemistry methods pioneered by Ralph Adams and co-workers[43] utilize a miniaturized version of such a detector placed directly into the brain region of interest. The active electrode usually consists of a carbon fiber, to the surface of which is applied a potential (vs. an Ag/AgCl reference electrode placed on the brain surface) sufficient to oxidize the compound of interest. The resultant current at the surface of the electrode is measured by an external potentiostat and its amplitude is proportional to the concentration of the compound surrounding the electrode. The potential waveform applied to the surface of the electrode may be in the form of a square wave (chronoamperometry), triangular wave (cyclic voltammetry), linear ramp, constant potential (amperometry), and variations thereof depending on the type of measurement being made. Technical information on these various methods can be found in several texts and recent reviews.[44–47]

The main advantage of this technique is its greater spatial and temporal resolution compared to the sampling methods described above. The fiber electrodes are of the order of 10 μm or less in diameter compared to 250 μm for microdialysis. Whereas microdialysis sampling is generally made at intervals of several minutes, electrochemical techniques offer resolution as low as several milliseconds both *in vivo*[48] and in tissue slices.[49] Moreover, using amperometry, individual quantal release events have been recorded from cells in culture.[50,51] The major limitation of this methodology is the relatively small number of compounds of biological interest that can be detected. Neuromodulators that are directly oxidizable within the potential window compatible with biological measurement are restricted to the biogenic amines:

dopamine,[52] norepinephrine,[49] serotonin,[53] and nitric oxide.[54] However, the ability to adsorb enzymes to the surface of platinum- or carbon-based electrodes offers the prospect of increasing the repertoire of compounds measurable with electrochemistry. For instance, glutamate oxidase has been adsorbed to the surface of both platinum filaments and carbon fibers enabling the measurement of glutamate *in vivo*.[55,56] The oxidation of glutamate catalyzed by the enzyme results in production of hydrogen peroxide, which can be oxidized at the surface of the platinum electrode.[55] Alternatively, by coupling glutamate oxidase to horseradish peroxidase and utilizing an osmium-containing polymer to facilitate electron transfer at carbon-fiber electrodes, it is possible to operate under reducing conditions, thereby avoiding cross-talk with biogenic amines.[56]

7.2.8 ANTIBODY MICROPROBES

This method was originally devised by Duggan and Hendry to measure release of substance P in the spinal cord *in vivo*.[57] Dissatisfied with the tissue damage produced by push–pull cannulae, Duggan and colleagues used glass microelectrodes for recording extracellular potentials in the spinal cord and treated these electrodes to immobilize a substance P antibody on the electrode surface. After removal from the spinal cord, the probes were analyzed for substance P content using a variant of radioimmunoassay.[58,59] Subsequent studies have utilized antibody microprobes to detect localized release of neurokinin A,[60] calcitonin gene-related peptide,[61] opioid peptides,[62,63] neuropeptide Y,[64] galanin,[65,66] and somatostatin[67] at various sites in the CNS. The ability to measure other peptides of interest with this technique is limited only by the availability of antibodies for a given peptide and the radiolabeled peptide for the radioimmunoassay. The technique has a couple of important advantages over microdialysis- and push–pull cannula-based techniques for peptide detection. One is the reduced tissue damage due to the relatively small diameters of the glass probes, and perhaps more importantly, reduced damage from the absence of dialysate or fluid at the site of sampling. The other main advantage of using antibody microprobes is the ability to anatomically localize release sites, since localized peptide release produces binding of the peptide at localized areas along the microprobe.[68] On the other hand, microdialysis and push–pull cannula techniques permit long-term continuous sampling from a focal area of interest, something that cannot be accomplished with the single-use antibody microprobes. Many articles and reviews describe the practical applications and detailed methodology of antibody microprobe manufacture and use.[69–71]

7.2.9 ION-SELECTIVE MICROELECTRODES (ISMs)

These electrochemical sensors allow for potentiometric determination of the activity (or free concentration) of a particular ion in the presence of other ions. ISMs allow for local measurement of ion activities either extracellularly or intracellularly, as well as *in vitro* or *in vivo*. In addition, multi-barreled microelectrodes can monitor activities of several ion species simultaneously. Not surprisingly, ISMs have contributed to advancing the knowledge of practically all biological systems where ion channels and/or ion transport are involved.

The methodology and techniques for manufacture of ISMs have undergone considerable changes in the last 4 decades. Microelectrodes with H^+-selective glass membranes were first used in 1954.[72] Different types of glass selective for various other ions were introduced in 1959,[73] but they never became popular, primarily due to difficulties in their manufacture and use.[74] Such glass microelectrodes were later replaced with ion-selective liquid-ion exchanger microelectrodes.[75–77] In the mid- to late 1970s, solutions based on neutral carriers became available as an alternative to ion-exchanger-based microelectrodes.[78,79] These quickly became very popular owing to their superior selectivity for various ions compared to the ion-exchanger-based microelectrodes. Further improvements were made to reduce the very high ohmic resistance of such neutral carrier-based liquid membranes.[78,80,81] Even with these improvements, ISMs require the use of high-input resistance amplifiers with low leak currents.[82,83] The high ohmic resistance of ISMs also places a limit on the speed of their responses to a change in ion activity, such that most ISMs have response times in the order of hundreds of milliseconds.

In order to prevent water from displacing the neutral carrier-based liquid membrane from the inner surface of the hydrophilic glass microelectrode tips, the tips are made to react with organic silicon compounds which makes the surface highly hydrophobic. This process of silanization was introduced by Walker[77] and has also been modified through the years.[84] Both vapor and liquid methods of silanization are used. Several improvements in the methods for filling and silanization of ISMs have been described.[85,86]

Measurement of extracellular ion activities has been a common application of ISMs. Thus, extracellular K^+, Na^+, Ca^{2+}, H^+, NH_4^+, Cl^-, and HCO_3^- activities have all been studied with ISMs under physiological and pathophysiological conditions in the CNS.[82] ISMs have also been used to estimate extracellular space volume by measuring the activity of impermeant extracellular marker ions.[82] Serum levels of Mg^{2+} have also been measured with ISMs in patients with migraine.[87]

For pain researchers interested in using ion-selective microelectrodes, the reviews and books on theory and practical applications by Thomas,[84] Ammann,[79] Carlini and Ransom,[82] and Schlue et al.,[83] are particularly recommended.

7.3 *IN VITRO* SAMPLING TECHNIQUES

7.3.1 OPEN FLOW PERFUSION

Perfusion of excised tissues of interest is a tremendously popular technique that has greatly advanced our knowledge of biological systems. Much of the popularity of this technique may be attributed to the ability to alter the extracellular milieu of perfused tissue at will, coupled with mechanical stability of such preparations which facilitates recording with various microelectrode electrophysiological techniques (see Spigelman, Light, and Gold, this volume). The absence of a blood-brain-barrier permits pre-loading of brain and spinal cord slice preparations with labeled (e.g., radioactive) precursors for studies of neurotransmitter release. During perfusion, brain or spinal cord slices may also be discretely stimulated to activate selective neuronal pathways and study neurotransmitter release from such pathways. The collected perfusate may then be subjected to almost any analytical technique for detecting chemicals of interest.

Another, less obvious advantage is the ability to greatly vary the perfusion flow rate depending on experimental requirements. Thus, slow perfusion rates may allow accumulation of detectable quantities of a substance released extracellularly in small amounts. Conversely, fast perfusion may remove a substance from its release sites before it is degraded by local extracellular, membrane-bound enzymes.

Other extracellular sampling methods may also be used in conjunction with collection and subsequent analysis of *in vitro* tissue perfusates. These include microdialysis,[88,89] electrochemistry,[46] antibody microprobes,[90] and ion-selective microelectrodes.[91]

Despite the many advantages of open flow perfusion, the excised tissues of interest are maintained post-mortem in rather artificial conditions that cannot truly represent the *in vivo* situation. The basal metabolism of brain slices is far lower than brain metabolism *in vivo*. Furthermore, the profound sensitivity of CNS tissue to anoxia almost invariably produces damage, compounding the mechanical damage inflicted by slicing procedures. Thus, intracellular organelles in slice preparations often appear abnormal under microscopic observation. Such differences between the *in vivo* and *in vitro* conditions may affect the extracellular concentrations of released neurotransmitters. For example, the evoked release of dopamine from brain slices appears to be greater than that observed *in vivo*.[46] Therefore, a balanced approach that utilizes both *in vivo* and *in vitro* sampling techniques may provide a more satisfactory appraisal of neurotransmitter release.

7.3.2 DETECTOR PATCHES

The concept of detecting substances of interest using biological membranes containing receptors for these substances is quite elegant in its simplicity. This became technically feasible only after the introduction of patch-clamp techniques for single-channel recording.[92,93] The detector patches may be obtained by excising an outside-out patch from a donor cell,[94] or by reconstituting channels in lipid bilayers formed at the tip of a patch pipette.[95–97] Once obtained, the detector patch pipette is placed close to the source of the chemical to be detected. The preparation may be chemically or electrically stimulated to release the chemical of interest. The level of channel activity in the detector patch then serves as an indicator of the extracellularly released chemical. Ligand-gated ion channels are activated by physiological concentrations of agonists on a millisecond to sub-millisecond time scale and with great amplification of the initiating chemical signal. These qualities impart the technique with both high speed and high sensitivity for detecting released substances of interest.

To date, the release of acetylcholine,[98,99] glutamate,[100] and adenosine triphosphate (ATP)[101,102] have been measured using detector patch methodology.

A variety of channels gated directly by released neurotransmitters can be used as detectors. However, for best results several factors need to be kept in mind. First, a detector patch containing a high density of ligand-gated channels is preferable, since this increases the signal-to-noise ratio of detection. Second, directly gated ion channels are preferable to channels that require generation of second messengers for activation, since second messenger systems are more likely to be disrupted in the dialyzed sub-membrane milieu of the detector patch. Third, the kinetics of the detector channels should be well characterized, if only to avoid false positives from the activation of other contaminating ligand-gated ion channels. Activation of contaminating channels may be reduced by introduction of selective channel blockers to the *in vitro* preparation. However, it should be recognized that a pharmacological channel blockade may also affect the release of the substance that is being detected. Fourth, channels with fast activation and deactivation kinetics allow for fast time resolution of release. Fifth, the concentration/response relationship for a particular ligand should optimally be centered around the range of physiological concentrations of the ligand encountered in the extracellular space. Such concentration–response relationships may be obtained by exposing the detector patch to different concentrations of a ligand. A perfusion system for fast application of different solutions to the detector patch is necessary for such experiments (e.g., Perfusion Fast-Step System from Warner Instruments Corp., Hamden, CT). One caveat here is that many ligand-gated ion channels are subject to considerable receptor desensitization/sensitization making them poorly suitable for rapid multiple or continuous exposures to an agonist.

7.3.3 ATOMIC FORCE MICROSCOPY (AFM)

This is a powerful and relatively new technique that allows investigation of the surface of living cells at ultra-high (nanometer scale) resolution.[103,104] Essentially, a functional tip/cantilever is moved over a given cell and reports back minute topological differences across the surface of a cell. However, when the functional tip/cantilever is appropriately treated, it may also be used to measure elasticity or enzyme function,[105,106] detect various chemicals,[107] or even measure receptor–ligand or antibody–antigen intermolecular forces.[108–110]

Recently, AFM was used to continuously detect extracellularly released ATP on the surfaces of living cells.[111] Here, the functional tip/cantilever was coated with the myosin subfragment S1, such that when the cantilever encountered ATP, it produced a deflection in the cantilever. Detectable deflections occurred in a concentration range of 10 to 500 nM ATP.[111] This method has several advantages over the traditional bioluminescence assay for ATP.[112–114] One major advantage is that it allows localization of changes in [ATP] along the surface of single cells. It also obviates the need for exposing the solution that is to be sampled to an exogenous substrate (luciferin) and O_2. In addition, the cantilever can be used repeatedly, whereas aqueous luciferase and luciferin must be replaced after each measurement in the traditional bioluminescence assay.

One of the few disadvantages of AFM for sampling the extracellular environment is its restriction to *in vitro* preparations. Furthermore, extreme care has to be taken to eliminate any sources of vibration, which include perfusion fluid movement. Also, the expense of such systems is currently prohibitive to all but a few well-established neuroscience laboratories.

ACKNOWLEDGMENTS

This work was supported by the NIH grants DE07212, DE00408, and the Whitehall Foundation.

REFERENCES

1. Kiistala, U. and Mustakallio, K.K., In vivo separation of epidermis by production of suction blisters, *Lancet,* I, (Abstr.) 1444, 1964.
2. Shelley, W.B. and Arthur, R.P., The neurohistology and neurophysiology of the itch sensation in man, *A.M.A. Arch. Derm.*, 76, 296, 1957.
3. Keele, C.A. and Armstrong, D., *Substances Producing Pain and Itch*, Williams & Wilkins, Baltimore, 1964.
4. Kennedy, W.R., Nolano, M., Wendelschafer-Crabb, G., Johnson, T.L., and Tamura, E., A skin blister method to study epidermal nerves in peripheral nerve disease, *Muscle Nerve*, 22, 360, 1999.
5. Aukland, K. and Fadnes, H.O., Protein concentration of interstitial fluid collected from rat skin by a wick method, *Acta Physiol. Scand.*, 88, 350, 1973.
6. Kramer, G.C., Sibley, L., Aukland, K., and Renkin, E.M., Wick sampling of interstitial fluid in rat skin: further analysis and modifications of the method, *Microvasc. Res.*, 32, 39, 1986.
7. Fadnes, H.O. and Aukland, K., Protein concentration and colloid osmotic pressure of interstitial fluid collected by the wick technique: analysis and evaluation of the method, *Microvasc. Res.*, 14, 11, 1977.
8. MacIntosh, F.C. and Oborin, P.E., Release of acetylcholine from intact cerebral cortex, *Abstr. Commun., XIX Int. Physiol. Congr., Montreal,* 580, 1953.
9. Myers, R.D., Methods for perfusing different structures of the brain, in *Methods in Psychobiology,* Myers, R.D., Ed., Academic Press, New York, 1972.
10. Moroni, F. and Pepeu, G., The cortical cup technique, in *Measurement of Neurotransmitter Release in Vivo,* Marsden, C.A., Ed., Wiley, New York, 1984.
11. Fox, R.H. and Hilton, S.M., Bradykinin formation in human skin as a factor in heat vasodilatation, *J. Physiol. (Lond.)*, 142, 219, 1958.
12. Gaddum, J.H., Push–pull cannulae, *J. Physiol. (Lond.),* 155, (Abstr.) 1P, 1961.
13. Myers, R.D., *Handbook of Drug and Chemical Stimulation*, Van Nostrand-Reinhold, New York, 1974.
14. Myers, R.D., Chronic methods: intraventricular infusion, cerebrospinal fluid sampling, and push–pull perfusion, in *Methods in Psychobiology,* Myers, R.D., Ed., Academic Press, New York, 1977.
15. Myers, R.D., Development of push-pull systems for perfusion of anatomically distinct regions of the brain of the awake animal, *Ann. N.Y. Acad. Sci.*, 473, 21, 1986.
16. Philippu, A., Use of push-pull cannulae to determine the release of endogenous neurotransmitters in distinct brain areas of anesthetized and freely moving animals, in *Measurement of Neurotransmitter Release in Vivo,* Marsden, C.A., Ed., Wiley, New York, 1984.
17. Yaksh, T.L. and Yamamura, H.I., Factors affecting performance of the push-pull cannula in brain, *J. Appl. Physiol.*, 37, 428, 1974.
18. Arancibia, S., Epelbaum, J., Boyer, R., and Assenmacher, I., In vivo release of somatostatin from rat median eminence after local K^+ infusion or delivery of nociceptive stress, *Neurosci. Lett.*, 50, 97, 1984.
19. Chase, T.N. and Kopin, I.J., Stimulus-induced release of substances from olfactory bulb using the push-pull cannula, *Nature*, 217, 466, 1968.
20. Kaehler, S.T., Sinner, C., and Philippu, A., Release of catecholamines in the locus coeruleus of freely moving and anaesthetized normotensive and spontaneously hypertensive rats: effects of cardiovascular changes and tail pinch, *Naunyn Schmiedeberg's Arch. Pharmacol.*, 361, 433, 2000.

21. Yonehara, N., Shibutani, T., Imai, Y., and Inoki, R., Involvement of descending monoaminergic systems in the transmission of dental pain in the trigeminal nucleus caudalis of the rabbit, *Brain Res.*, 508, 234, 1990.

22. McCarson, K.E. and Goldstein, B.D., Release of substance P into the superficial dorsal horn following nociceptive activation of the hind paw of the rat, *Brain Res.*, 568, 109, 1991.

23. Takagi, H., Experimental pain and neuropeptides, *Clin. Ther.*, 7, 35, 1984.

24. Zachariou, V. and Goldstein, B.D., Delta-opioid receptor modulation of the release of substance P-like immunoreactivity in the dorsal horn of the rat following mechanical or thermal noxious stimulation, *Brain Res.*, 736, 305, 1996.

25. Alstergren, P., Kopp, S., and Theodorsson, E., Synovial fluid sampling from the temporomandibular joint: sample quality criteria and levels of interleukin-1 beta and serotonin, *Acta Odontol. Scand.*, 57, 16, 1999.

26. Yaksh, T.L. and Rudy, T.A., Chronic catheterization of the spinal subarachnoid space, *Physiol. Behav.*, 17, 1031, 1976.

27. Yaksh, T.L., Spinal superfusion in the rat and cat, in *Measurement of Neurotransmitter Release in Vivo,* Marsden, C.A., Ed., Wiley, Chichester, 1984, chap. 5.

28. Ungerstedt, U., Measurement of neurotransmitter release by intracranial dialysis, in *Measurement of Neurotransmitter Release in Vivo,* Marsden, C.A., Ed., Wiley, Chichester, 1984, chap. 4.

29. Clapp-Lilly, K.L., Roberts, R.C., Duffy, L.K., Irons, K.P., Hu, Y., and Drew, K.L., An ultrastructural analysis of tissue surrounding a microdialysis probe, *J. Neurosci. Meth.*, 90, 129, 1999.

30. Cadoni, C., Solinas, M., and Di, C.G., Psychostimulant sensitization: differential changes in accumbal shell and core dopamine, *Eur. J. Pharmacol.*, 388, 69, 2000.

31. Fuentealba, J.A., Forray, M.I., and Gysling, K., Chronic morphine treatment and withdrawal increase extracellular levels of norepinephrine in the rat bed nucleus of the stria terminalis, *J. Neurochem.*, 75, 741, 2000.

32. Tao, R., Ma, Z., and Auerbach, S.B., Differential effect of local infusion of serotonin reuptake inhibitors in the raphe versus forebrain and the role of depolarization-induced release in increased extracellular serotonin, *J. Pharmacol. Exp. Ther.*, 294, 571, 2000.

33. Matsuno, K., Matsunaga, K., Senda, T., and Mita, S., Increase in extracellular acetylcholine level by sigma ligands in rat frontal cortex, *J. Pharmacol. Exp. Ther.*, 265, 851, 1993.

34. Wolf, M.E. and Xue, C.J., Amphetamine-induced glutamate efflux in the rat ventral tegmental area is prevented by MK-801, SCH 23390, and ibotenic acid lesions of the prefrontal cortex, *J. Neurochem.*, 73, 1529, 1999.

35. Timmerman, W., Bouma, M., De, V.J., Davis, M., and Westerink, B.H., A microdialysis study on the mechanism of action of gabapentin, *Eur. J. Pharmacol.*, 398, 53, 2000.

36. Berman, R.F., Fredholm, B.B., Aden, U., and O'Connor, W.T., Evidence for increased dorsal hippocampal adenosine release and metabolism during pharmacologically induced seizures in rats, *Brain Res.*, 872, 44, 2000.

37. Olive, M.F. and Maidment, N.T., Opioid regulation of pallidal enkephalin release: bimodal effects of locally administered mu and delta opioid agonists in freely moving rats, *J. Pharmacol. Exp. Ther.*, 285, 1310, 1998.

38. Zangen, A., Herzberg, U., Vogel, Z., and Yadid, G., Nociceptive stimulus induces release of endogenous β-endorphin in the rat brain, *Neuroscience*, 85, 659, 1998.

39. Xin, L., Geller, E.B., Liu-Chen, L.Y., Chen, C., and Adler, M.W., Substance P release in the rat periaqueductal gray and preoptic anterior hypothalamus after noxious cold stimulation: effect of selective mu and kappa opioid agonists, *J. Pharmacol. Exp. Ther.*, 282, 1055, 1997.

40. Calcutt, N.A., Stiller, C., Gustafsson, H., and Malmberg, A.B., Elevated substance-P-like immunoreactivity levels in spinal dialysates during the formalin test in normal and diabetic rats, *Brain Res.*, 856, 20, 2000.

41. Neubert, J.K., Maidment, N.T., Matsuka, Y., Adelson, D.W., Kruger, L., and Spigelman, I., Inflammation-induced changes in primary afferent-evoked release of substance P within trigeminal ganglia in vivo, *Brain. Res.*, 871, 181, 2000.

42. Robinson, T.E. and Justice, Jr., J.B., *Microdialysis in the Neurosciences,* Robinson, T.E. and Justice, Jr., J.B., Eds., Elsevier, Amsterdam, 1991.

43. Adams, R.N., Conti, J., Marsden, C.A., and Strope, E., The measurement of dopamine and 5-hydroxytryptamine release in CNS of freely moving unanaesthetised rats, *Proc. Br. J. Pharmacol.*, 64, 473P, 1978.

44. Justice, J.B. Jr., *Voltammetry in the Neurosciences,* Justice, J.B. Jr., Ed., Humana Press, Clifton, 1987.

45. Marsden, C.A., *Measurement of Neurotransmitter Release in Vivo,* Marsden, C.A., Ed., Wiley, Chichester, 1984.

46. Michael, D.J. and Wightman, R.M., Electrochemical monitoring of biogenic amine neurotransmission in real time, *J. Pharm. Biomed. Anal.*, 19, 33, 1999.

47. Kruk, Z.L. and O'Connor, J.J., Fast electrochemical studies in isolated tissues, *Trends Pharmacol. Sci.*, 16, 145, 1995.

48. Kilpatrick, M.R., Rooney, M.B., Michael, D.J., and Wightman, R.M., Extracellular dopamine dynamics in rat caudate-putamen during experimenter-delivered and intracranial self-stimulation, *Neuroscience*, 96, 697, 2000.

49. Callado, L.F. and Stamford, J.A., Spatiotemporal interaction of α_2 autoreceptors and noradrenaline transporters in the rat locus coeruleus: implications for volume transmission, *J. Neurochem.*, 74, 2350, 2000.

50. Pothos, E.N., Przedborski, S., Davila, V., Schmitz, Y., and Sulzer, D., D2-like dopamine autoreceptor activation reduces quantal size in PC12 cells, *J. Neurosci.*, 18, 5575, 1998.

51. Hochstetler, S.E., Puopolo, M., Gustincich, S., Raviola, E., and Wightman, R.M., Real-time amperometric measurements of zeptomole quantities of dopamine released from neurons, *Anal. Chem.*, 72, 489, 2000.

52. Brake, W.G., Sullivan, R.M., and Gratton, A., Perinatal distress leads to lateralized medial prefrontal cortical dopamine hypofunction in adult rats, *J. Neurosci.*, 20, 5538, 2000.

53. Hentall, I.D., Kurle, P.J., and White, T.R., Correlations between serotonin level and single-cell firing in the rat's nucleus raphe magnus, *Neuroscience*, 95, 1081, 2000.

54. Friedemann, M.N., Robinson, S.W., and Gerhardt, G.A., o-Phenylenediamine-modified carbon fiber electrodes for the detection of nitric oxide, *Anal. Chem.*, 68, 2621, 1996.

55. Hu, Y., Mitchell, K.M., Albahadily, F.N., Michaelis, E.K., and Wilson, G.S., Direct measurement of glutamate release in the brain using a dual enzyme-based electrochemical sensor, *Brain Res.*, 659, 117, 1994.

56. Kulagina, N.V., Shankar, L., and Michael, A.C., Monitoring glutamate and ascorbate in the extracellular space of brain tissue with electrochemical microsensors, *Anal. Chem.*, 71, 5093, 1999.

57. Duggan, A.W. and Hendry, I.A., Laminar localization of the sites of release of immunoreactive substance P in the dorsal horn with antibody coated microelectrodes, *Neurosci. Lett.*, 68, 134, 1986.

58. Duggan, A.W., Morton, C.R., Zhao, Z.Q., and Hendry, I.A., Noxious heating of the skin releases immunoreactive substance P in the substantia gelatinosa of the cat: a study with antibody microprobes, *Brain Res.*, 403, 345, 1987.

59. Duggan, A.W., Hendry, I.A., Morton, C.R., Hutchison, W.D., and Zhao, Z.Q., Cutaneous stimuli releasing immunoreactive substance P in the dorsal horn of the cat, *Brain Res.*, 451, 261, 1988.

60. Duggan, A.W., Hope, P.J., Jarrott, B., Schaible, H.G., and Fleetwood-Walker, S.M., Release, spread and persistence of immunoreactive neurokinin A in the dorsal horn of the cat following noxious cutaneous stimulation. Studies with antibody microprobes, *Neuroscience*, 35, 195, 1990.

61. Morton, C.R., Hutchison, W.D., and Hendry, I.A., Intraspinal release of substance P and calcitonin gene-related peptide during opiate dependence and withdrawal, *Neuroscience*, 43, 593, 1991.

62. Williams, C.A., Holtsclaw, L.I., and Chiverton, J.A., Release of immunoreactive enkephalinergic substances in the periaqueductal grey of the cat during fatiguing isometric contractions, *Neurosci. Lett.*, 139, 19, 1992.

63. Duggan, A.W., Hope, P.J., Lang, C.W., and Bjelke, B., Noxious mechanical stimulation of the hind paws of the anaesthetized rat fails to elicit release of immunoreactive beta-endorphin in the periaqueductal grey matter, *Neurosci. Lett.*, 149, 205, 1993.

64. Williams, C.A., Holtsclaw, L.I., and Chiverton, J.A., Release of immunoreactive neuropeptide Y from brainstem sites in the cat during isometric contractions, *Neuropeptides*, 24, 53, 1993.

65. Morton, C.R. and Hutchison, W.D., Release of sensory neuropeptides in the spinal cord: studies with calcitonin gene-related peptide and galanin, *Neuroscience*, 31, 807, 1989.

66. Colvin, L.A. and Duggan, A.W., Primary afferent-evoked release of immunoreactive galanin in the spinal cord of the neuropathic rat, *Br. J. Anaesth.*, 81, 436, 1998.

67. Morton, C.R., Hutchison, W.D., Hendry, I.A., and Duggan, A.W., Somatostatin: evidence for a role in thermal nociception, *Brain Res.*, 488, 89, 1989.

68. Duggan, A.W., Detection of neuropeptide release in the central nervous system with antibody microprobes, *J. Neurosci. Meth.*, 34, 47, 1990.

69. Duggan, A.W., Hendry, I.A., Green, J.L., Morton, C.R., and Hutchison, W.D., The preparation and use of antibody microprobes, *J. Neurosci. Meth.*, 23, 241, 1988.

70. Duggan, A.W. and Furmidge, L.J., Probing the brain and spinal cord with neuropeptides in pathways related to pain and other functions, *Front. Neuroendocrinol.*, 15, 275, 1994.

71. Hendry, I.A., Morton, C.R., and Duggan, A.W., Analysis of antibody microprobe autoradiographs by computerized image processing, *J. Neurosci. Meth.*, 23, 249, 1988.

72. Caldwell, P.C., An investigation of intracellular pH of crab muscle fibers by means of micro-glass and micro-tungsten electrodes, *J. Physiol. (Lond.)*, 126, 169, 1954.

73. Hinke, J.A.M., Glass microelectrodes for measuring intracellular activities of sodium and potassium, *Nature*, 184, 1257, 1959.

74. Eisenmann, G., *Glass Electrodes for Hydrogen and Other Cations*, Dekker, New York, 1967.

75. Orme, F.W., Liquid ion-exchanger microelectrodes, in *Glass Microelectrodes*, Lavallée, M., Schanne, O., and Hebert, N.C., Eds., Wiley, New York, 1969.

76. Cornwall, M.C., Peterson, D.F., Kunze, D.L., Walker, J.L., and Brown, A.M., Intracellular potassium and chloride activities measured with liquid ion exchanger microelectrodes, *Brain Res.*, 23, 433, 1970.

77. Walker, J.L., Ion specific liquid ion exchanger microelectrodes, *Anal. Chem.*, 43, 89A, 1971.
78. Oehme, M. and Simon, W., Microelectrode based on potassium ions based on a neutral carrier and comparison of its characteristics with a cation exchanger sensor, *Anal. Chem.*, 86, 21, 1976.
79. Ammann, D., *Ion-Selective Microelectrodes: Principles, Design and Application*, Springer, Berlin, 1986.
80. Wuhrmann, P., Ineichen, H., Riesen-Willi, U., and Lezzi, M., Change in nuclear potassium electrochemical activity and puffing of potassium-sensitive salivary chromosome regions during Chironomus development, *Proc. Natl. Acad. Sci. USA*, 76, 806, 1979.
81. Ammann, D., Chao, P.S., and Simon, W., Valinomycin-based K^+ selective microelectrodes with low electrical membrane resistance, *Neurosci. Lett.*, 74, 221, 1987.
82. Carlini, W.G. and Ransom, B.R., Fabrication and implementation of ion-selective microelectrodes, in *Neurophysiological Techniques: Basic Methods and Concepts*, Boulton, A.A., Baker, G.B., and Vanderwolf, C.H., Eds., Humana Press, Clifton, 1990.
83. Schlue, W.R., Kilb, W., and Günzel, D., Ultramicroelectrodes for membrane research, *Electrochim. Acta*, 42, 3197, 1997.
84. Thomas, R.C., *Ion-Sensitive Intracellular Microelectrodes*, Academic Press, London, 1978.
85. Semb, S.O., Amundsen, B., and Sejersted, O.M., A new improved way of making double-barrelled ion-selective micro-electrodes, *Acta Physiol. Scand.*, 161, 1, 1997.
86. Woodruff, R.I., Fabrication of ion-selective microelectrodes by a centrifugation/suction method, *Biotechniques*, 23, 100, 1997.
87. Mauskop, A., Altura, B.T., Cracco, R.Q., and Altura, B.M., Intravenous magnesium sulfate rapidly alleviates headaches of various types, *Headache*, 36, 154, 1996.
88. Bradberry, C.W., Sprouse, J.S., Sheldon, P.W., Aghajanian, G.K., and Roth, R.H., *In vitro* microdialysis: a novel technique for stimulated neurotransmitter release measurements, *J. Neurosci. Meth.*, 36, 85, 1991.
89. Duport, S., Robert, F., Muller, D., Grau, G., Parisi, L., and Stoppini, L., An *in vitro* blood-brain barrier model: cocultures between endothelial cells and organotypic brain slice cultures, *Proc. Natl. Acad. Sci. USA*, 95, 1840, 1998.
90. Dun, N.J., Dun, S.L., Wu, S.Y., Williams, C.A., and Kwok, E.H., Endomorphins: localization, release and action on rat dorsal horn neurons, *J. Biomed. Sci.*, 7, 213, 2000.
91. Croning, M.D. and Haddad, G.G., Comparison of brain slice chamber designs for investigations of oxygen deprivation in vitro, *J. Neurosci. Meth.*, 81, 103, 1998.
92. Neher, E. and Sakmann, B., Single-channel currents recorded from membrane of denervated frog muscle fibres, *Nature*, 260, 799, 1976.
93. Sakmann, B. and Neher, E., *Single Channel Recording*, 2nd ed., Plenum, New York, 1995.
94. Hamill, O.P., Marty, A., Neher, E., Sackmann, B., and Sigworth, F.J., Improved patch-clamp techniques for high resolution current from cells and cell-free membrane patches, *Pflugers Arch.*, 391, 85, 1981.
95. Hanke, W., Boheim, G., Barhanin, J., Pauron, D., and Lazdunski, M., Reconstitution of highly purified saxitoxin-sensitive Na^+-channels into planar lipid bilayers, *EMBO J.*, 3, 509, 1984.
96. Boheim, G., Hanke, W., Barrantes, F.J., Eibl, H., Sakmann, B., Fels, G., and Maelicke, A., Agonist-activated ionic channels in acetylcholine receptor reconstituted into planar lipid bilayers, *Proc. Natl. Acad. Sci. USA*, 78, 3586, 1981.

97. Coyne, M.D., Dagan, D., and Levitan, I.B., Calcium and barium permeable channels from aplysia nervous system reconstituted in lipid bilayers, *J. Membr. Biol.*, 97, 205, 1987.

98. Hume, R.I., Role, L.W., and Fischbach, G.D., Acetylcholine release from growth cones detected with patches of acetylcholine receptor-rich membranes, *Nature*, 305, 632, 1983.

99. Young, S.H. and Poo, M.M., Spontaneous release of transmitter from growth cones of embryonic neurones, *Nature*, 305, 634, 1983.

100. Copenhagen, D.R. and Jahr, C.E., Release of endogenous excitatory amino acids from turtle photoreceptors, *Nature*, 341, 536, 1989.

101. Silinsky, E.M. and Gerzanich, V., On the excitatory effects of ATP and its role as a neurotransmitter in coeliac neurons of the guinea-pig, *J. Physiol. (Lond.)*, 464, 197, 1993.

102. Silinsky, E.M. and Redman, R.S., Synchronous release of ATP and neurotransmitter within milliseconds of a motor nerve impulse in the frog, *J. Physiol. (Lond.)*, 492, 815, 1996.

103. Binnig, G. and Quate, C.F., Atomic force microscope, *Phys. Rev. Lett.*, 56, 930, 1986.

104. Henderson, E., Haydon, P.G., and Sakaguchi, D.S., Actin filament dynamics in living glial cells imaged by atomic force microscopy, *Science*, 257, 1944, 1992.

105. Radmacher, M., Fritz, M., Hansma, H.G., and Hansma, P.K., Direct observation of enzyme activity with the atomic force microscope, *Science*, 265, 1577, 1994.

106. Radmacher, M., Cleveland, J.P., Fritz, M., Hansma, H.G., and Hansma, P.K., Mapping interaction forces with the atomic force microscope, *Biophys. J.*, 66, 2159, 1994.

107. Frisbie, C.D., Rozsnyai, L.F., Noy, A., Wrighton, M.S., and Lieber, C.M., Functional group imaging by chemical force microscopy, *Science*, 265, 2071, 1994.

108. Moy, V.T., Florin, E.L., and Gaub, H.E., Intermolecular forces and energies between ligands and receptors, *Science*, 266, 257, 1994.

109. Florin, E.L., Moy, V.T., and Gaub, H.E., Adhesion forces between individual ligand-receptor pairs, *Science*, 264, 415, 1994.

110. Dammer, U., Hegner, M., Anselmetti, D., Wagner, P., Dreier, M., Huber, W., and Guntherodt, H.J., Specific antigen/antibody interactions measured by force microscopy, *Biophys. J.*, 70, 2437, 1996.

111. Schneider, S.W., Egan, M.E., Jena, B.P., Guggino, W.B., Oberleithner, H., and Geibel, J.P., Continuous detection of extracellular ATP on living cells by using atomic force microscopy, *Proc. Natl. Acad. Sci. USA*, 96, 12180, 1999.

112. Strehler, B.L. and Totter, J.R., Firefly luminescence in the study of energy transfer mechanisms. I. Substrate and enzyme determination, *Arch. Biochem.*, 40, 28, 1952.

113. Seliger, H.H. and McElroy, W.D., Spectral emission and quantum yield of firefly bioluminescence, *Arch. Biochem. Biophys.*, 88, 136, 1960.

114. Silinsky, E.M., A simple, rapid method for detecting the efflux of small quantities of adenosine triphosphate from biological tissues, *Comp. Biochem. Physiol. A.*, 48, 561, 1974.

8 Electrophysiological Recording Techniques in Pain Research

Igor Spigelman, Michael S. Gold, and Alan R. Light

CONTENTS

0-8493-0035-5/01/$0.00+$1.50
© 2001 by CRC Press LLC

8.1 INTRODUCTION

Studies that utilize electrophysiological techniques are of paramount importance in pain research because they currently represent the only means by which we can directly measure the transduction, propagation, and encoding of innocuous and noxious stimuli. The electrophysiological approach is also highly relevant to studies focusing on the processing of encoded sensory signals by the central nervous system (CNS). Many injury-induced changes in sensory neurons are manifest at the site of injury, most often at the afferent terminal, making this structure an ideal site to investigate the cellular mechanisms underlying these changes. Unfortunately, the morphological complexity, small size, and relative inaccessibility of afferent terminals preclude direct cellular studies with conventional microelectrophysiological techniques. Studying the consequences of peripheral injury on central neurons carries an additional set of difficulties due to the complexity of CNS morphology and function. Nevertheless, researchers have devised and employed methods for studying these systems. Specifically, there are five alternative general experimental approaches that may be employed to identify injury-induced changes in the electrophysiological properties of peripheral and central neurons. These are

1. Extracellular recording *in vivo* from axon tracts, individual axons, or somata of neurons
2. Intracellular recording *in vivo*
3. Intracellular recording from neurons in intact ganglia or tissue slices *in vitro*
4. Intracellular recording from dissociated neurons *in vitro*
5. Patch clamp recording *in vitro* and *in vivo*

This chapter discusses the relative merits and provides some practical suggestions for studies that may employ such experimental approaches.

8.2 EXTRACELLULAR RECORDING *IN VIVO*

The majority of studies relevant to pain research have focused on extracellular recordings from primary afferent neurons and to a lesser extent on second-order neurons in the spinal cord and brainstem. Other studies have targeted the CNS areas considered important for perception and modulation of pain sensation. Advantages of this approach include (1) access to primary afferents *in vivo*, enabling a complete characterization of receptive fields, response properties, and conduction velocity; (2) minimization of the amount of tissue injury required to gain access to the afferents, while leaving terminals intact; (3) the ability to study changes in peripheral terminals of sensory neurons; and (4) the ability to activate brain regions that connect with primary pain relay regions to assess the CNS modulation of inputs. With this approach, it may also be possible to identify injury-induced changes in mechanical, thermal, and chemosensitivity. Disadvantages of this approach are that it is impossible to determine with certainty whether a putative change in signal processing or sensitivity occurred primarily within the afferent or a neighboring cell (i.e., direct vs. indirect). More importantly, it is rarely possible to determine which ion channel(s) are responsible for changes in excitability due to the lack of electrical control over, or intracellular access to, the soma, axon, or terminals. Nevertheless, this approach is critical to test hypotheses (generated with more reductionist approaches) related to the identification of the mechanisms underlying injury-induced changes in primary afferent excitability and the corresponding changes in central neurons.

Some of the most important discoveries about the nature of pain and nociception were initially determined with extracellular recordings. One of these was the discovery of the importance of the most superficial laminae of the spinal cord for nociception. In a very careful series of experiments, Christensen and Perl[1] definitively demonstrated the presence of spinal cord neurons with specific responses to mechanical and thermal nociceptive inputs in lamina I, and then Kumazawa and Perl[2] extended this to lamina II as well. These findings were confirmed with intracellular recordings that clearly demonstrated only nociceptor activation of some superficial dorsal horn cells, while others were activated only by innocuous thermal inputs.[3] These studies also labeled the neurons, clearly placing their cell bodies and dendrites in laminae I and II.

Another very important finding using these methods was the plasticity of neuronal responses of neurons that are located deeper in the dorsal horn to putative nociceptive inputs. In these experiments, extracellular recordings from neurons demonstrated enhanced responses when peripheral nerves were stimulated at C-fiber intensities.[4] This property was labeled "windup" and its mechanisms have been further determined recently with intracellular recordings.[5] This review also emphasizes that windup is not identical to central sensitization that has long been associated with pain and hyperalgesia.

Later extracellular recording studies traced the impact of noxious stimuli on the nervous system to the thalamus.[6] Most recently, various imaging techniques have revealed sites of nociceptive processing in various parts of the cerebral cortex.[7,8]

8.2.1 COMPOUND ACTION POTENTIAL (CAP) RECORDING FROM NERVES AND FIBER TRACTS

Early recordings with bipolar electrodes established that the various peaks in the recorded signals were related to the conduction velocity of various axon populations in the peripheral nerve.[9,10] Electrical stimulation and recording of nerve CAPs in humans were crucial to determining that impulse conduction in the slowest fibers was correlated with the sensation of pain.[11-13] Field potential analysis of primary afferent activity in CNS tracts has also proven quite valuable in determining the rostrocaudal distribution of nociceptive primary afferent axons and their terminal arborizations.[14,15] However, field potential recordings offer only qualitative information on the amplitude of the CAP. Suction electrodes provide a better means for quantitative recording of propagated electrical activity from peripheral nerves and CNS fiber tracts. For example, dorsal root CAP recordings may be used in combination with intracellular recordings from the dorsal root ganglion (DRG) neurons and stimulation of the spinal nerve (Fig. 8.1A). However, recording and stimulating with suction electrodes necessarily involve cutting the nerve fibers of interest. This limits their use to terminal *in vivo* or *in vitro* exper-

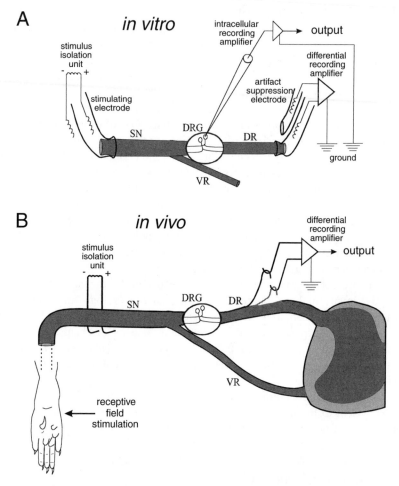

FIGURE 8.1 Schematic diagram of the recording arrangements from mammalian sensory neurons *in vitro* (A) and *in vivo* (B). SN, spinal nerve; DR, dorsal root; VR, ventral root; DRG, dorsal root ganglion.

iments, as well as complicates the interpretation of studies on injury-induced changes in primary afferent fibers. Detailed methodological approaches to the uses of suction electrodes have been described.[16,17]

8.2.2 MICRONEUROGRAPHY IN HUMANS

Pain is a subjective experience and thus it is only through experiments in humans that an accurate description of both the quality and intensity of pain may be obtained. Introduction of the microneurography recording technique[18] led to many human studies that clearly defined the involvement of unmyelinated C-fibers in pain sensation, as well as the association of pathology in human nerves with abnormal spontaneous activity.[19–23]

A major advantage of microneurography is that recorded axons can be stimulated relatively selectively by applying small pulses of current to the microelectrode following isolation of a single unit recording.[24,25] Although somewhat controversial, this has been used to directly relate activation of single primary afferent nerve fibers with specific modalities of cutaneous and muscle sensation.[25–27] Thus, cutaneous C-polymodal primary afferents have been associated with dull, burning pain; group IV muscle receptors with dull, cramping pain; cutaneous slowly adapting receptors with pressure; and rapidly adapting receptors with tap and flutter sensations. Most recently, microneurography in humans has also been used to identify the elusive itch receptors that would have been extremely difficult to identify in animal models.[28]

Since its introduction, microneurography has undergone relatively minor modifications related primarily to the size and composition of the recording electrodes, as well as analysis of the recorded signals. The reader is directed to excellent publications that provide detailed instructions on the construction and various uses of recording electrodes.[29–31]

8.2.3 SINGLE UNIT RECORDING FROM AXON FIBER STRANDS

While performed in the 1950s by brave pioneering neuroscientists on their own peripheral nerves,[32] single unit recording from peripheral nerve filaments is generally reserved for experiments in experimental animals. Such recordings have become an important part of the arsenal of electrophysiological techniques available to pain researchers. The technique is relatively straightforward: Once the epineurial connective tissue and perineurial sheath are removed to open a "picking window," small bundles of nerve fibers (10 to 20 μm diameter) are then picked using sharpened forceps, teased free of the nerve, and wrapped around one terminal of a bipolar recording electrode, while a filament of connective tissue is wrapped around the other (indifferent) terminal (Fig. 8.1B). The arrangement works equally well for recording *in vivo* or *in vitro*. Platinum/iridium electrodes are preferable to Ag-AgCl electrodes because the latter are known to be toxic to excitable tissue over long-term exposure.[33,34]

Single unit recordings from sensory fibers *in vivo* coupled with surgical or pharmacological blockade of impulse conduction were crucial in determining that much of the spontaneous discharge in sensory fibers after peripheral nerve injury originates within sensory ganglia.[35,36] Furthermore, it was shown that this ectopic spike generation may be augmented by stimulating the axons of nearby neurons, suggesting the presence of non-synaptic communication between sensory neurons.[37] These studies provided the necessary impetus to the use of intracellular recording techniques to address the cellular mechanisms by which spontaneous discharge might arise within sensory neurons.

8.3 INTRACELLULAR RECORDING *IN VIVO*

This approach has all the advantages described for the first approach, in addition to a degree of electrical control over the neuron studied. If recording is performed in the cell body, it may also be possible to identify injury-induced changes that occur in the cell body, as well as changes that occur in the peripheral terminals. Partial control over the intracellular milieu is obtained through contact with the contents of the intracellular recording microelectrode, limited by the small diameter and corresponding high resistance of the electrode tip. Similarly, it is possible to label the neuron studied so that morphological and histological characterization may be performed in conjunction with the electrophysiological characterization. However, this approach suffers from the same disadvantages as those associated with the first approach, with the added disadvantage that the recording itself may damage the neuron either by mechanical puncture via the electrode, or by diffusion of important substances out of or into the recording electrode.

Intracellular recordings *in vivo* have allowed important advances in the classi-fication of sensory neurons based on their action potential characteristics and the receptive properties of their peripheral terminations.[38–41] Most of the current studies conducted on dissociated sensory neurons *in vitro* rely on the classification data obtained from such *in vivo* recordings. The *in vivo* recordings also contributed to our understanding of the changes in sensory neuron membrane properties as a consequence of peripheral inflammation or nerve injury.[42,43]

Extracellular and intracellular recordings *in vivo* have demonstrated many of the inputs to large neurons located deep in the dorsal horn in a number of species, including primates. These studies have defined the inputs to identified spinothalamic neurons that have been postulated to have a major role in the transmission of nociceptive information that can lead to pain.[44,45]

However, the neurons in the superficial dorsal horn, including the substantia gelatinosa, were technically difficult to record from, and only a few intracellular recording studies have been published. These demonstrated that innocuous inputs were sometimes subliminal in what appeared to be nociceptive specific neurons, but that other neurons only had inputs from nociceptors. Other intracellular recording experiments demonstrated substantial modulation of substantia gelatinosa neurons by brainstem structures.[3,46–48] However, all of these studies were limited by the stability of the recordings, and by the destruction of the cells by the impalement with sharp electrodes. The smaller neurons in the rat were especially difficult to record from intracellularly, and a survey of these neurons was done only recently with whole-cell patch recording methods (see below).

8.4   INTRACELLULAR RECORDING FROM NEURONS IN THE INTACT GANGLIA OR TISSUE SLICES *IN VITRO*

Advantages of this approach are (1) better control of the extracellular milieu than that provided by the *in vivo* approaches, including the absence of a blood-brain or blood-nerve barrier; (2) some degree of electrical control of the soma membrane; (3) the possibility of identifying primary afferents according to their conduction velocity (if enough peripheral axon is retained), and in some instances according to their receptive field properties;[49] (4) the possibility of observing an injury-induced increase in excitability that does not involve a change in an intrinsic property of the particular neuron studied; and (5) no enzymatic or mechanical treatment of neurons is necessary prior to recording. Clearly there are many advantages to this approach, and it has been used effectively to characterize electrophysiological properties of primary afferent and central neurons. Often, a combination of extracellular and intracellular recordings is utilized within such *in vitro* preparations (see Fig. 8.1). However, this approach also suffers from several disadvantages.

1. When the somata of primary afferents are used as the recording site, several of the proteins necessary for the transduction of stimuli appear to be absent, or expressed at such low levels in the uninjured non-dissociated afferent cell body that stimulus transduction is often undetectable. Indeed, mechanical and thermal sensitivity develops in the cut ends of primary afferents, but only after more than 2 h following transection.[50–52] Thus, characterization of the ion channels involved in these transduction processes is not possible with this approach.
2. It may not be possible to determine whether an injury-induced change in excitability reflects a direct change in a particular neuron studied or an indirect change in a cell associated with the neuron studied (e.g., glial or neuronal network interactions).
3. The presence of the axonal process and dendrites (on CNS cells) influences the electrical control over the neuronal membrane and therefore the ability to characterize the ion channels present in these neurons.

However, there are ways to mitigate the loss of clamp control associated with the presence of an axonal process or dendrites emerging from the cell body.[53,54] Also, the introduction of whole-cell patch recording in brain slices has dramatically improved our ability to control the intracellular milieu of recorded neurons, while maintaining synaptic connections.[55] The recent addition of infrared video microscopy further extended the technique to allow recording from neuronal somata and dendrites under visual guidance.[56]

8.4.1 SURGERY AND TISSUE ISOLATION

The isolation of tissues for electrophysiological studies requires surgical intervention that will depend on the tissue being studied. For example, recordings from neurons in dorsal root ganglia or spinal cord slices require that a laminectomy be performed under general anesthesia in order to maintain an adequate blood supply up to the moment of tissue excision. This is not surprising given that neurons, especially CNS neurons, are quite vulnerable to anoxia. Further dissection in chilled, oxygenated artificial cerebrospinal fluid (ACSF) is often required to remove excess debris or to make tissue slices. For example, it is usually necessary to remove or tear the tough sheath of connective tissue that surrounds sensory ganglia to allow for impalement of sensory neurons with fragile sharp microelectrodes. In some instances, it is advantageous to perform the dissection in modified ACSF containing reduced $[Na^+]$ and $[Ca^{2+}]$, in order to enhance neuronal survival during the course of the *in vitro* experiment.[57] This can be extended to actual intravenous perfusion of the experimental animal with similarly modified ACSF prior to the dissection.[58] The procedure appears to greatly increase the viability of CNS slice preparations, especially when working with adult tissue (unpublished observations).

Many investigators utilize commercially available tissue slicers to obtain viable sections of interest. The most common models are those where the horizontally vibrating blade is advanced through the tissue (e.g., Oxford Vibratome), or the tissue is advanced through the vibrating blade (e.g., Campden Instruments Vibroslicer). Prior to cutting, tissue is attached to a plastic block with cyanoacrylate-based glue and then covered with oxygenated cold ACSF. Tissue slicers differ primarily in whether the blade/tissue is advanced and retracted manually or automatically, and also in the mode of blade vibration. Excessive vertical displacement of the blade during horizontal vibration can easily damage the top layers of cells in a tissue slice, making them useless for intracellular or patch-clamp recordings. This represents a problem when combining infrared imaging with patch-clamp recording that relies on the viability of neurons within 100 μm of the slice surface. Some manufacturers claim minimal vertical blade displacement during horizontal blade movement (e.g., Leica VT1000S), thus minimizing damage to cells near the slice surface.

8.4.2 RECORDING CHAMBERS

The choice of recording chamber design is also dependent on the type of tissue that is being studied. Most commercially available recording chambers are of interphase or submersion design allowing for superfusion of the tissue with pre-oxygenated and pH-buffered ACSF. Interphase chambers allow for a humidified and oxygenated atmosphere to blow over the partially submerged tissue, enhancing the mechanical stability of the preparation. Temperature control of the recording chamber is achieved through cooling/heating of the perfusing fluid (e.g., Medical Systems Corp. model TC-202) and/or by heating the chamber only with Peltier-type devices (e.g., Warner Instruments Co.). Custom-made recording chambers may offer some advantages, especially if one is interested in multicompartment chambers that allow separate access and superfusion of primary afferent terminals, sensory ganglion, and spinal cord.[49,59,60] Application of drugs may be achieved by superfusion following appropriate dilution in ACSF, by picospritzer (pressure) ejection, and/or by iontophoresis of charged molecules from glass micropipettes.[61] A relatively new addition to drug application methodology is flash photolysis of caged compounds which, in combination with infrared illumination, allows for visual identification of cells targeted for drug action.[62,63]

8.5 INTRACELLULAR RECORDING FROM DISSOCIATED NEURONS *IN VITRO*

This approach offers several distinct advantages over the other three approaches:

1. Dissociated neurons are completely isolated from one another making it possible to obtain virtually complete control of the extracellular milieu.
2. Patch-clamp techniques employed for electrophysiological recording make it possible to control the intracellular milieu, as well as extract the intracellular contents for analysis of genes and gene products.

3. Isolating neurons *in vitro* makes it possible to determine whether an injury-induced change in neuronal excitability reflects a direct change in the neuron studied.

4. It is possible to record from many more neurons from a single animal than with the other methods.

5. The acutely isolated neuron soma possesses an optimal geometry for electrophysiological characterization of voltage- and ligand-gated currents, because space-clamp errors are minimized when voltage-clamping, especially in the case of a nearly spherical soma of a primary afferent neuron.

6. Dissociation of sensory neurons appears to facilitate the redistribution of proteins to the plasma membrane that are normally present in afferent terminals. For example, a proton receptor/ion channel complex appears to be present in afferent terminals[64] but not the cell body *in vivo*;[65] yet a proton receptor/ion channel complex is present in the cell body *in vitro*.[66,67] Similarly, the ionic mechanisms underlying mechanical and thermal transduction appear to be present in the cell bodies of dissociated sensory neurons.

There also are several disadvantages associated with the use of this approach. These include

1. The inability to identify primary afferents with respect to conduction velocity or receptive field properties.

2. The potential for damage to membrane properties associated with the use of enzymes required to dissociate the neurons.

3. The act of placing neurons *in vitro* necessarily involves injury.

4. The potential for alteration of neuron properties by the lack of unknown important factors both in the bath and in the electrodes is maximized.

5. Neurons can only be studied for a limited time in culture.

6. Because neurons in living animals never operate in the absence of supporting cells and other neurons, the results obtained from such studies cannot be directly related to the conditions in the behaving animal, without confirmation using some of the preparations described above.

8.6 PATCH-CLAMP RECORDING *IN VITRO*

The patch-clamp technique is an extremely powerful method for studying electrophysiological properties of biological membranes.[68] First described by Neher and Sakmann in 1976,[69] it is now routinely used in laboratories around the world. The power of the patch-clamp technique comes from the fidelity with which electrophysiological signals may be recorded. The technique was not widely used by pain researchers until the beginning of this last decade, at which point researchers began to use it to characterize electrophysiological properties and chemosensitivity of neurons involved in the transduction and transmission of nociceptive stimuli.

Patch-clamp techniques have been applied most widely in the pain field to the study of primary afferent neurons. Consequently, application of this technique to the study of primary afferent neurons will be the focus of this section. However, the technique has been successfully employed for the study of neurons of the CNS in dissociated neurons,[70,71] slice preparations,[72–78] and more recently, *in vivo*.[79] Two approaches utilizing the patch-clamp technique to study CNS neurons will be discussed at the end of this chapter.

8.6.1 CELL-ATTACHED PATCH-CLAMP RECORDING FROM AFFERENT TERMINALS

The closest any investigator has come to cell-attached patch recording of an afferent terminal was recently reported by Brock and colleagues, who recorded extracellular activity from the peripheral terminals of corneal afferents.[80] However, this approach has been used to record ion channel activity from C-fiber axons as well as the afferent cell body.[81,82] If single-channel recordings are obtained, this approach may enable generation of the most detailed picture of the biophysical properties of the ion channel under study. Furthermore, because it is possible to record from specific sites on a neuron, it is possible to obtain information concerning the relative distribution of ion channels. However, this approach is probably the most technically difficult and labor intensive of all the approaches discussed here.

Unmyelinated axons are small and delicate, so access to isolated axons is not achieved easily. Ion channels in myelinated axons have to be studied at nodes of Ranvier unless axons are demyelinated to allow access to the membrane-bound ion channels normally covered by myelin; a process that may by itself modify the behavior of these ion channels. Furthermore, given the propensity of neurons to selectively distribute ion channels throughout the plasma membrane, the probability of obtaining single-channel recordings is low. Finally, a lot of data are required to generate a representative picture of the behavior of any given channel, and therefore, using this approach to identify injury-induced changes in channel activity is even more difficult.

8.6.2 PATCH-CLAMP RECORDING FROM SOMATA OF PRIMARY AFFERENT NEURONS

Since many investigators will choose to study the cell body of sensory neurons *in vitro*, it is worth considering the extent to which it is reasonable to extrapolate findings obtained in the cell body to the afferent terminal. While there will always be legitimate concerns about such extrapolation, several lines of evidence suggest that the cell body of acutely isolated sensory neurons *in vitro* is a valid model for the afferent terminal *in vivo*. First, with few exceptions, receptors and ion channels that are transported to the peripheral or central terminals of sensory neurons are present and functional in the plasma membrane of the cell body *in vitro*. Moreover, many of the receptors for agents present on the cell body of sensory neurons *in vitro*[83–86] appear pharmacologically similar to those near the peripheral and central arbors.[87–92] Second, it is possible to induce changes in the excitability of the cell body *in vitro* with the same manipulations that induce changes in the peripheral terminal excitability *in vivo*. For example, the cell body *in vitro* is sensitized by inflammatory mediators such as PGE_2.[93–98] This observation indicates that the cellular components (i.e., receptors, second messengers, ionic currents, etc.) necessary to achieve inflammatory mediator-induced increases in nociceptor excitability are present in the cell body *in vitro*. Third, the sensory neuron cell body *in vitro* can be induced to release neurotransmitters.[99] While it has yet to be determined whether the mechanisms underlying transmitter release from the cell body are analogous to those mediating transmitter release from central and peripheral terminals, it is noteworthy that neurotransmitter release from primary afferent terminals and the cell body are Ca^{2+}-dependent.[99]

Since a detailed discussion of the theory and practice of patch-clamp electrophysiological recording may be found elsewhere,[68,100] this section will address some of the practical issues that must be considered before patch-clamp recording may be performed.

8.6.3 DISSOCIATION PROTOCOLS FOR STUDYING ISOLATED SENSORY NEURONS

8.6.3.1 Choice of Enzymes

While there are a few investigators who employ a non-enzymatic approach to obtaining isolated neurons, the vast majority of protocols involve at least one enzyme (most commonly collagenase) and often a combination of enzymes.[101] The trade-off in the choice of enzymatic treatments is between speed (minimizing the time between tissue harvest and recording) and preservation of plasma membrane proteins. More aggressive enzymatic treatments that produce more rapid dissociation require shorter exposure time, but they increase the likelihood that plasma membrane proteins (i.e., the ion channels to be studied) may be altered or destroyed. Investigators who are new to the purchase of enzymes should not be surprised by the number of different enzyme preparations available. Companies that specialize in enzymes will prepare them within a range of specific activity (specific activity is defined in several different ways depending on the substrate and reaction endpoint). While these ranges in activity are useful in narrowing the choice of enzyme, there may be a lot of variability between preparations of the same enzyme, so it is necessary to test the enzyme under the specific conditions in which it will be used. Thus, it may be helpful to order several small aliquots of an enzyme with different lot numbers, test these, and then order a large quantity of the lot that gave the best results.

Whatever enzyme combination is employed, some mechanical perturbation of the ganglia is usually necessary, particularly when adult animals are used, to facilitate enzyme access to neurons. Thus, manual removal of the connective tissue surrounding the ganglia with sharpened jeweler's forceps under a dissecting microscope is recommended, prior to exposing ganglia to enzymes. Since sensory neurons within the trigeminal ganglia are somatotopically organized[101] and some of the neurons of interest may be located very superficially, it may be necessary to avoid removal of the connective tissue from these ganglia. Alternatively, ganglia may be cut into small pieces with scissors prior to enzyme exposure.

Several different approaches have been employed to further decrease the time that the ganglia must be exposed to digestive enzymes. Shaking water baths maintained at 37°C are typically used; however, it is possible to further decrease incubation times by employing a more aggressive shaking system such as that provided by the Nutator™ or the Belly-Dancer™ shaker. We have also had success bubbling the ganglia/enzyme mix with carbogen (5% CO_2, 95% O_2), which serves the dual purpose of agitating the ganglia and maintaining pH (if a bicarbonate-buffered medium is used with the enzymes). Serum is omitted from the enzyme solution if bubbling is employed to avoid excessive frothing.

8.6.3.2 Choice of Medium

The choice of medium depends on the nature of the specific questions to be addressed and the nature of the experimental protocol. Many factors can influence the properties of the neurons to be studied. Which factors are necessary depends on the age of the animal from which the cells were obtained and the time period over which the neurons will be studied after they are harvested. For example, nerve growth factor is necessary for the survival of sensory neurons obtained from embryos,[102] but is also involved in the regulation of phenotypic properties of sensory neurons from the adult.[103] Similarly, Delree and colleagues[104] have demonstrated that even brief exposure of sensory neurons to serum profoundly influences the properties of sensory neurons maintained in culture. The simplest solution to this concern is to study the neurons for as short a time as possible after removal from the animal. For short-term culture, investigators will often choose to provide for only the most minimal of survival needs. Thus, a minimal essential medium (MEM) is often sufficient to maintain cells for several hours. Investigators have also employed Ham's F-12, F-14, or a combination of MEM and F-12 with success in maintaining sensory neurons in culture for short periods of time. The addition of serum or serum substitutes (see below) to these media enables the maintenance of sensory neuronal culture for weeks. Neurons will also survive in an even more restrictive medium, as several groups routinely maintain neurons in a physiological saline solution with glucose prior to recording.[105,106] To further limit the possibility of phenotypic changes associated with dissociation, sensory neurons may be stored at room temperature or lower.[101,107]

8.6.3.3 The Use of Serum

Many cell types are maintained in culture in a medium containing 10% serum. Sensory neurons will also thrive in medium containing 10% serum (generally fetal bovine serum, but other sera have also been shown to work well). Our own experience is that if it is necessary to maintain sensory neurons in culture for longer than 24 h, the use of serum or a serum substitute becomes increasingly important to keep the cells healthy. The problem with serum is that its contents are partially unknown, the contents of different lots may vary, and as noted above, even short exposure to serum may result in phenotypic changes.[104] We routinely use heat-inactivated serum in an attempt to mitigate some of these concerns, but this is clearly not a perfect solution. Consequently, investigators have turned to serum substitutes. Several of these are commercially available and appear to enable the maintenance of healthy sensory neurons.[108] The other two solutions that have been employed to address the serum problem are the use of a completely defined medium,[104] or omitting serum and maintaining neurons for a short period of time in serum-free media.[109]

8.6.3.4 Plating Substrates

To study several neurons in a single field of view under conditions where the extracellular solution is continually exchanged, it is often useful to have neurons adhere to the bottom of a recording chamber. To facilitate neuronal adhesion, the plating surface is often coated with a charged molecule. Because of its ease of use, poly-d-lysine is probably used most commonly. Typically, the plating surface is soaked in a 0.1-mg/ml solution for 10s of minutes to hours, washed with water, and then air dried. Because poly-lysine may be toxic, care must be taken to thoroughly wash excess poly-lysine from the surface prior to drying. Due to the potential toxicity of poly-lysine, we have routinely employed poly-L-ornithine (0.1% solution) for 30 min.[95] Surfaces coated with poly-L-ornithine need only be air dried. We add laminin (5 µg/ml), an extracellular matrix protein to the poly-L-ornithine prior to coating, to further promote adhesion to poly-L-ornithine-coated surfaces (the drawback of the laminin is that it promotes neurite extension, thereby limiting the time during which neurons may be adequately voltage clamped after plating). Other charged molecules such as conconavalin-A are also employed with success.[109]

8.6.3.5 Obtaining a "Pure" Neuronal Culture

Sensory ganglia contain several different cell types in addition to neurons, and thus, obtaining a "pure" neuronal culture may be difficult. We have routinely avoided attempting to obtain a pure neuronal culture because non-neuronal cells appear to provide substances (such as nerve growth factor, NGF) that help maintain healthy neuronal cultures. However, if it is necessary to obtain a pure culture, there are several techniques that may be useful to clean up the neurons. Probably the most effective technique we have found is pre-plating the neurons. That is, following dissociation, cells are plated onto poly-L-ornithine-coated tissue culture plates and left to adhere for 2 to 4 h. The plating medium is then collected along with medium used to wash plates several times. This medium is centrifuged and the pellet is re-suspended and plated again onto fresh poly-L-ornithine/laminin-coated tissue culture plates. This second plating will contain primarily neurons, as the majority of non-neuronal cells will be left adhering to the first plates. Neuronal cultures may be further purified over time if mitotic inhibitors are included in the solution, as this will preferentially lead to the loss of non-neuronal cells. Of note for sensory neurons, it is preferable to use compounds other than cytosine arabinoside as a mitotic inhibitor, because this compound has been shown to inhibit the actions of NGF.[110]

8.6.3.6 Use of Trophic Factors

Whether or not to add trophic factors to the plating medium is becoming a more complicated question as the number of trophic factors increases along with our understanding of the various roles these compounds play in the physiology and pathophysiology of the nervous system. Historically, investigators have routinely added NGF to culture medium in concentrations ranging between 5 and 5000 ng/ml. Given our current knowledge of the role NGF plays in mediating phenotypic changes in sensory neurons,[39,111–117] the amount of NGF added to the culture medium may clearly influence the electrophysiological properties of the neurons studied. More recently, glial-cell-derived neurotrophic factor (GDNF) has been shown to control expression of several ion channels in a subpopulation of sensory neurons.[118,119] Thus, at this time, the questions of which trophic factors to add and at what concentrations have yet to be clearly answered. If neurons are studied soon after dissociation it is probably fair to argue that the impact of trophic factors (whether added or omitted) is minimal. However, if longer culture times are to be used, the impact of specific culture conditions, including the presence or absence of trophic factors, will have to be determined with respect to the specific properties of interest.

Many of the issues discussed above with respect to the dissociation and plating of sensory neurons also apply to the dissociation and plating of CNS neurons. However, there are several additional issues that must be considered.

8.6.3.7 Time in Culture

Many CNS neurons do not survive when plated on typical substrates such as lysine or ornithine. Consequently, investigators who wish to study isolated dorsal horn or brain-stem neurons must study neurons within hours after dissociation and plating,[70,71] or plate such neurons on support cells.[101,120–122] Astrocytes grown to confluency appear to provide a suitable substrate to maintain CNS neurons in culture.[101] Eckert and colleagues provide a detailed approach to the isolation and maintenance of astrocytes.[101]

8.6.4 Unique Applications

MacDermott and colleagues[120–123] have studied dissociated sensory and dorsal horn neurons in combination through an approach that has enabled them to characterize, at an electrophysiological level, events controlling the release of transmitter from the central terminals of primary afferent neurons. To create a situation *in vitro* where one or two sensory neurons form synapses on one or two dorsal horn neurons, these investigators establish micro-islands of dorsal horn neurons. This technique was originally developed by Segal and Furshpan[124] for the study of hippocampal neurons and was subsequently adapted by MacDermott and colleagues.[121] To obtain micro-islands, cover-slips are first coated with poly-D-lysine and then dipped in 0.5% agarose and air dried. Dry cover-slips are then sprayed with rat tail collagen (2 mg/ml in acetic acid) with an atomizer. Cover-slips are sterilized and then plated with astrocytes. The astrocytes will adhere to the collagen, thus forming the basis of the micro-island. From 3 to 7 days after plating the astrocytes, dorsal horn and DRG neurons are plated on top of the astrocytes at a density of 10 to 30 K dorsal horn neurons and 30 to 50 K DRG neurons per dish. Once established, it is possible to record from pairs of dorsal horn and sensory neurons using the electrophysiolog-ical responses of the dorsal horn neuron as an assay with which to assess events controlling release of transmitter from the sensory neuron.[121–123]

8.7 PATCH-CLAMP RECORDING *IN VIVO*

Recently, because of discrepancies between results obtained with intracellular recording techniques *in vivo,* and whole-cell patch-clamp techniques in culture and slices, and because of reports of success using whole-cell techniques *in vivo* in other systems (bat inferior colliculus[125] and rat cortex[126]), it was determined that whole-cell recording techniques could also be applied to the nociceptive systems of the spinal cord of the rat *in vivo.*[79] All the cell types that have previously been observed in the superficial dorsal horn by extracellular recording techniques may be found with whole-cell patch techniques in the superficial dorsal horn of the rat.[79]

The advantages of this technique have been outlined above in conjunction with intracellular recording techniques *in vivo.* This technique has added advantages:

1. Better control over the electrical properties of the recorded neuron because of the very low resistance of the patch pipette and the resultant very high resolution of the recording of electrical events within the neuron.
2. Small cells can be sampled even in the difficult to stabilize rat spinal cord, presumably because the whole-cell technique is more robust than sharp electrode intracellular recording techniques.
3. Cells can easily be labeled by placing .05% biocytin in the patch pipette.
4. Intracellular medium can be controlled by placing drugs or ions in the patch pipette which has easy access to the inside of the patched neuron.
5. The extracellular medium can also be controlled to a certain degree by superfusion with artificial cerebrospinal fluid, much like superfusion of slices in a chamber. Superfusion is actually very effective because the cells of the superficial dorsal horn are very close to the surface (30 to 200 μm deep).

Some other potential advantages of this technique include on-cell patching *in vivo* that would allow the observation of single channels in their native milieu, and the ability to apply drugs to the inside of a single, physiologically identified cell to determine the effects on nociceptive throughput. The major disadvantages of this technique are

1. Difficulties in proper stabilization of experimental animals *in vivo.*
2. Interactions of the necessary anesthetics with the nociceptive functions of the neurons.
3. Difficulties in obtaining adequate seals because of movement, and covering astrocytes (in adult rats, up to 80% of blind seals form on astrocytes, not neurons, after which the pipette must be discarded).
4. Possible alteration of the response of the recorded neuron because of dialysis of factors not present within the patch pipette.
5. Not knowing the nature of the recorded neuron until after the experiment, since cells cannot be visualized until then (although, the projection targets of the recorded neuron may be determined by antidromic stimulation).

In practice, it has been much more difficult to achieve whole-cell recordings in *in vivo* than in *in vitro* slices, which in turn is more difficult than achieving whole-cell recordings in culture. The *in vivo* whole-cell technique is not the best method for obtaining large samples of patch-clamped neurons to study. Recently, Yoshimura's lab has published a study on substantia gelatinosa neurons recorded *in vivo* in the whole-cell mode, in which they failed to find nociceptive thermal inputs that excited cells.[127] While we have found substantia gelatinosa neurons with noxious thermal inputs in the rat,[79] it is possible that sampling problems can selectively limit the population of neurons recorded with this technique.

In summary, each experimental electrophysiological technique described here has advantages and disadvantages. Which approach is best depends on the question being asked. Furthermore, given that there are disadvantages to each approach that may limit the interpretation of results obtained, it is useful whenever possible to employ several different approaches to the characterization of ion channels involved in the control of afferent excitability and subsequent CNS processing of stimuli perceived to be painful.

ACKNOWLEDGEMENTS

This work was supported by the Whitehall Foundation (I.S.), NINDS grants NS36929 (M.S.G), NS39420 and NS16433 (A.R.L).

REFERENCES

1. Christensen, B.N. and Perl, E.R., Spinal neurons specifically excited by noxious or thermal stimuli: marginal zone of the dorsal horn., *J. Neurophysiol.*, 33, 293, 1970.
2. Kumazawa, T. and Perl, E.R., Excitation of marginal and substantia gelatinosa neurons in the primate spinal cord: indications of their place in dorsal horn functional organization, *J. Comp. Neurol.*, 177, 417, 1978.
3. Light, A.R., Trevino, D.L., and Perl, E.R., Morphological features of functionally defined neurons in the marginal zone and substantia gelatinosa of the spinal dorsal horn, *J. Comp. Neurol.*, 186, 151, 1979.
4. Mendell, L.M. and Wall, P.D., Responses of single dorsal cord cells to peripheral cutaneous unmyelinated fibres, *Nature*, 206, 97, 1965.
5. Woolf, C.J., Windup and central sensitization are not equivalent [editorial], *Pain*, 66, 105, 1996.
6. Craig, A.D., Bushnell, M.C., Zhang, E.T., and Blomqvist, A., A thalamic nucleus specific for pain and temperature sensation, *Nature*, 372, 770, 1994.
7. Blomqvist, A., Zhang, E.T., and Craig, A.D., Cytoarchitectonic and immunohistochemical characterization of a specific pain and temperature relay, the posterior portion of the ventral medial nucleus, in the human thalamus, *Brain*, 123, 601, 2000.
8. Craig, A.D., Chen, K., Bandy, D., and Reiman, E.M., Thermosensory activation of insular cortex, *Nat. Neurosci.*, 3, 184, 2000.
9. Clark, D., Hughes, J., and Gasser, H.S., Afferent function in the group of nerve fibers of slowest conduction velocity, *Am. J. Physiol.*, 114, 69, 1935.
10. Gasser, H.S., The classification of nerve fibers, *Ohio J. Sci.*, 41, 145, 1941.

11. Heinbecker, P., Bishop, G.H., and O'Leary, J., Pain and touch fibers in peripheral nerves, *Arch. Neurol. Psychiatr.*, 153, 113, 1933.
12. Heinbecker, P., Bishop, G.H., and O'Leary, J., Analysis of sensation in terms of the nerve impulse, *Arch. Neurol. Psychiatr.*, 31, 34, 1934.
13. Collins, W.F., Jr., Nulsen, F.E., and Randt, C.T., Relation of peripheral nerve fiber size and sensation in man, *Arch. Neurol. (Chicago)*, 3, 381, 1960.
14. Traub, R.J. and Mendell, L.M., The spinal projection of individual identified A-delta- and C-fibers, *J. Neurophysiol.*, 59, 41, 1988.
15. Traub, R.J., Sedivec, M.J., and Mendell, L.M., The rostral projection of small diameter primary afferents in Lissauer's tract, *Brain Res.*, 399, 185, 1986.
16. Stys, P.K., Suction electrode recording from nerves and fiber tracts, in *Practical Electrophysiological Methods. A Guide for In Vitro Studies in Vertebrate Neurobiology,* Kettenmann, H. and Grantyn, R., Eds., Wiley-Liss, New York, 1992.
17. Stys, P.K. and Kocsis, J.D., Electrophysiological approaches to the study of axons, in *The Axon: Structure, Function and Pathophysiology,* Waxman, S. G., Kocsis, J.D., and Stys, P. K., Eds., Oxford University Press, New York, 1995, chap. 17.
18. Hagbarth, K.-E. and Vallbo, A.B., Mechanoreceptor activity recorded percutaneously with semi-microelectrodes in human peripheral nerves, *Acta Physiol. Scand.*, 69, 121, 1967.
19. Torebjörk, H.E. and Hallin, R.G., C-fibre units recorded from human sensory nerve fascicles in situ. A preliminary report, *Acta Soc. Med. Ups.*, 75, 81, 1970.
20. Van, H.J. and Gybels, J.M., Pain related to single afferent C fibers from human skin, *Brain Res.*, 48, 397, 1972.
21. Torebjörk, H.E., Ochoa, J.L., and McCann, F.V., Paresthesiae: abnormal impulse generation in sensory nerve fibres in man, *Acta Physiol. Scand.*, 105, 518, 1979.
22. Ochoa, J., Torebjörk, H.E., Culp, W.J., and Schady, W., Abnormal spontaneous activity in single sensory nerve fibers in humans, *Muscle Nerve*, 5, S74-S77, 1982.
23. Ochoa, J.L. and Torebjörk, H.E., Paraesthesiae from ectopic impulse generation in human sensory nerves, *Brain*, 103, 835, 1980.
24. Simone, D.A., Marchettini, P., Caputi, G., and Ochoa, J.L., Identification of muscle afferents subserving sensation of deep pain in humans, *J. Neurophysiol.*, 72, 883, 1994.
25. Torebjörk, H.E., Vallbo, A.B., and Ochoa, J.L., Intraneural microstimulation in man. Its relation to specificity of tactile sensations, *Brain*, 110, 1509, 1987.
26. Ochoa, J. and Torebjörk, H.E., Sensations evoked by intraneural microstimulation of C nociceptor fibres in human skin nerves, *J. Physiol. (Lond.)*, 415, 583, 1989.
27. Marchettini, P., Simone, D.A., Caputi, G., and Ochoa, J.L., Pain from excitation of identified muscle nociceptors in humans, *Brain Res.*, 740, 109, 1996.
28. Schmelz, M., Schmidt, R., Bickel, A., Handwerker, H.O., and Torebjörk, H.E., Specific C-receptors for itch in human skin, *J. Neurosci.*, 17, 8003, 1997.
29. Hallin, R.G. and Wu, G., Protocol for microneurography with concentric needle electrodes, *Brain Res. Protoc.*, 2, 120, 1998.
30. Hagbarth, K.E., Exteroceptive, proprioceptive, and sympathetic activity recorded with microelectrodes from human peripheral nerves, *Proc. Mayo Clin.*, 54, 353, 1979.
31. Vallbo, A.B., Hagbarth, K.E., Torebjörk, H.E., and Wallin, B.G., Somatosensory, proprioceptive, and sympathetic activity in human peripheral nerves, *Physiol. Rev.*, 59, 919, 1979.
32. Hensel, H. and Boman, K.K.A., Afferent impulses in cutaneous sensory nerves in human subjects, *J. Neurophysiol.*, 23, 564, 1960.

33. Fisher, G., Sayre, R.G., and Bickford, R.G., Histological changes in cat's brain after introduction of metalic and coated wire used in encephalography, *Proc. Mayo Clin.*, 32, 14, 1957.

34. Jackson, W.F. and Tuling, B.R., Toxic effects of silver/silver chloride electrodes on vascular smooth muscle, *Circ. Res.*, 53, 105, 1983.

35. Kajander, K.C., Wakisaka, S., and Bennett, G.J., Spontaneous discharge originates in the dorsal root ganglion at the onset of a painful peripheral neuropathy in the rat, *Neurosci. Lett.*, 138, 225, 1992.

36. Xie, Y., Zhang, J., Petersen, M., and LaMotte, R.H., Functional changes in dorsal root ganglion cells after chronic nerve constriction in the rat, *J. Neurophysiol.*, 73, 1811, 1995.

37. Devor, M. and Wall, P.D., Cross-excitation in dorsal root ganglia of nerve-injured and intact rats, *J. Neurophysiol.*, 64, 1733, 1990.

38. Koerber, H.R., Druzinsky, R.E., and Mendell, L.M., Properties of somata of dorsal root ganglion cells differ according to peripheral receptor innervated, *J. Neurophysiol.*, 60, 1584, 1988.

39. Ritter, A.M. and Mendell, L.M., Somal membrane properties of physiologically identified sensory neurons in the rat: effects of nerve growth factor, *J. Neurophysiol.*, 68, 2033, 1992.

40. Lawson, S.N., Crepps, B.A., and Perl, E.R., Relationship of substance P to afferent characteristics of dorsal root ganglion neurones in guinea-pig, *J. Physiol. (Lond.)*, 505, 177, 1997.

41. Djouhri, L., Bleazard, L., and Lawson, S.N., Association of somatic action potential shape with sensory receptive properties in guinea-pig dorsal root ganglion neurones, *J. Physiol. (Lond.)*, 513, 857, 1998.

42. Czeh, G., Kudo, N., and Kuno, M., Membrane properties and conduction velocity in sensory neurones following central or peripheral axotomy, *J. Physiol. (Lond.)*, 270, 165, 1977.

43. Djouhri, L. and Lawson, S.N., Changes in somatic action potential shape in guinea-pig nociceptive primary afferent neurones during inflammation in vivo, *J. Physiol. (Lond.)*, 520, 565, 1999.

44. Willis, W.D., Trevino, D.L., Coulter, J.D., and Maunz, R.A., Responses of primate spinothalamic tract neurons to natural stimulation of hindlimb, *J. Neurophysiol.*, 37, 358, 1974.

45. Giesler, G.J.J., Gerhart, K.D., Yezierski, R.P., Wilcox, T.K., and Willis, W.D., Post-synaptic inhibition of primate spinothalamic neurons by stimulation in nucleus raphe magnus, *Brain Res.*, 204, 184, 1981.

46. Bennett, G.J., Hayashi, H., Abdelmoumene, M., and Dubner, R., Physiological properties of stalked cells of the substantia gelatinosa intracellularly stained with horse-radish peroxidase, *Brain Res.*, 164, 285, 1979.

47. Steedman, W.M., Molony, V., and Iggo, A., Nociceptive neurones in the superficial dorsal horn of cat lumbar spinal cord and their primary afferent inputs, *Exp. Brain Res.*, 58, 171, 1985.

48. Light, A.R., Casale, E.J., and Menetrey, D.M., The effects of focal stimulation in nucleus raphe magnus and periaqueductal gray on intracellularly recorded neurons in spinal laminae I and II, *J. Neurophysiol.*, 56, 555, 1986.

49. Schneider, S.P., Functional properties and axon terminations of interneurons in laminae III-V of the mammalian spinal dorsal horn in vitro, *J. Neurophysiol.*, 68, 1746, 1992.

50. Michaelis, M., Blenk, K.H., Janig, W., and Vogel, C., Development of spontaneous activity and mechanosensitivity in axotomized afferent nerve fibers during the first hours after nerve transection in rats, *J. Neurophysiol.*, 74, 1020, 1995.

51. Blenk, K.H., Michaelis, M., Vogel, C., and Janig, W., Thermosensitivity of acutely axotomized sensory nerve fibers, *J. Neurophysiol.*, 76, 743, 1996.

52. Michaelis, M., Vogel, C., Blenk, K.H., Arnarson, A., and Janig, W., Inflammatory mediators sensitize acutely axotomized nerve fibers to mechanical stimulation in the rat, *J. Neurosci.*, 18, 7581, 1998.

53. Johnston, D. and Brown, T.H., Interpretation of voltage-clamp measurements in hippocampal neurons, *J. Neurophysiol.*, 50, 464, 1983.

54. Villiere, V. and McLachlan, E.M., Electrophysiological properties of neurons in intact dorsal root ganglia classified by conduction velocity and action potential duration, *J. Neurophysiol.*, 76, 1924, 1996.

55. Edwards, F. A., Konnerth, A., Sakmann, B., and Takahashi, T., A thin slice preparation for patch clamp recordings from neurones of the mammalian central nervous system, *Pflügers Arch.*, 414, 600, 1989.

56. Stuart, G.J., Dodt, H.-U., and Sakmann, B., Patch-clamp recordings from the soma and dendrites of neurons in brain slices using infrared video microscopy, *Pflügers Arch.*, 423, 511, 1993.

57. Agajanian, G.K. and Rasmussen, K., Intracellular studies in the facial nucleus illustrating a simple new method for obtaining viable motoneurons in adult rat brain slices, *Synapse*, 3, 331, 1989.

58. Colbert, C.M., Magee, J.C., Hoffman, D.A., and Johnston, D., Slow recovery from inactivation of Na^+ channels underlies the activity-dependent attenuation of dendritic action potentials in hippocampal CA1 pyramidal neurons, *J. Neurosci.*, 17, 6512, 1997.

59. Yanagisawa, M., Hosoki, R., and Otsuka, M., The isolated spinal cord-skin preparation of the newborn rat and effects of some algogenic and analgesic substances, *Eur. J. Pharmacol.*, 220, 111, 1992.

60. Jeftinija, S., Urban, L., and Kojic, L., The selective activation of dorsal horn neurons by potassium stimulation of high threshold primary afferent neurons *in vitro*, *Neuroscience*, 56, 473, 1993.

61. Curtis, D.R., Microelectrophoresis, in *Physical Techniques in Biological Research*, Nastuk, W.L., Ed., Academic Press, New York, 1964, chap. 4.

62. Katz, L.C. and Dalva, M.B., Scanning laser photostimulation: a new approach for analyzing brain circuits, *J. Neurosci. Methods*, 54, 205, 1994.

63. Kotter, R., Staiger, J.F., Zilles, K., and Luhmann, H.J., Analysing functional connectivity in brain slices by a combination of infrared video microscopy, flash photolysis of caged compounds and scanning methods, *Neuroscience*, 86, 265, 1998.

64. Steen, K.H., Issberner, U., and Reeh, P.W., Pain due to experimental acidosis in human skin: evidence for non-adapting nociceptor excitation, *Neurosci. Lett.*, 199, 29, 1995.

65. Tegeder, C. and Reeh, P.W., Acid pH has different effects on somata of primary afferent neurons than on nociceptive nerve endings, *7th World Congr. Pain,* 7, 142, 1993.

66. Bevan, S. and Yeats, J., Protons activate a cation conductance in a sub-population of rat dorsal root ganglion neurones, *J. Physiol. (Lond.)*, 433, 145, 1991.

67. Lingueglia, E., de Weillen, R., Bassilana, F., Heurteaux, C., Sakai, H., Waldmann, R., and Lazdunski, M., A modulatory subunit of acid sensing ion channels in brain and dorsal root ganglion cells, *J. Biol. Chem.*, 272, 29778, 1997.

68. Hamill, O.P., Marty, A., Neher, E., Sackmann, B., and Sigworth, F.J., Improved patch-clamp techniques for high resolution current from cells and cell-free membrane patches, *Pflügers Arch.*, 391, 85, 1981.

69. Neher, E. and Sakmann, B., Single-channel currents recorded from membrane of denervated frog muscle fibres, *Nature*, 260, 799, 1976.

70. Reichling, D.B., Kyrozis, A., Wang, J., and MacDermott, A.B., Mechanisms of GABA and glycine depolarization-induced calcium transients in rat dorsal horn neurons, *J. Physiol. (Lond.)*, 476, 411, 1994.

71. Rusin, K.I., Jiang, M.C., Cerne, R., and Randic, M., Interactions between excitatory amino acids and tachykinins in the rat spinal dorsal horn, *Brain Res. Bull.*, 30, 329, 1993.

72. Baba, H., Kohno, T., Okamoto, M., Goldstein, P.A., Shimoji, K., and Yoshimura, M., Muscarinic facilitation of GABA release in substantia gelatinosa of the rat spinal dorsal horn, *J. Physiol. (Lond.)*, 508, 83, 1998.

73. Bao, J., Li, J.J., and Perl, E.R., Differences in Ca^{2+} channels governing generation of miniature and evoked excitatory synaptic currents in spinal laminae I and II, *J. Neurosci.*, 18, 8740, 1998.

74. Pan, Z.Z. and Fields, H.L., Endogenous opioid-mediated inhibition of putative pain-modulating neurons in rat rostral ventromedial medulla, *Neuroscience*, 74, 855, 1996.

75. Pan, Z.Z., Tershner, S.A., and Fields, H.L., Cellular mechanism for anti-analgesic action of agonists of the kappa-opioid receptor, *Nature*, 389, 382, 1997.

76. Pan, Z.Z., μ-Opposing actions of the κ-opioid receptor, *Trends Pharmacol. Sci.*, 19, 94, 1998.

77. Yoshimura, M. and Nishi, S., Blind patch-clamp recordings from substantia gelatinosa neurons in adult rat spinal cord slices: pharmacological properties of synaptic currents, *Neuroscience*, 53, 519, 1993.

78. Schneider, S.P., Eckert, W.A., and Light, A.R., Opioid-activated postsynaptic, inward rectifying potassium currents in whole cell recordings in substantia gelatinosa neurons, *J. Neurophysiol.*, 80, 2954, 1998.

79. Light, A.R. and Willcockson, H.H., Spinal laminae I-II neurons in rat recorded in vivo in whole cell, tight seal configuration: properties and opioid responses, *J. Neurophysiol.*, 82, 3316, 1999.

80. Brock, J.A., McLachlan, E.M., and Belmonte, C., Tetrodotoxin-resistant impulses in single nociceptor nerve terminals in guinea-pig cornea, *J. Physiol. (Lond.)*, 512, 211, 1998.

81. Reid, G., Scholz, A., Bostock, H., and Vogel, W., Human axons contain at least five types of voltage-dependent potassium channel, *J. Physiol. (Lond.)*, 518, 681, 1999.

82. Scholz, A., Reid, G., Vogel, W., and Bostock, H., Ion channels in human axons, *J. Neurophysiol.*, 70, 1274, 1993.

83. Liu, L. and Simon, S.A., Capsaicin and nicotine both activate a subset of rat trigeminal ganglion neurons, *Am. J. Physiol.*, 270, C1807, 1996.

84. Lovinger, D.M. and Weight, F.F., Glutamate induces a depolarization of adult rat dorsal root ganglion neurons that is mediated predominantly by NMDA receptors, *Neurosci. Lett.*, 94, 314, 1988.

85. Pidoplichko, V.I., Ammonia and proton gated channel populations in trigeminal ganglion neurons, *Gen. Physiol. Biophys.*, 11, 39, 1992.

86. Todorovic, S. and Anderson, E.G., $5-HT_2$ and $5-HT_3$ receptors mediate two distinct depolarizing responses in rat dorsal root ganglion neurons, *Brain Res.*, 511, 71, 1990.

87. Carlton, S.M. and Coggeshall, R.E., Immunohistochemical localization of $5-HT_{2A}$ receptors in peripheral sensory axons in rat glabrous skin, *Brain Res.*, 763, 271, 1997.

88. Carlton, S.M., Zhou, S., and Coggeshall, R.E., Peripheral GABA(A) receptors: evidence for peripheral primary afferent depolarization, *Neuroscience*, 93, 713, 1999.

89. Chen, X., Belmonte, C., and Rang, H.P., Capsaicin and carbon dioxide act by distinct mechanisms on sensory nerve terminals in the cat cornea, *Pain*, 70, 23, 1997.

90. Chen, X., Gallar, J., and Belmonte, C., Reduction by antiinflammatory drugs of the response of corneal sensory nerve fibers to chemical irritation, *Invest. Ophthalmol. Vis. Sci.*, 38, 1944, 1997.

91. Coggeshall, R.E. and Carlton, S.M., Ultrastructural analysis of NMDA, AMPA, and kainate receptors on unmyelinated and myelinated axons in the periphery, *J. Comp. Neurol.*, 391, 78, 1998.

92. Liu, H., Wang, H., Sheng, M., Jan, L.Y., Jan, Y.N., and Basbaum, A.I., Evidence for presynaptic N-methyl-D-aspartate autoreceptors in the spinal cord dorsal horn, *Proc. Natl. Acad. Sci. USA*, 91, 8383, 1994.

93. Baccaglini, P.I. and Hogan, P.G., Some rat sensory neurons in culture express characteristics of differentiated pain sensory cells, *Proc. Natl. Acad. Sci. USA*, 80, 594, 1983.

94. Fowler, J.C., Wonderlin, W.F., and Weinreich, D., Prostaglandins block a Ca^{2+}-dependent slow spike afterhyperpolarization independent of effects on Ca^{2+} influx in visceral afferent neurons, *Brain Res.*, 345, 345, 1985.

95. Gold, M.S., Dastmalchi, S., and Levine, J.D., Co-expression of nociceptor properties in dorsal root ganglion neurons from the adult rat *in vitro*, *Neuroscience*, 71, 265, 1996.

96. Nicol, G.D. and Cui, M., Enhancement by prostaglandin E2 of bradykinin activation of embryonic rat sensory neurones, *J. Physiol. (Lond.)*, 480, 485, 1994.

97. Vasko, M.R., Campbell, W.B., and Waite, K.J., Prostaglandin E2 enhances bradykinin-stimulated release of neuropeptides from rat sensory neurons in culture, *J. Neurosci.*, 14, 4987, 1994.

98. Weinreich, D. and Wonderlin, W.F., Inhibition of calcium-dependent spike after-hyperpolarization increases excitability of rabbit visceral sensory neurones, *J. Physiol. (Lond.)*, 394, 415, 1987.

99. Huang, L.-Y.M. and Neher, E., Ca^{2+}-dependent exocytosis in the somata of dorsal root ganglion neurons, *Neuron*, 17, 135, 1996.

100. Sakmann, B. and Neher, E., *Single Channel Recording*, 2nd ed., Plenum, New York, 1995.

101. Eckert, S.P., Taddese, A., and McCleskey, E.W., Isolation and culture of rat sensory neurons having distinct sensory modalities, *J. Neurosci. Meth.*, 77, 183, 1997.

102. Liuzzi, A., Angeletti, P.U., and Levi-Montalcini, R., Metabolic effects of a specific nerve growth factor (NGF) on sensory and sympathetic ganglia: enhancement of lipid biosynthesis, *J. Neurochem.*, 12, 705, 1965.

103. Woolf, C.J., Phenotypic modification of primary sensory neurons: the role of nerve growth factor in the production of persistent pain, *Philos. Trans. Roy. Soc. Lond. B. Biol. Sci.*, 351, 441, 1996.

104. Delree, P., Ribbens, C., Martin, D., Rogister, B., Lefebvre, P.P., Rigo, J.M., Leprince, P., Schoenen, J., and Moonen, G., Plasticity of developing and adult dorsal root ganglion neurons as revealed in vitro, *Brain Res. Bull.*, 30, 231, 1993.

105. Cardenas, C.G., Del, M.L., and Scroggs, R.S., Variation in serotonergic inhibition of calcium channel currents in four types of rat sensory neurons differentiated by membrane properties, *J. Neurophysiol.*, 74, 1870, 1995.

106. Scroggs, R.S., Todorovic, S.M., Anderson, E.G., and Fox, A.P., Variation in IH, IIR, and ILEAK between acutely isolated adult rat dorsal root ganglion neurons of different size, *J. Neurophysiol.*, 71, 271, 1994.

107. White, G., Lovinger, D.M., and Weight, F.F., Transient low-threshold Ca^{2+} current triggers burst firing through an afterdepolarizing potential in an adult mammalian neuron, *Proc. Natl. Acad. Sci. USA*, 86, 6802, 1989.

108. Winter, J., Brain derived neurotrophic factor, but not nerve growth factor, regulates capsaicin sensitivity of rat vagal ganglion neurones, *Neurosci. Lett.*, 241, 21, 1998.

109. Scroggs, R.S. and Fox, A.P., Calcium current variation between acutely isolated adult rat dorsal root ganglion neurons of different size, *J. Physiol. (Lond.)*, 445, 639, 1992.

110. Tomkins, C.E., Edwards, S.N., and Tolkovsky, A.M., Apoptosis is induced in post-mitotic rat sympathetic neurons by arabinosides and topoisomerase II inhibitors in the presence of NGF, *J. Cell Sci.*, 107, 1499, 1994.

111. Aguayo, L.G. and White, G., Effects of nerve growth factor on TTX- and capsaicin-sensitivity in adult rat sensory neurons, *Brain Res.*, 570, 61, 1992.

112. Black, J.A., Langworthy, K., Hinson, A.W., Dib-Hajj, S.D., and Waxman, S.G., NGF has opposing effects on Na+ channel III and SNS gene expression in spinal sensory neurons, *Neuroreport*, 8, 2331, 1997.

113. Dib-Hajj, S.D., Black, J.A., Cummins, T.R., Kenney, A.M., Kocsis, J.D., and Waxman, S.G., Rescue of α-SNS sodium channel expression in small dorsal root ganglion neurons after axotomy by nerve growth factor *in vivo*, *J. Neurophysiol.*, 79, 2668, 1998.

114. Fjell, J., Cummins, T.R., Fried, K., Black, J.A., and Waxman, S.G., *In vivo* NGF deprivation reduces SNS expression and TTX-R sodium currents in IB4-negative DRG neurons, *J. Neurophysiol.*, 81, 803, 1999.

115. Helliwell, R.J.A., Winter, J., McIntyre, P., and Bevan, S., NGF regulates the expression of a tetrodotoxin-resistant sodium channel in cultured sensory neurones, *Soc. Neurosci. Abstr.*, 23, 911, 1997.

116. Oyelese, A.A., Rizzo, M.A., Waxman, S.G., and Kocsis, J.D., Differential effects of NGF and BDNF on axotomy-induced changes in GABA(A)-receptor-mediated conductance and sodium currents in cutaneous afferent neurons, *J. Neurophysiol.*, 78, 31, 1997.

117. Steers, W.D., Kolbeck, S., Creedon, D., and Tuttle, J.B., Nerve growth factor in the urinary bladder of the adult regulates neuronal form and function, *J. Clin. Invest.*, 88, 1709, 1991.

118. Fjell, J., Cummins, T.R., Dib-Hajj, S.D., Fried, K., Black, J.A., and Waxman, S.G., Differential role of GDNF and NGF in the maintenance of two TTX-resistant sodium channels in adult DRG neurons, *Mol. Brain Res.*, 67, 267, 1999.

119. Molliver, D.C., Wright, D.E., Leitner, M.L., Parsadanian, A.S., Doster, K., Wen, D., Yan, Q., and Snider, W.D., IB4-binding DRG neurons switch from NGF to GDNF dependence in early postnatal life, *Neuron*, 19, 849, 1997.

120. Gu, J.G., Albuquerque, C., Lee, C.J., and MacDermott, A.B., Synaptic strengthening through activation of Ca2+-permeable AMPA receptors, *Nature*, 381, 793, 1996.

121. Gu, J.G. and MacDermott, A.B., Activation of ATP P2X receptors elicits glutamate release from sensory neuron synapses, *Nature*, 389, 749, 1997.

122. MacDermott, A.B., Role, L.W., and Siegelbaum, S.A., Presynaptic ionotropic receptors and the control of transmitter release, *Ann. Rev. Neurosci.*, 22, 443, 1999.

123. Lee, C.J., Engelman, H.S., and MacDermott, A.B., Activation of kainate receptors on rat sensory neurons evokes action potential firing and may modulate transmitter release, *Ann. N.Y. Acad. Sci.*, 868, 546, 1999.

124. Segal, M.M. and Furshpan, E.J., Epileptiform activity in microcultures containing small numbers of hippocampal neurons, *J. Neurophysiol.*, 64, 1390, 1990.

125. Covey, E., Kauer, J.A., and Casseday, J.H., Whole-cell patch-clamp recording reveals subthreshold sound-evoked postsynaptic currents in the inferior colliculus of awake bats, *J. Neurosci.*, 16, 3009, 1996.

126. Moore, C.I. and Nelson, S.B., Spatio-temporal subthreshold receptive fields in the vibrissa representation of rat primary somatosensory cortex, *J. Neurophysiol.*, 80, 2882, 1998.

127. Furue, H., Narikawa, K., Kumamoto, E., and Yoshimura, M., Responsiveness of rat substantia gelatinosa neurones to mechanical but not thermal stimuli revealed by *in vivo* patch-clamp recording, *J. Physiol. (Lond.)*, 521, 529, 1999.

9 Membrane Properties: Ion Channels

Michael S. Gold

CONTENTS

0-8493-0035-5/01/$0.00+$1.50

9.1 INTRODUCTION

Tissue injury is associated with ongoing pain and/or an increased sensitivity to noxious (hyperalgesia) and innocuous (allodynia) stimuli. These changes in sensation reflect, at least in part, changes in the excitability of primary afferent neurons. Under normal conditions, the biophysical properties, density, and distribution of the ion channels present in the plasma membrane of a neuron determine various aspects of its excitability such as resting membrane potential, action potential threshold, and inter-spike-interval. Therefore, injury-induced changes in the excitability of primary afferent neurons are likely to reflect changes in the ion channels in these neurons. Injury-induced changes in the biophysical properties,[1,2] density,[2,3] and distribution[4] of ion channels present in primary afferent neurons have been documented. Furthermore, similar changes have been documented for ion channels present in neurons of the central nervous system. However, because a role in nociceptive processing has been suggested for many of the ion channels present in primary afferent neurons and because the methods used to study ion channels in primary afferent neurons are essentially identical to those used to study ion channels in other neurons, the focus of the present chapter is on the methods involved in the electrophysiological characterization of the biophysical properties of ion channels in primary afferent neurons. A proper characterization of the biophysical properties of ion channels is important because even small changes in one or more properties of a single population of channels can have a profound influence on the excitability of a neuron.[5,6]

9.2 A MODEL SYSTEM

Because many injury-induced changes in sensory neurons are manifest at the site of injury, often at the afferent terminal, this structure would be the ideal site to investigate the cellular mechanisms underlying these changes. Unfortunately, the morphological complexity, small size, and relative inaccessibility of afferent terminals preclude direct cellular studies with conventional microelectrophysiological methodology. One approach that has been used to circumvent these obstacles is to utilize the isolated cell body *in vitro* as a model for the afferent terminal *in vivo*.

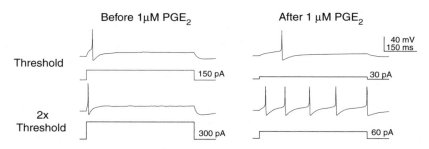

FIGURE 9.1 Prostaglandin E_2 (PGE$_2$) sensitizes a subpopulation of sensory neurons *in vitro*. A small diameter DRG neuron was studied in current clamp mode, 3 h after dissociation and plating. The resting membrane potential was –57 mV. Depolarizing current injection (500 msec square pulse) was used to evoke action potentials. Prior to the application of PGE$_2$ (Before 1 μM PGE$_2$), the minimum amount of depolarizing current necessary to evoke an action potential was 150 pA (Threshold). Current injection two times threshold (2× Threshold) still resulted in the generation of a single action potential. However, 60 s after the application of PGE$_2$ (After PGE$_2$) the threshold for action potential generation was decreased to 30 pA and current injection two times larger resulted in the generation of multiple action potentials. Changes in several voltage- and Ca^{2+}-dependent ion channels may contribute to this increase in excitability.

There are several methodological reasons for using the cell body as a model of the afferent terminal. First, because dissociated neurons are completely isolated from one another, it is possible to obtain virtually complete control of the extracellular milieu. Second, because patch-clamp techniques may be employed for electrophysiological recording, it is possible to control the intracellular milieu, as well. Third, because the neurons are isolated *in vitro*, it is possible to determine whether an injury-induced change in neuronal excitability reflects a direct change in the neuron studied (Fig. 9.1). Fourth, it is possible to record from many neurons from a single animal. Fifth, the acutely isolated sensory neuron cell body possesses an optimal geometry for electrophysiological characterization of voltage- and ligand-gated currents, because space-clamp errors are minimized when voltage-clamping the nearly spherical afferent cell body. And sixth, dissociation of sensory neurons appears to facilitate the redistribution of proteins to the plasma membrane that are normally present in afferent terminals. For example, a proton receptor/ion channel complex appears to be present in afferent terminals[7] but not the cell body *in vivo*;[8] while a proton receptor/ion channel complex is present in the cell body *in vitro*.[9] Similarly, the ionic mechanisms underlying mechanical[10,11] and thermal transduction[12,13] appear to be present in the cell bodies of dissociated sensory neurons.

While there are clearly methodological advantages of using the sensory neuron cell body *in vitro* as a model of the afferent terminal, there are several disadvantages associated with the use of this approach that should be kept in mind. These include

1. The inability to identify afferents with respect to conduction velocity or receptive field properties.
2. The potential for damage to membrane properties associated with the use of enzymes required to dissociate the neurons.

3. The act of placing sensory neurons *in vitro* necessarily involves injury.
4. Neurons can only be studied for a limited time in culture.
5. Change in phenotype to multipolar neurons.

9.3 ION CHANNELS IN SENSORY NEURONS

Ion channels present in primary afferent neurons may be divided into two main classes: those involved in signal transduction and those involved in the control of excitability. Ion channels involved in signal transduction may be further subclassified on the basis of the stimulus modality(s) to which they are responsive. That is, ion channels have been identified that are opened or closed in response to mechanical,[14,15] thermal,[16] and chemical stimuli.[9,17–19] This final subclass of signal transducing ion channels includes the prototypical ligand-gated ion channels.[18,20–22] The second main subclass of ion channels includes background or leak channels, voltage-gated channels, and Ca^{2+}-dependent channels.

9.3.1 TRANSDUCERS

9.3.1.1 Changes in Transducer Properties Influence Nociceptor Processing

In the presence of injury, inflammatory mediators are released at the site of injury. Many of these mediators are direct acting, resulting in the activation (a depolarization of sufficient magnitude for action potential generation) and/or sensitization (an increase in afferent sensitivity to subsequent stimuli) of primary afferent neurons. At least one of the mechanisms underlying the actions of inflammatory mediators is a change in an ion channel responsible for the transduction of thermal stimuli, specifically a heat transducer. That is, inflammatory mediators such as bradykinin,[12] prostaglandin E_2,[13] and protons,[16,23] induce changes in a heat transducer such that the channel is activated at lower temperatures. Such a change could explain those in thermal sensitivity observed in the presence of inflammation. Inflammation also results in changes in the mechanosensitivity of primary afferent neurons.[24–26] Similarly, injury results in changes in chemosensitivity.[27–30] All of these observations point to potential therapeutic interventions for the treatment of pain associated with injury as well as underscore the importance of the identification and characterization of ion channels involved in sensory transduction.

   While a detailed description of the methods used in the isolation and characterization of ion channels involved in sensory transduction is beyond the scope of the present chapter, the following is a brief discussion of several issues that investigators new to this field may wish to keep in mind.

9.3.1.2 Characterization of Transducers

First, and probably most importantly, the study of ion channels involved in transduction will likely require specialized equipment above and beyond the equipment ordinarily required for electrophysiological recording; that is, able to apply adequate and appropriate stimuli. Investigators have employed several distinct stimuli in an effort to identify and characterize mechanical transducers. The application of positive and negative pressure to a patch pipette in the "cell-attached patch" configuration has long been known to activate stretch-activated channels in virtually every cell studied. However, upon establishing the "whole-cell" configuration, it was not been possible to demonstrate whole cell currents evoked with mechanical deformation of the plasma membrane. McBride and Hamill[31] theorized that the reason for this failure was because the large volume of the patch electrode served to attenuate changes in pressure at the cell. Consequently, these investigators devised a system to clamp the pressure within the patch pipette.[31–33] Recently, however, McCarter and colleagues[10] reported whole-cell currents evoked with various mechanical stimuli while employing no special equipment to clamp the pipette pressure.

9.3.1.2.1 Mechanical Transducers

Several different mechanical stimuli have been employed in the study of mechanotransduction. One involves the application of hypo- and hyperosmotic extracellular solutions causing cells to swell or shrink in response to changes in osmotic pressure.[10] Another involves a micromanipulator to apply a probe directly to the neuron studied.[34] A third involves a hydraulic stimulus in which an extracellular solution is puffed onto a cell.[10] Finally, Maingret and colleagues[14] have employed trinitrophenol to produce crenation of the plasma membrane. While all of these stimuli produce electrophysiological changes, whether any are capable of utilizing the ion channels involved in the mechanotransduction that occurs in the peripheral terminals of sensory neurons has yet to be determined.

9.3.1.2.2 Thermal Transducers

As is to be expected, characterization of the ionic mechanisms underlying the transduction of heat requires an apparatus to heat and cool extracellular solutions. The simplest approach to this problem is to apply solutions warmed to a specific temperature directly to the recording chamber.[16] Minimally, one should be able to position a calibrated thermister close to the cell being studied so that the temperature "seen" by the neuron may be determined. This approach becomes cumbersome if one wishes to test a series of different temperatures and/or hold the temperature at a specific value. Consequently, investigators have turned to the use of Peltier devices to heat and cool solutions.[13] Use of a water-cooled Peltier system enables relatively rapid changes in temperature with the generation of virtually no electrical noise. Resistor-based heating systems can also raise temperatures relatively rapidly. Unlike Peltier systems, however, it is not possible to actively cool solutions heated in this way. Implementing a feedback control circuit to a Peltier- or resistor-based heating system enables one to hold the temperature at a specific value. Several of these devices are commercially available.

9.3.1.2.3 Chemical Transducers

The ionic mechanisms underlying chemical transduction may require the least sophisticated equipment of the modalities. That is, many compounds will produce large sustained responses in neurons and so they may be applied via the recording chamber perfusion system.[9] The advantage of this approach is that the concentration of a test agent may be determined with certainty. However, there are two major disadvantages of this approach. First, it necessary to use relatively large volumes of test agents, thereby prohibiting the study of relatively scarce or expensive compounds. Second, there are many ligand-gated responses that are rapidly desensitizing.[35,36] Therefore, if agonists are applied through the relatively slow chamber perfusion system, the channels may desensitize before they mediate a detectable response. For example, bradykinin-evoked responses are virtually undetectable when applied through the chamber perfusion system, while they are clearly detectable when applied rapidly and briefly (unpublished observation).

Investigators have employed several different approaches to the problem of rapid drug application. One relatively inexpensive approach is the use of a "puffer" pipette. In this approach, a microelectrode (tip diameter 30 to 100 μm) is positioned near (< 100 μm) the cell under study. This microelectrode, which may be filled with the test agent(s) of choice, is attached to a device that can apply a variable amount of pressure for a variable duration. The magnitude of applied pressure should be under experimental control so that adequate pressure may be applied to pipettes of different resistances to ensure ejection of the test agent. Furthermore, the time of application should be controllable for obvious reasons. The limitation of the puffer approach is that it is not possible to determine the concentration of drug seen by the cell (because it is mixed with normal extracellular solution), and the rate of drug removal depends on the flow of the extracellular solution through the recording chamber. A more sophisticated version of the puffer pipette involves a multi-barrel pipette and a switching valve, such that drug application is terminated rapidly by the application of a control solution in addition to the termination of the test solution. An alternate approach is the use of a "sewer-pipe." This approach utilizes several large-bore pipettes that are attached to a stepping motor. The motor positions the pipette of choice in front of the cell, enabling a rapid exchange of pipettes. Because both the flow through the pipettes and the exchange of pipettes is rapid, solution exchanges may be achieved within milliseconds. Many of these devices are commercially available.

9.3.2 IONIC CURRENTS

9.3.2.1 Voltage-Gated K+ Currents

9.3.2.1.1 Voltage-Gated K+ Currents and Nociceptive Processing

Voltage-activated K+ currents (VGKCs) have been shown to control action potential thresholds, resting potentials, and firing patterns in other neuronal preparations.[37] In primary afferent neurons, at least six VGKCs have been characterized,[38-43] including three inactivating currents and three non-inactivating currents;[38] five of these currents are likely to be present in nociceptors.[38] Two of the non-inactivating currents have low thresholds for activation and are likely, therefore, to contribute to the control of the action potential threshold. The inactivating K+ currents may also contribute to neuronal excitability as it has been demonstrated that a selective block of a slowly inactivating K^+ current decreases both action potential threshold and accommodation.[43] Recent evidence suggests VGKCs contribute to injury-induced increases in afferent excitability. For example, K^+ current density is decreased by inflammatory mediators[44] and nerve injury.[3]

9.3.2.1.2 Characterization of Voltage-Gated K^+ Currents

The separation of VGKCs from other voltage-gated ionic currents simply requires replacing Na^+ and Ca^{2+} in the extracellular solution with non-permeant, non-blocking ionic species. I replace Na^+ with choline[+] and Ca^{2+} with Co^{2+}. Because the voltage-sensitivity of some VGKCs is influenced by the concentration and species of divalent ions in the extracellular solution,[38,40] it is also possible to eliminate voltage-gated Ca^{2+} currents (VGCCs) with blockers. However, given that a single sensory neuron may express as many as five VGCCs,[45,46] the use of selective VGCC blockers can become expensive. Consequently, many investigators employ non-selective VGCC blockers such as cadmium (50 μM).

While isolating VGKCs from other voltage-gated channels is relatively easy, distinguishing VGKCs from one another can be quite difficult. There are at least three reasons for this. First, several different VGKCs are often expressed in the same neuron.[38] Second, the biophysical properties of VGKCs often overlap.[38] Third, the pharmacological tools available for the isolation of VGKCs are relatively non-specific.[37] Consequently, it is often difficult to determine with certainty the relative contribution of various VGKCs to the total outward current observed in a neuron. As a result, an experimental design utilizing each neuron as its own control may provide the most unequivocal results in terms of specific VGKCs influenced by specific experimental manipulations; for example, a VGKC is inhibited by prosta-glandin E_2 (PGE_2).[44] It is possible to determine the biophysical properties of the PGE_2-sensitive VGKC by digitally subtracting current traces obtained after the application of PGE_2 from those obtained before its application.

Because it is often not possible to use each neuron as its own control, it is worth considering some issues that may facilitate a more detailed characterization of the VGKC(s) influenced by a specific experimental manipulation. First, both transient and sustained VGKCs may be subject to steady-state inactivation over similar voltage ranges.[38,39] Consequently, while studying currents evoked from two different conditioning potentials may provide valuable information, such a manip-ulation may be insufficient to separate transient and sustained currents. Second, sensory neurons express at least three transient currents that inactivate with mark-edly different time constants.[38] Consequently, evoked currents must be sampled rapidly enough to resolve a current that inactivates within 10 msec, yet for long enough to resolve a current that inactivates with a time constant of more than 100 msec. Third, tail currents may be very informative. Various VGKCs may deactivate (i.e., a term used to describe the voltage-dependent transition of ion channels from open to the closed states) with markedly different time constants. Thus, the rate and complexity of current decay following a test potential may suggest the presence of one or more VGKCs contributing to the total outward current at the end of a test pulse.

9.3.2.2 Voltage-Gated Na⁺ Currents

9.3.2.2.1 Voltage-Gated Na⁺ Currents and Nociceptive Processing

At least seven distinct voltage-gated sodium currents (VGSCs) have been described in sensory neurons on the basis of distinct biophysical and pharmacological properties.[47–50] These currents are generally separated into two groups on the basis of their sensitivity to tetrodotoxin (TTX). One group is blocked by nanomolar concentrations of TTX, while the other is resistant to TTX at concentrations greater than 1 μM. Of the currents that have been described in sensory neurons, three are TTX sensitive and four (possibly five) are TTX-resistant. Molecular biological studies support the suggestion that sensory neurons express many VGSCs,[51] although the association between cloned channels and macroscopic currents is a question that is still being actively investigated. Furthermore, while expression of these currents appears to be tightly regulated, researchers are only just beginning to identify the factors underlying this regulation. For example, at least one TTX-sensitive current is expressed only during development or following nerve injury,[52] and the expression of this current appears to be regulated by access to nerve growth factor.[53]

A detailed characterization of the biophysical properties of VGSCs present in sensory neurons has important implications for our understanding of the underlying mechanisms of pain for several reasons. First, activation of VGSCs is critical for the generation and propagation of action potentials. Second, clinical and basic research indicates that compounds known to block Na⁺ channels may be effective for the treatment of hyperalgesia and pain.[54–59] Third, the biophysical properties of VGSCs have a profound impact on neuronal excitability. For example, the ratio of TTX-resistant to TTX-sensitive VGSCs available for activation influences excitability[5] and can change rapidly in response to changes in resting membrane potential. Rapid changes in the ratio of these two currents reflect differences in steady-state inactivation properties and rates of recovery from inactivation of these two currents. Fourth, changes in the expression and/or biophysical properties of VGSCs are associated with the development and maintenance of pain, hyperalgesia, and allodynia associated with several distinct forms of injury including inflammation,[60–64] nerve injury,[1,4,64] and diabetes.[65]

9.3.2.2.2 Characterization of Voltage-Gated Na⁺ Currents

VGSCs may be studied in isolation by blocking K⁺ and Ca²⁺ currents as described above. Eliminating VGKCs is generally achieved by replacing K⁺ with Cs⁺ in the electrode solution if whole-cell patch configuration is utilized. However, because a relatively large subpopulation of rat sensory neurons expresses a VGKC permeable to Cs⁺,[38] a complete block of outward currents with this approach may be difficult. Furthermore, this current is also incompletely blocked with very high (~90 mM) concentrations of tetraethylammonium (TEA) in bath solutions.[38] However, I have found that including 30 to 40 mM TEA in the electrode solution is quite effective at eliminating residual outward currents in sensory neurons (Fig. 9.2). Voltage-gated Ca²⁺ currents (VGCCs) may be eliminated by decreasing the concentration of extracellular Ca²⁺ (≤ 0.1 mM) and/or adding non-selective Ca²⁺ channel blockers such as Cd²⁺ (50 μM) to the extracellular solution.

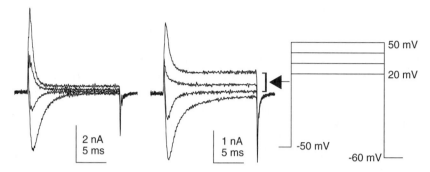

FIGURE 9.2 Voltage-gated sodium currents studied in isolation following elimination of voltage-gated potassium and calcium currents. Two small diameter DRG neurons were studied in voltage-clamp with the whole-cell patch configuration. The extracellular solution contained 35 mM NaCl (to mitigate voltage errors associated with large inward current), 65 mM choline (a non-blocking, non-permeant cation used as a substitute for extracellular NaCl), 40 mM tetraethylammonium (TEA) (to block some voltage-gated potassium currents), 0.1 mM CaCl$_2$ (to limit the size of voltage-gated calcium currents and calcium-dependent chloride and/or potassium currents), 5 mM MgCl$_2$ (to help stabilize the plasma membrane in the presence of a low concentration of extracellular calcium), 10 mM glucose (to help the cells with their energy requirements) and 10 mM HEPES (to buffer the pH to 7.4). The intracellular solution contained 1 mM CaCl$_2$, 2 mM MgCl$_2$, and 11 mM EGTA (resulting in a free calcium concentration of ~40 nM), 2 mM Mg-ATP (to mitigate run-down of currents and provide additional substrate for phosphorylation reactions), 1 mM Li-GTP (to facilitate G-protein-mediated reactions), and 10 mM HEPES (to buffer the pH to 7.2). The only difference between the solutions used to record the traces on the left and those in the middle was that 40 mM TEA was included in the electrode solution (the CsCl concentration was reduced to 100 mM). Including TEA in the electrode solution largely eliminated the sustained outward current indicated by the bracket and arrow. The voltage-protocol used to evoke the currents is shown on the right.

Because VGSCs activate rapidly and generally carry inward current, they generate depolarizing voltage errors that can lead to loss of clamp control. Loss of clamp control can be mitigated through the use of relatively low resistance patch electrodes (< 2 MΩ), use of amplifier series resistance compensation circuitry, and decreasing the driving force on Na$^+$ by lowering the concentration of extracellular Na$^+$. I generally use 35 mM Na$^+$ in the bath solution (with choline making up the bulk of the cation in the bath solution) in order to obtain peak whole-cell currents that are generally less than 5 nA; 5-nA current with a 2-MΩ electrode and 80% series-resistance compensation will generate a 2-mV voltage error.

There are several other issues to keep in mind when designing voltage-clamp protocols for the characterization of VGSCs present in sensory neurons. First, these channels may undergo a process referred to as slow-inactivation. This is a voltage-dependent process similar to the more typical inactivation described with a steady-state inactivation protocol, except that it occurs over many seconds, rather than tens to hundreds of milliseconds.[66] The result is that the fraction of available channels may be dramatically influenced by the duration that the membrane is held at a given potential. Ogata and Tatebayashi[66] described the slow inactivation of TTX-resistant

VGSCs in sensory neurons from neonatal rats that results in a dramatic reduction in peak current over tens to hundreds of seconds. Second, to achieve full recovery from steady-state inactivation of TTX-sensitive VGSCs, it is often necessary to hyperpolarize the membrane less than –120 mV. In my experience, sensory neurons do not tolerate hyperpolarized potentials well, often becoming unstable and leaky. However, I have found that increasing the concentration of divalent ions in solution can attenuate this problem. Third, it is generally possible to isolate TTX-resistant VGSCs from TTX-sensitive VGSCs with a 500-msec depolarizing voltage-step to –50 mV (a potential at which the steady-state inactivation of TTX-sensitive currents is complete). However, there does appear to be a TTX-sensitive current that requires membrane depolarizations of –40 mV or greater to fully inactivate (unpublished observation).

9.3.2.3   Voltage-Gated Ca²⁺ Currents

9.3.2.3.1   Voltage-Gated Ca²⁺ Currents and Nociceptive Processing

Four voltage-activated Ca^{2+} currents have been described in mammalian sensory neurons including the T, L, N, and P currents specifically blocked by amiloride, dihydropyridines, ω-conotoxin (ω-CTx), and ω-agatoxin (ω-AgTx), respectively.[45,46] A fifth Ca^{2+} current may also exist in DRG that is resistant to these blockers.[45,46] High-threshold Ca^{2+} currents (N, L, and P) are generally thought to control the efferent function of sensory neurons (i.e., by controlling Ca^{2+} entry and the subsequent Ca^{2+}-dependent release of transmitter from neuronal terminals).[67] However, these currents may contribute indirectly to nociceptor excitability through effects on Ca^{2+}-dependent currents,[68] and/or through control of the release of transmitters from the primary afferent that may modulate nociceptor excitability directly and/or indirectly.[69] In contrast, low-threshold T-type currents have been shown to underlie action potential bursts in sensory neurons,[70] although the magnitude of this current is relatively small in putative nociceptors.[45] Recent evidence suggests that voltage-gated Ca^{2+} currents may also contribute to injury-induced increases in afferent excitability as nerve injury results in changes in both T- and N-type currents.[71,72]

9.3.2.3.2   Characterization of Voltage-Gated Ca²⁺ Currents

VGCCs are studied in isolation following elimination of VGKCs and voltage-gated Na^+ currents (VGSCs). Eliminating VGSCs may be achieved by substituting a choline for Na^+ in the extracellular solution. Eliminating VGKCs is generally achieved by replacing K^+ with Cs^+ in the electrode solution if whole-cell patch-configuration is utilized. Because outward Cs^+ current (passing through voltage-gated K^+ channels) may contaminate VGCCs as well as VGSCs, 30 to 40 mM TEA in the electrode solution may be used to eliminate residual outward currents in sensory neurons (see Fig. 9.2).

While there are differences between VGCCs with respect to their biophysical properties, with the exception of T-type currents, these differences do not enable the isolation of one current from another with the use of unique voltage-clamp protocols and digital subtraction. Rather, the pharmacological differences between VGCCs provide the basis by which one current is distinguished from another. Because of this, the role of a particular VGCC is generally implicated with an occlusion experiment in which one assesses the fraction of total VGCC sensitive to a specific VGCC blocker (e.g., dihydropyridine), before and after application of a test agent. The application of a saturating concentration of VGCC blocker will occlude the effects of a test agent if the compounds are acting on the same channel.[46,73]

The greatest difficulty in recording VGCCs is that these currents exhibit an exponential decay or "run-down" following establishment of the whole-cell config-uration. All of the mechanisms underlying VGCC run-down have yet to be elucidated, but appear to involve Ca^{2+}-dependent desensitization of channels, phosphorylation/ dephosphorylation events, and changes in the interaction with cytoskeletal pro-teins.[67,74] Thus, VGCC run-down may be attenuated by substituting Ba^{2+} for Ca^{2+}, as this ion permeates Ca^{2+} channels, yet does not appear to be as effective at mediating Ca^{2+}-dependent desensitization. That Ba^{2+} may also limit the number of Ca^{2+}-depen-dent cellular processes influenced by an increase in intracellular Ca^{2+} concentration is another reason to use Ba^{2+} instead of Ca^{2+} in the extracellular solution. Other investigators have also had some success with the inclusion of ATP and/or com-pounds that facilitate the regeneration of ATP in the electrode solution.[75,76] Finally, because the time course of run-down is such that the majority of current decay occurs within the first 5 min after establishing whole-cell access, one simple way to mitigate problems associated with a changing baseline is to wait 5 to 10 min after establishing whole-cell access prior to performing any experimental manipulations.

9.3.2.4 Inward-Rectifying Currents

9.3.2.4.1 Inward-Rectifying Currents and Nociceptive Processing

Sensory neurons express at least two types of inward rectifying current. One is K^+ current referred to as I_{IR}, present in a subpopulation of sensory neurons that has yet to be studied in detail.[77] Such currents are thought to be involved in the determination of the inter-spike-interval as they slow membrane depolarization following an action potential. A second inward-rectifying non-selective cationic current, often referred to as I_h (for hyperpolarization-activated current), has been studied in somewhat greater detail.[78,79] This current is also thought to contribute to determining the inter-spike-interval. However, because I_h is a non-selective cationic current, it increases the rate of depolarization following an action potential. There is at least one report suggesting that I_h may be involved in the control of action potential conduction velocity in slowly conducting afferents.[80] Evidence also exists that modulation of the biophysical properties of this current can influence excitability.[79] However, I_h may not be present in putative nociceptors.[79,81]

9.3.2.4.2 Characterization of Inward-Rectifying Currents

Because I_{IR} and I_h are activated with membrane hyperpolarization, these currents are easily isolated from other voltage-gated currents, and are typically activated with a hyperpolarizing voltage-step to –80 or –100 mV from a potential of between –60 and –40 mV. The two currents may be distinguished from one another on the basis of their rates of activation and inactivation, in addition to their ionic selectivity. The two currents also display a differential pharmacological profile: I_h and I_{IR} are blocked by extracellular Cs^+ (5 mM); I_{IR} is blocked by extracellular Ba^{2+} (100 μM).[77]

9.3.3 Ca^{2+}-Dependent Ion Channels

9.3.3.1 Ca^{2+}-Dependent K$^+$ Currents

9.3.3.1.1 Ca^{2+}-Dependent K$^+$ Currents and Nociception

There is compelling evidence to suggest that at least two of the three Ca^{2+}-dependent K$^+$ currents that have been described in mammalian visceral sensory neurons (in the nodose ganglion) may be involved in nociceptive processing. These currents include a slowly activating – slowly decaying current (I_{sAHP}),[82] a charybdotoxin- (ChTx) and iberiotoxin-sensitive current (I_C),[83] and a ChTx- and apamin-resistant current (I_{CA-R}).[84] Both I_{sAHP} and I_C are present in putative nociceptors. Because I_{sAHP} is a current driving a slow afterhyperpolarization (sAHP) that can last for many seconds follow-ing action potential generation, it limits action potential burst duration.[82,85,86] I_{sAHP} is selectively blocked by several hyperalgesic inflammatory mediators resulting in a dramatic increase in nociceptor excitability.[85,86] While there is some evidence sug-gesting I_{sAHP} may be selectively blocked by apamin,[82] this property may be species specific as I_{sAHP} present in nodose ganglion neurons from the rabbit, ferret, and guinea pig is apamin insensitive.[87] I_C has been shown to be involved in action potential repolarization, and in the early phase of the AHP following action potentials in other cell types.[83,88,89] Thus, this current may regulate spike frequency early in a burst of action potentials;[37] I_C has only been studied in sensory neurons with respect to its role in action potential repolarization.[83] The distribution of I_{CA-R} among sensory neurons and its functional role has yet to be determined.

9.3.3.1.2 Characterization of Ca^{2+}-Dependent K$^+$ Currents

Isolation and characterization of Ca^{2+}-dependent currents are relatively difficult for several reasons. One reason is that these currents require an increase in intracellular Ca^{2+}. Because this is most easily accomplished via activation of VGCCs, it is possible that Ca^{2+}-dependent outward and inward currents are active simultaneously. This is the case for I_C, which activates rapidly. As a result, I_C must be studied as a difference current (i.e., current obtained by subtracting current evoked in the presence of blocker from that evoked in its absence), taking advantage of the fact that there are selective blockers for I_C. In contrast, inward and outward currents are separated temporally with the activation of I_{sAHP}, because of the relatively slow activation of this current. The rate of activation of I_{CA-R} at the whole-cell level has yet to be determined, likely reflecting the fact that a selective blocker for this channel has not been identified.

A second reason these currents are difficult to study is because their activation depends on, or in the case of I_C is modulated by, increases in intracellular Ca^{2+}. As a result, it must be determined whether the kinetics of current activation and decay reflect a biophysical property of the underlying channels or the time course of changes in intracellular Ca^{2+}. Cell-free patch recording is one way to address this question electrophysiologically. Alternatively, it is possible to utilize fluorimetry in combination with electrophysiological recording, so that current and changes in intracellular Ca^{2+} may be recorded simultaneously.[87]

A third difficulty in characterizing Ca^{2+}-dependent K$^+$ currents is that the acti-vation of these currents may be influenced by both Ca^{2+} and voltage. For example, large conductance K$^+$ channels referred to as "maxi-K$^+$" or "BK" channels were originally described as Ca^{2+}-dependent K^+ channels because of the dramatic increase in the open probability of these channels observed in association with increases in the concentration of intracellular Ca^{2+}. Based on more recent evidence it is now clear that Ca^{2+} is better described as a modulator, rather than activator of maxi-K^+ channels.[90] That is, these channels are actually voltage dependent, yet this voltage dependence is exquisitely sensitive to changes in intracellular Ca^{2+}.[90] Consequently, small changes in the resting free Ca^{2+} concentration may have profound effects on the activation of these channels. That maxi-K^+ channels underlie I_C is suggested by similarities between the two with respect to pharmacological sensitivity and biophysical properties. Thus, the characterization of I_C is complicated by the fact that the biophysical properties of this current depend on and are influenced by an interaction between voltage and Ca^{2+}. Whether a similar complication impacts the characterization of I_{CA-R} has yet to be determined. If patch electrodes are utilized, this potential problem may be mitigated through the judicious use of Ca^{2+} chelators. If sharp electrodes or the perforated-patch configuration is utilized, direct measurement of resting intracellular Ca^{2+} concentration may be necessary.

A fourth reason Ca^{2+}-dependent K^+ currents are difficult to study is that they may be labile during whole-cell recording. This is particularly true for I_{sAHP}, which can run-down to undetectable levels within minutes of establishing whole-cell access.[91] Fortunately, it is possible to mitigate this problem by utilizing sharp electrodes[92] or the perforated-patch configuration of the whole-cell patch clamp.[91]

9.3.3.2 Ca^{2+}-Dependent Cl^- Currents

9.3.3.2.1 Ca^{2+}-Dependent Cl^- Currents and Nociception

A Ca^{2+}-activated Cl^- current has been described in dorsal root ganglion neurons.[93,94] In most neurons, Cl^- is passively distributed across the membrane, so activation of Cl^- current is inhibitory. However, because some dorsal root ganglion neurons accumulate Cl^- and have depolarizing Cl^- currents that reverse tens of mV positive to the resting potential,[95] it is difficult to predict the net effect of Cl^- currents on dorsal root ganglion neuron excitability. That said, evidence suggests that a Ca^{2+}-activated Cl^- current is upregulated in the cell bodies of sympathetic post-ganglionic neurons following injury to distal axons.[96] In these cells, the current drives a membrane depolarization following an action potential. However, this depolarization is inhibitory to subsequent activation. The distribution of this current among sensory neurons remains to be determined.

9.3.3.2.2   Characterization of Ca^{2+}-Dependent Cl^- Currents

Ca^{2+}-activated Cl^- currents are generally studied as tail currents evoked at hyperpolarized membrane potentials following a membrane depolarization of sufficient magnitude to enable the influx of Ca^{2+} through VGCCs.[93] Nevertheless, it should be noted that the current appears to activate during the membrane depolarization.[96] There are several relatively selective blockers of these currents including niflumic acid and flufenemic acid.[96]

Investigators wishing to study Ca^{2+}-activated Cl^- currents will face many of the same problems described above that are associated with the study of Ca^{2+}-dependent

K$^+$ currents. In addition, use of Ag/AgCl in electrodes necessitates the inclusion of several mM Cl$^-$ in electrode solutions rendering it impossible to study the current in the absence of intracellular Cl$^-$.

9.4 SUMMARY AND CONCLUSIONS

Activation of primary afferent nociceptors is the first step leading to the perception of pain following noxious stimulation of peripheral tissue. This activation reflects a change in membrane potential resulting from the opening and closing of ion channels present in the plasma membrane of this subpopulation of sensory neurons. These ion channels are opened and closed in response to specific stimuli (mechanical, thermal, and chemical) changes in membrane potential, and/or changes in concentration of intracellular Ca^{2+}. Because the ongoing pain and tenderness associated with tissue injury reflects changes in the excitability of primary afferent nociceptors, which in turn reflect changes in the properties of ion channels, the detailed characterization of the mechanisms controlling ion channels in this population of afferents may lead to novel therapeutic interventions for the treatment of pain.

REFERENCES

1. Kral, M.G. et al., Alteration of Na$^+$ currents in dorsal root ganglion neurons from rats with a painful neuropathy, *Pain*, 81, 15, 1999.
2. Cummins, T.R. and Waxman, S.G., Downregulation of tetrodotoxin-resistant sodium currents and upregulation of a rapidly repriming tetrodotoxin-sensitive sodium current in small spinal sensory neurons after nerve injury, *J. Neurosci.*, 17, 3503, 1997.
3. Everill, B. and Kocsis, J.D., Reduction in potassium currents in identified cutaneous afferent dorsal root ganglion neurons after axotomy, *J. Neurophysiol.*, 82, 700, 1999.
4. Novakovic, S.D. et al., Distribution of the tetrodotoxin-resistant sodium channel PN3 in rat sensory neurons in normal and neuropathic conditions, *J. Neurosci.*, 18, 2174, 1998.
5. Schild, J.H. and Kunze, D.L., Experimental and modeling study of Na$^+$ current heterogeneity in rat nodose neurons and its impact on neuronal discharge, *J. Neurophysiol.*, 78, 3198, 1997.
6. Matzner, O. and Devor, M., Na$^+$ conductance and the threshold for repetitive neuronal firing, *Brain Res.*, 597, 92, 1992.
7. Steen, K.H. et al., A dominant role of acid pH in inflammatory excitation and sensitization of nociceptors in rat skin, in vitro, *J. Neurosci.*, 15, 3982, 1995.
8. Tegeder, C. and Reeh, P.W., Acid pH has different effects on somata of primary afferent neurons than on nociceptive nerve endings, *Abstr. 7th World Congr. Pain*, 142, 1993.
9. Bevan, S. and Yeats, J., Protons activate a cation conductance in a sub-population of rat dorsal root ganglion neurones, *J. Physiol. (Lond.)*, 433, 145, 1991.
10. McCarter, G.C. et al., Mechanical transduction by rat dorsal root ganglion neurons in vitro, *Neurosci. Lett.*, 273, 179, 1999.
11. Young, S.H. et al., Calcium waves in colonic myocytes produced by mechanical and receptor-mediated stimulation, *Am. J. Physiol.*, 276, G1204, 1999.

12. Cesare, P. and McNaughton, P., A novel heat-activated current in nociceptive neurons and its sensitization by bradykinin, *Proc. Natl. Acad. Sci. USA*, 93, 15435, 1996.

13. Reichling, D.B. and Levine, J.D., Heat transduction in rat sensory neurons by calcium-dependent activation of a cation channel, *Proc. Natl. Acad. Sci. USA*, 94, 7006, 1997.

14. Maingret, F. et al., TRAAK is a mammalian neuronal mechano-gated K^+ channel, *J. Biol. Chem.*, 274, 1381, 1999.

15. Maingret, F. et al., Mechano- or acid stimulation, two interactive modes of activation of the TREK-1 potassium channel, *J. Biol. Chem.*, 274, 26691, 1999.

16. Caterina, M.J. et al., The capsaicin receptor: a heat-activated ion channel in the pain pathway, *Nature*, 389, 816, 1997.

17. Baccaglini, P.I. and Hogan, P.G., Some rat sensory neurons in culture express characteristics of differentiated pain sensory cells, *Proc. Natl. Acad. Sci. USA*, 80, 594, 1983.

18. Cook, S.P. et al., Distinct ATP receptors on pain-sensing and stretch-sensing neurons, *Nature*, 387, 505, 1997.

19. Chen, C.C. et al., A sensory neuron-specific, proton-gated ion channel, *Proc. Natl. Acad. Sci. USA*, 95, 10240, 1998.

20. Todorovic, S. and Anderson, E.G., 5-HT2 and 5-HT3 receptors mediate two distinct depolarizing responses in rat dorsal root ganglion neurons, *Brain Res.*, 511, 71, 1990.

21. Oyelese, A.A. et al., Enhancement of GABAA receptor-mediated conductances induced by nerve injury in a subclass of sensory neurons, *J. Neurophysiol.*, 74, 673, 1995.

22. Lovinger, D.M. and Weight, F.F., Glutamate induces a depolarization of adult rat dorsal root ganglion neurons that is mediated predominantly by NMDA receptors, *Neurosci. Lett.*, 94, 314, 1988.

23. Tominaga, M. et al., The cloned capsaicin receptor integrates multiple pain-producing stimuli, *Neuron*, 21, 531, 1998.

24. Martin, H.A. et al., Leukotriene and prostaglandin sensitization of cutaneous high-threshold C- and A-delta mechanonociceptors in the hairy skin of rat hindlimb, *Neuroscience*, 22, 651, 1987.

25. Davis, K.D. et al., Chemosensitivity and sensitization of nociceptive afferents that innervate the hairy skin of monkey, *J. Neurophysiol.*, 69, 1071, 1993.

26. Strassman, A.M. and Raymond, S.A., Electrophysiological evidence for tetrodotoxin-resistant sodium channels in slowly conducting dural sensory fibers, *J. Neurophysiol.*, 81, 413, 1999.

27. Ji, R.R. et al., Expression of mu-, delta-, and kappa-opioid receptor-like immuno-reactivities in rat dorsal root ganglia after carrageenan-induced inflammation, *J. Neurosci.*, 15, 8156, 1995.

28. Nicol, G.D. and Cui, M., Enhancement by prostaglandin E2 of bradykinin activation of embryonic rat sensory neurones, *J. Physiol. (Lond.)*, 480, 485, 1994.

29. Lopshire, J.C. and Nicol, G.D., Activation and recovery of the PGE2-mediated sensitization of the capsaicin response in rat sensory neurons, *J. Neurophysiol.*, 78, 3154, 1997.

30. Khasar, S.G. et al., Inflammation modulates the contribution of receptor-subtypes to bradykinin-induced hyperalgesia in the rat, *Neuroscience*, 69, 685, 1995.

31. McBride, Jr., D.W. and Hamill, O.P., Pressure-clamp: a method for rapid step perturbation of mechanosensitive channels, *Pflügers Arch.*, 421, 606, 1992.

32. McBride, Jr., D.W. and Hamill, O.P., Simplified fast pressure-clamp technique for studying mechanically gated channels, *Meth. Enzymol.*, 294, 482, 1999.

33. McBride, Jr., D.W. and Hamill, O.P., Pressure-clamp technique for measurement of the relaxation kinetics of mechanosensitive channels, *Trends Neurosci.*, 16, 341, 1993.

34. Gschossmann, J.M. et al., Mechanical activation of dorsal root ganglion cells in vitro: comparison with capsaicin and modulation by kappa-opioids, *Brain Res.*, 856, 101, 2000.

35. Cook, S.P. and McCleskey, E.W., Desensitization, recovery and Ca(2+)-dependent modulation of ATP-gated P2X receptors in nociceptors, *Neuropharmacology*, 36, 1303, 1997.

36. Cholewinski, A. et al., The role of calcium in capsaicin-induced desensitization in rat cultured dorsal root ganglion neurons, *Neuroscience*, 55, 1015, 1993.

37. Rudy, B., Diversity and ubiquity of K channels, *Neuroscience*, 25, 729, 1988.

38. Gold, M.S. et al., Characterization of six voltage-gated K+ currents in adult rat sensory neurons, *J. Neurophysiol.*, 75, 2629, 1996.

39. Akins, P.T. and McCleskey, E.W., Characterization of potassium currents in adult rat sensory neurons and modulation by opioids and cyclic AMP, *Neuroscience*, 56, 759, 1993.

40. Mayer, M.L. and Sugiyama, K., A modulatory action of divalent cations on transient outward currents in cultured rat sensory neurones, *J. Physiol. (Lond.)*, 396, 417, 1988.

41. McFarlane, S. and Cooper, E., Kinetics and voltage dependence of A-type currents on neonatal rat sensory neurons, *J. Neurophysiol.*, 66, 1380, 1991.

42. Kostyuk, P.G. et al., Ionic currents in the somatic membrane of rat dorsal root ganglion neurons-III. Potassium currents, *Neuroscience*, 6, 2439, 1981.

43. Stansfeld, C.E. et al., Mast cell degranulating peptide and dendrotoxin selectively inhibit a fast-activating potassium current and bind to common neuronal proteins, *Neuroscience*, 23, 893, 1987.

44. Nicol, G.D. et al., Prostaglandins suppress an outward potassium current in embryonic rat sensory neurons, *J. Neurophysiol.*, 77, 167, 1997.

45. Scroggs, R.S. and Fox, A.P., Calcium current variation between acutely isolated adult rat dorsal root ganglion neurons of different size, *J. Physiol. (Lond.)*, 445, 639, 1992.

46. Rusin, K.I. and Moises, H.C., mu-Opioid receptor activation reduces multiple components of high-threshold calcium current in rat sensory neurons, *J. Neurosci.*, 15, 4315, 1995.

47. Caffrey, J.M. et al., Three types of sodium channels in adult rat dorsal root ganglion neurons, *Brain Res.*, 592, 283, 1992.

48. Rush, A.M. et al., Electrophysiological properties of sodium current subtypes in small cells from adult rat dorsal root ganglia, *J. Physiol. (Lond.)*, 511, 771, 1998.

49. Kostyuk, P.G. et al., Ionic currents in the somatic membrane of rat dorsal root ganglion neurons - I. Sodium currents, *Neuroscience*, 6, 2424, 1981.

50. Scholz, A. et al., Two types of TTX-resistant and one TTX-sensitive Na+ channel in rat dorsal root ganglion neurons and their blockade by halothane, *Suppl. Eur. J. Neurosci.*, 10, 2547, 1998.

51. Black, J.A. et al., Spinal sensory neurons express multiple sodium channel alpha-subunit mRNAs, *Brain Res. Mol. Brain Res.*, 43, 117, 1996.

52. Waxman, S.G. et al., Type III sodium channel mRNA is expressed in embryonic but not adult spinal sensory neurons, and is reexpressed following axotomy, *J. Neurophysiol.*, 72, 466, 1994.

53. Black, J.A. et al., NGF has opposing effects on Na+ channel III and SNS gene expression in spinal sensory neurons, *Neuroreport*, 8, 2331, 1997.

54. Abram, S.E. and Yaksh, T.L., Systemic lidocaine blocks nerve injury-induced hyperalgesia and nociceptor-driven spinal sensitization in the rat, *Anesthesiology*, 80, 383, 1994.

55. Chabal, C. et al., The effect of intravenous lidocaine, tocainide, and mexiletine on spontaneously active fibers originating in rat sciatic neuromas, *Pain*, 38, 333, 1989.
56. Puig, S. and Sorkin, L.S., Formalin-evoked activity in identified primary afferent fibers: systemic lidocaine suppresses phase-2 activity, *Pain*, 64, 345, 1996.
57. Devor, M. et al., Systemic lidocaine silences ectopic neuroma and DRG discharge without blocking nerve conduction, *Pain*, 48, 261, 1992.
58. Chabal, C. et al., The use of oral mexiletine for the treatment of pain after peripheral nerve injury, *Anesthesiology*, 76, 513, 1992.
59. Rizzo, M.A., Successful treatment of painful traumatic mononeuropathy with carbamazepine: insights into a possible molecular pain mechanism, *J. Neurol. Sci.*, 152, 103, 1997.
60. Gold, M.S. et al., Hyperalgesic agents increase a tetrodotoxin-resistant Na^+ current in nociceptors, *Proc. Natl. Acad. Sci. USA*, 93, 1108, 1996.
61. Khasar, S.G. et al., A tetrodotoxin-resistant sodium current mediates inflammatory pain in the rat, *Neurosci. Lett.*, 256, 17, 1998.
62. Cardenas, C.G. et al., 5HT4 receptors couple positively to tetrodotoxin-insensitive sodium channels in a subpopulation of capsaicin-sensitive rat sensory neurons, *J. Neurosci.*, 17, 7181, 1997.
63. England, S. et al., PGE2 modulates the tetrodotoxin-resistant sodium current in neonatal rat dorsal root ganglion neurons via the cyclic AMP-protein kinase A cascade, *J. Physiol. (Lond.)*, 495, 429, 1996.
64. Porreca, F. et al., A comparison of the potential role of the tetrodotoxin-insensitive sodium channels, PN3/SNS and NaN/SNS2, in rat models of chronic pain, *Proc. Natl. Acad. Sci. USA*, 96, 7640, 1999.
65. Hirade, M. et al., Tetrodotoxin-resistant sodium channels of dorsal root ganglion neurons are readily activated in diabetic rats, *Neuroscience*, 90, 933, 1999.
66. Ogata, N. and Tatebayashi, H., Slow inactivation of tetrodotoxin-insensitive Na^+ channels in neurons of rat dorsal root ganglia, *J. Membr. Biol.*, 129, 71, 1992.
67. McCleskey, E.W., Calcium channels: cellular roles and molecular mechanisms, *Curr. Opin. Neurobiol.*, 4, 304, 1994.
68. Blatz, A.L. and Magleby, K.L., Calcium-activated potassium channels, *Trends in Neurosci.*, 10, 463, 1987.
69. Sluka, K.A. et al., The role of dorsal root reflexes in neurogenic inflammation, *Pain Forum*, 4, 141, 1995.
70. White, G. et al., Transient low-threshold Ca^{2+} current triggers burst firing through an afterdepolarizing potential in an adult mammalian neuron, *Proc. Natl. Acad. Sci. USA*, 86, 6802, 1989.
71. Baccei, M.L. and Kocsis, J.D., Voltage-gated calcium currents in axotomized adult rat cutaneous afferent neurons, *J. Neurophysiol.*, 83, 2227, 2000.
72. Hogan, Q.H. et al., Painful neuropathy decreases membrane calcium current in mammalian primary afferent neurons, *Pain*, 86, 43, 2000.
73. Schroeder, J.E. and McCleskey, E.W., Inhibition of Ca^{2+} currents by a mu-opioid in a defined subset of rat sensory neurons, *J. Neurosci.*, 13, 867, 1993.
74. Schroeder, J.E. et al., Two components of high-threshold Ca^{2+} current inactivate by different mechanisms, *Neuron*, 5, 445, 1990.
75. Thayer, S.A. and Miller, R.J., Regulation of the intracellular free calcium concentration in single rat dorsal root ganglion neurones in vitro, *J. Physiol. (Lond.)*, 425, 85, 1990.
76. Forscher, P. et al., Noradrenaline modulates calcium channels in avian dorsal root ganglion cells through tight receptor-channel coupling, *J. Physiol. (Lond.)*, 379, 131, 1986.

77. Scroggs, R.S. et al., Variation in IH, IIR, and ILEAK between acutely isolated adult rat dorsal root ganglion neurons of different size, *J. Neurophysiol.*, 71, 271, 1994.

78. Cardenas, C.G. et al., Serotonergic modulation of hyperpolarization-activated current in acutely isolated rat dorsal root ganglion neurons, *J. Physiol. (Lond.)*, 518, 507, 1999.

79. Ingram, S.L. and Williams, J.T., Opioid inhibition of Ih via adenylyl cyclase, *Neuron*, 13, 179, 1994.

80. Djouhri, L. and Lawson, S.N., Changes in somatic action potential shape in guinea-pig nociceptive primary afferent neurones during inflammation in vivo, *J. Physiol. (Lond.)*, 520, 565, 1999.

81. Cardenas, C.G. et al., Variation in serotonergic inhibition of calcium channel currents in four types of rat sensory neurons differentiated by membrane properties, *J. Neurophysiol.*, 74, 1870, 1995.

82. Morita, K. and Katayama, Y., Calcium-dependent slow outward current in visceral primary afferent neurones of the rabbit, *Pflügers Arch.*, 414, 171, 1989.

83. Christian, E.P. et al., Guinea pig visceral C-fiber neurons are diverse with respect to the K^+ currents involved in action-potential repolarization, *J. Neurophysiol.*, 71, 561, 1994.

84. Hay, M. and Kunze, D.L., An intermediate conductance calcium-activated potassium channel in rat visceral sensory afferent neurons, *Neurosci. Lett.*, 167, 179, 1994.

85. Weinreich, D. and Wonderlin, W.F., Inhibition of calcium-dependent spike afterhyperpolarization increases excitability of rabbit visceral sensory neurones, *J. Physiol.*, 394, 415, 1987.

86. Fowler, J.C. et al., Prostaglandins block Ca^{2+}-dependent slow afterhyperpolarization independent of effects on Ca^{2+} influx in visceral afferent neurons, *Brain Res.*, 345, 345, 1985.

87. Cordoba-Rodriguez, R. et al., Calcium regulation of a slow post-spike hyperpolarization in vagal afferent neurons, *Proc. Natl. Acad. Sci. USA*, 96, 7650, 1999.

88. Pineda, J.C. et al., Charybdotoxin and apamin sensitivity of the calcium-dependent repolarization and the afterhyperpolarization in neostriatal neurons, *J. Neurophysiol.*, 68, 287, 1992.

89. Sah, P. and McLachlan, E.M., Potassium currents contributing to action potential repolarization and the afterhyperpolarization in rat vagal motoneurons, *J. Neurophysiol.*, 68, 1834, 1992.

90. Wallner, M. et al., Ca^{2+}-dependent K^+ channels in muscle and brain: molecular structure, function and diseases, in *Potassium Ion Channels: Molecular Structure, Function and Diseases*, Fambrough, D. et al., Eds., Academic Press, New York, 1998.

91. Gold, M.S. et al., Role of a slow Ca^{2+}-dependent slow afterhyperpolarization in prostaglandin E2-induced sensitization of cultured rat sensory neurons, *Neurosci. Lett.*, 205, 161, 1996.

92. Weinreich, D., Cellular mechanisms of inflammatory mediators acting on vagal sensory nerve excitability, *Pulm. Pharmacol.*, 8, 173, 1995.

93. Mayer, M., A calcium-activated chloride current generates the afterdepolarization of rat sensory neurons in culture, *J. Physiol. (Lond.)*, 364, 217, 1985.

94. Scott, R.H. et al., Modulation of divalent cation-activated chloride ion currents, *Br. J. Pharmacol.*, 94, 653, 1988.

95. Gallagher, J.P. et al., Characterization and ionic basis of GABA-induced depolarizations recorded in vitro from cat primary afferent neurones, *J. Physiol. (Lond.)*, 275, 263, 1978.

96. Sanchez-Vives, M.V. and Gallego, R., Calcium-dependent chloride current induced by axotomy in rat sympathetic neurons, *J. Physiol. (Lond.)*, 475, 391, 1994.

10 Anatomical Methods in Pain Research

Susan M. Carlton and Andrew Todd

CONTENTS

0-8493-0035-5/01/$0.00+$1.50
© 2001 by CRC Press LLC

10.1   INTRODUCTION

A variety of anatomical techniques are used in pain research. Immunocytochemistry has been used for decades, for example, to map the distribution of neurotransmitters, and more recently their receptors in both the central and peripheral nervous systems. Likewise, the visualization of long-distance pathways by "tract tracing" in the brain and spinal cord has provided valuable information about the neuronal circuits that underlie nociceptive processing. Techniques such as *in situ* hybridization and TUNEL allow us to visualize the responses of neurons to noxious input or to damaging stimuli. This chapter describes some of the most commonly used anatomical techniques employed by pain researchers in their quest to understand nociceptive processing in the nervous system.

10.2   IMMUNOCYTOCHEMISTRY AT THE LIGHT MICROSCOPIC LEVEL

10.2.1   GENERAL PRINCIPLES

Immunocytochemistry involves the use of antibodies to detect particular chemicals (antigens) in tissue sections that are subsequently examined with light or electron microscopy (LM or EM). The simplest approach is to examine the distribution of a single compound, but recently methods have been developed that allow two or even three different compounds to be revealed in the same section (multiple immunolabelling). Many types of antigens can be detected with immunocytochemistry, and those most relevant for pain research include neurotransmitters (e.g., glutamate, γ-aminobutyric acid (GABA), glycine, monoamines) and their synthetic enzymes (e.g., choline acetyltransferase, glutamate decarboxylase), neuropeptides (e.g., substance P, enkephalins), neurotransmitter and neuropeptide receptors (e.g., α-amino-3-hydroxy-5-methyl-4-isoxazolepropionic acid (AMPA), N-methyl-D-aspartate (NMDA), γ-aminobutyric acid A ($GABA_A$), neurokinin and opiate receptors), enzymes involved in signalling pathways (e.g., protein kinases), immediate early genes (e.g., c-fos), and neuronal tracer substances.

It is not possible in this chapter to describe all of the methods that are available for light and electron microscopic immunocytochemistry, and we have been necessarily selective. For a more detailed account of the subject the reader is referred to Côté et al.[1]

10.2.2 Tissue Preparation

10.2.2.1 Fixation

Sections to be examined with either LM or EM require chemical fixation to provide mechanical stability, and in some cases to retain antigens within the tissue. Various fixatives can be used for immunocytochemistry. The one most commonly used for LM consists of 4% formaldehyde in phosphate buffer. This is prepared from the polymer paraformaldehyde (see Protocol 1). For EM, fixatives containing both glutaraldehyde and formaldehyde are commonly used[2] (see 10.3.2.1), and glutaraldehyde should also be used in LM studies when amino acids (e.g., glutamate, GABA, glycine) are to be detected, because formaldehyde (which has only one aldehyde group per molecule) provides very limited retention of amino acids in tissue.

10.2.2.2 Methods of Fixation

There are various ways in which fixation can be carried out, and the choice depends on the species from which tissue is obtained as well as the antigen that is to be investigated.

10.2.2.3 Perfusion Fixation

Perfusion of fixative through the vascular system is usually the method of choice for animal tissues, because it provides very rapid and thorough fixation. The animal is deeply anesthetized (e.g., with i.p. pentobarbitone) and the thoracic cavity is then opened by cutting through the attachment of the diaphragm to the ribs, cutting vertically through the ribcage on either side and lifting the ribs and sternum. The right atrium is opened immediately before the perfusion starts, and fluids are then perfused under pressure into either the left ventricle, or directly into the ascending aorta, through a wide-bore (e.g., 16G) needle or cannula. Initially, a buffer (e.g., Ringer's solution) is perfused to wash out the blood, and then this is replaced with fixative. The most convenient method is to use a "Y" connector with two "fluid-administration sets," each consisting of a reservoir and a suitable length of tubing with a tap. The perfusion pressure should be as close as possible to the systolic pressure of the animal. The volume of fixative required will depend on the species: for a rat, 300 to 500 mls is sufficient, depending upon the weight of the animal. For immunocytochemical detection of amino acids and certain other antigens (e.g., some receptors), the speed of fixation appears to be particularly important, presumably because of the rapid diffusion or breakdown of these molecules. In these cases, keeping the duration of the rinse short (approximately 10 s) is strongly recommended.

After the perfusion is complete, the relevant pieces of tissue are dissected out and stored in the same fixative for a further period, usually 4 to 24 h, and then rinsed in buffer prior to sectioning. With certain antibodies (for example, some of those directed against neurotransmitter receptors), post-fixation appears to be highly detrimental to the immunoreaction, and this process should be kept to a minimum (2 h or less). For this reason, when a new antibody fails to provide satisfactory staining it is worth testing it on tissue which has had minimal post-fixation.

10.2.2.4 Immersion Fixation

Where vascular perfusion is not possible, for example, in studies of human material, pieces of tissue cut to a suitable size (ideally with at least one dimension smaller than 5 mm) can be immersed overnight in fixative. If visualizing receptors is the goal, smaller pieces of tissue and a limited time in the fixative (2 h) are optimal.

10.2.2.5 Post-Fixation

For certain antigens, the prolonged fixation which results from perfusion or immersion does not permit satisfactory immunostaining, presumably because the fixation process alters the conformation or accessibility of the antigen, making it unrecognizable by the antibody. In these cases, a much briefer period of fixation can be achieved by rapidly freezing fresh, unfixed tissue (e.g., over liquid nitrogen) and then cutting sections with a cryostat. The cryostat sections are collected on gelatinized slides and allowed to dry for up to 1 h at room temperature. The slides are immersed briefly (e.g., for 5 min) in 4% formaldehyde before the immunocytochemical procedure is carried out.

10.2.3 Sectioning

Because of the limited penetration of reagents into the tissue, immunocytochemistry is generally carried out on histological sections (<100 μm thick). Several methods are available for cutting sections from blocks of tissue for immunocytochemistry, and the choice of method obviously depends on the equipment available, as well as the section thickness required and the extent to which antigens of interest will tolerate the different cutting methods. If a stereological method is to be used to quantify numbers of labelled cells, then a sectioning strategy must be designed at this point (see Chapter 11 on stereological procedures).

10.2.3.1 Vibratome Sections

We routinely use a Vibratome to cut sections for light and electron microscopic immunocytochemistry. A Vibratome (or similar tissue-sectioning device) can be used to cut sections of fixed tissue between 25 and 100 μm thick. Blocks of tissue (up to 5 mm long) are glued onto the chuck of the Vibratome with cyanoacrylate adhesive and then cut into buffer at an appropriate thickness (e.g., 60 to 70 μm). We generally immerse sections in 50% ethanol in water for 30 min immediately after cutting, because this dramatically enhances penetration of antibodies during the immunocytochemical procedures.[3] If the fixative contained glutaraldehyde, sections are treated for 30 min in 1% sodium borohydride,[4] because this can reduce the deleterious effects of glutaraldehyde on antigenicity. Considerable care is needed in handling sodium borohydride, as it is potentially explosive. Borohydride treatment should be followed by extensive rinsing of the sections in phosphate buffer. Vibratome sections are reacted "free-floating," which has the advantage of exposing both cut surfaces to the reagents throughout the procedure.

10.2.3.2 Frozen Sections

Fixed tissue that is to be frozen prior to sectioning should be cryoprotected in order to minimize the structural damage caused by development of ice crystals. Blocks of tissue can be cryoprotected by immersing them in 10 to 30% sucrose in phosphate buffer for 24 h (or until the blocks sink).

Frozen sections can be cut with either a cryostat or with a freezing microtome. A cryostat is particularly useful for cutting thin (4 to 20 µm) sections. Since these are very fragile, they are mounted directly onto gelatinized slides and the subsequent immunocytochemical reaction is carried out by applying droplets of reagents onto the slide-mounted sections. Both cryostat and freezing microtome also can be used to cut thicker sections (e.g., 40 to 100 µm) that are processed free-floating.

Blocks of tissue are placed on the chuck of the cryostat (or freezing microtome) and frozen rapidly either by immersing the chuck in isopentane at −50°C, or by suspending it above liquid nitrogen. OCT compound (BDH) can be used to "glue" the block to the chuck, and this also provides support to the tissue while it is being sectioned. Thin ($\leq$ 20 µm) sections are picked off the cryostat knife by carefully applying the gelatinized slide to the knife surface. They are allowed to thaw and dry on the slide for around 30 min. Thicker sections can be picked up by touching them with the blade of a fine forceps, and they are then immersed in buffer.

10.2.3.3 Sections from Embedded Blocks

Immunocytochemical detection of certain antigens can be performed on conventional histological sections cut from paraffin-embedded blocks. The main advantages of this approach are that relatively thin (< 10 µm) sections can be cut and kept in serial order, and the structural preservation of the tissue can be very good. In addition, it allows the study of material that has been prepared and stored in this way, for example, pathological specimens. Many antigens are unrecognizable in paraffin-embedded material; however, there are methods for unmasking antigenic sites in these sections, either involving treatment with proteolytic enzymes[1] or microwaves.[5]

For light microscopic detection of certain antigens (e.g., GABA and glycine), post-embedding immunocytochemistry can be carried out on semi-thin (0.5 to 1 µm thick) sections cut from blocks of tissue that have been osmicated and embedded in epoxy resin.[6] These sections are cut on an ultramicrotome with a glass knife, and are mounted onto gelatinized slides. For details of post-embedding immuno-cytochemical protocols at the light microscopic level, see Todd and Sullivan[6] and Somogyi et al.[7]

10.2.4 IMMUNOSTAINING METHODS

Although in principle it is possible to use labelled antibodies directed against the molecules to be detected in the tissue, this approach is seldom used. In virtually all cases, unlabelled primary antibodies that recognize the antigen of interest are first applied to the section, and then these are revealed by means of labelled secondary antibodies (anti-immunoglobulins) that bind to the primary antibodies. The secondary antibody is raised in a different species than that used for generating the primary antibody. Various types of labels can be used to reveal the location of the antigen(s) in the tissue section. Here we will describe the two types of method that are most commonly used: immunoperoxidase and immunofluorescence staining.

10.2.4.1 Immunoperoxidase Methods

In these methods the enzyme horseradish peroxidase (HRP) is used to convert a soluble reagent (usually 3,3′-diaminobenzidine, DAB) into an insoluble, colored precipitate in the presence of hydrogen peroxide. Although HRP can be attached to the secondary antibody directly, far greater sensitivity is obtained by using biotin-labelled (biotinylated) secondary antibodies, and then applying HRP linked to avidin. Avidin and biotin bind to each other with extremely high affinity, which means that HRP molecules become located at sites where the primary and secondary antibodies have attached to antigens in the tissue. Two forms of HRP-avidin are available: the avidin–biotin complex (ABC, e.g., Vectastain, Vector Laboratories) and avidin–HRP conjugates (e.g., ExtrAvidin peroxidase, Sigma). ABC complexes are large structures each of which contains several HRP molecules, and for this reason they are generally thought to provide greater sensitivity than avidin–HRP conjugates, although we have found little difference in sensitivity between the two reagents. Avidin–HRP conjugates have two advantages over ABC: first, they are considerably cheaper, and second, they penetrate better into sections than the large ABC complexes. A description of the method applied to free-floating sections is given in Protocol 2, and examples of immunoperoxidase staining are shown in Fig. 10.1.

10.2.4.2 Immunofluorescence

Fluorescent dyes (fluorochromes) attached to secondary antibodies can be used as labels for both conventional epi-fluorescence microscopy, and also for confocal laser scanning microscopy. Many different fluorochromes are available, and these have characteristic absorption (excitation) and emission spectra. The most commonly used fluorochromes are fluorescein (fluorescein isothiocyanate, FITC), which absorbs blue and emits green light, and various forms of rhodamine (e.g., lissamine rhodamine sulphonyl chloride, LRSC), which absorb yellow and emit red light. However, several other dyes with different spectral characteristics can also be used, including 7-amino-4-methylcomarin-3-acetic acid (AMCA), which is excited by ultraviolet and emits blue light, and cyanine 5.18 (Cy5), which absorbs red and emits infrared radiation. The choice of fluorescent dye obviously depends on the wavelengths emitted by the bulb (for a fluorescence microscope) or laser (for a confocal microscope) and the filter sets that are available. However, there are certain advantages and disadvantages associated with each fluorochrome. With epi-fluorescence microscopy, FITC is more readily detected than rhodamine dyes, because the human eye is more sensitive to green than red light; however, FITC shows much more tendency to fading (photo-bleaching) after prolonged illumination. With confocal microscopy these differences may be less significant; rhodamine dyes are just as easily detected as FITC, and we have found that fading of fluorochromes is seldom a problem when sections are scanned with the confocal microscope.

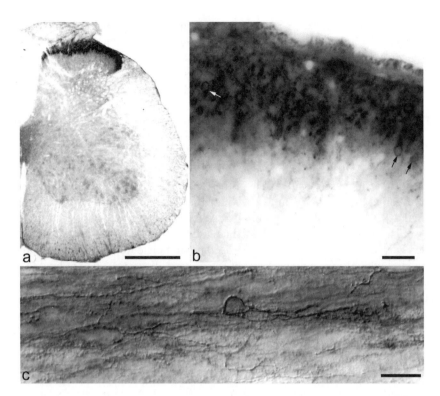

FIGURE 10.1 Immunocytochemical staining with an antibody directed against the μ-opioid receptor (MOR1) in rat spinal cord at the light microscope level. Vibratome sections were reacted with an immunoperoxidase method. (a) A transverse section of the spinal cord at low magnification reveals a dense band of immunostaining for the receptor in the superficial part of the dorsal horn (laminae I and II), with little staining elsewhere. (b) At higher magnification it is possible to distinguish the cell bodies of MOR1-immunoreactive neurons in lamina II. Three of these are marked with arrows. (c) A parasagittal section viewed with Nomarski optics at even higher magnification shows a single MOR1-immunoreactive neuron. The primary dendrites can be followed from rostral and caudal ends of the soma. Scale bars: a = 500 μm, b = 30 μm, c = 20 μm. (From Kemp, T., Spike, R.C., Watt, C., and Todd, A.J., The μ-opioid receptor (MOR1) is mainly restricted to neurons that do not contain GABA or glycine in the superficial dorsal horn of the rat spinal cord, *Neuroscience*, 75, 1231, 1996. With permission from Elsevier Science.)

10.2.4.3 Multiple Immunolabelling at the Light Microscopic Level

There are many situations in which it is necessary to identify two (or even three) different antigens in the same tissue section. This approach is particularly useful in studies of neuronal circuitry. If two populations of neurons can be revealed with different antibodies, then immunocytochemistry can be used to determine whether, for example, the axons of one population form contacts with the cell bodies or dendrites of the other. It is sometimes possible to define a particular neuronal population by the presence of two different antigens; for example, substance P-containing primary afferents in the spinal cord can be distinguished from other substance P-containing axons because they also contain calcitonin gene-related peptide (CGRP).[8] In this case, the ability to detect a third antigen can allow the post-synaptic targets of these afferents to be recognized. We have been able to demonstrate that substance P-containing afferents (labelled with antibodies directed against substance P and CGRP) make numerous contacts on large neurons in lamina III or IV that express the neurokinin 1 (NK1) receptor.[9]

Because secondary antibodies can be made specific for the immunoglobulin of a particular species (e.g., rabbit, mouse, rat), it is possible to detect two or more antigens in a single section by using primary antibodies raised in different species, and revealing each primary antibody with an appropriate secondary antibody that is used to deposit a particular label in the section. This approach has become easier over the last few years because of the increased availability of primary antibodies raised in a variety of different species, together with the development of secondary antibodies which show a high degree of species specificity (i.e., do not bind to the immunoglobulins of other species). Although it is possible to detect two different antibodies with brightfield microscopy (by using different colored dyes), multiple-labelling is far more straightforward with immunofluorescence methods, because it is possible to excite two or three fluorochromes selectively by means of appropriate filter sets, which means that each of them can be viewed independently of the others. This is particularly important when two different antigens are contained within the same structure. A method for carrying out multiple-labelling immuno-fluorescence staining is given in Protocol 3 and an example of triple-staining is illustrated in Color Figure 1.*

10.2.4.4 Advantages of Confocal Microscopy

Sections reacted by an immunofluorescence method can be viewed with a conventional epi-fluorescence microscope; however, there are several advantages to using a confocal microscope. The confocal microscope provides extremely good spectral separation of different fluorescent dyes with minimal bleed-through fluorescence, which means that in multiple-labelling immunofluorescence studies, each antigen can be reliably identified with little interference from the others. With conventional fluorescence microscopy, fluorescent-labelled structures which are out of the plane of focus frequently cause glare, which makes visualization difficult, and for this reason, it is best to use thin (e.g., slide-mounted cryostat) sections with this approach. Because structures that are out of the plane of focus are largely excluded from view with confocal microscopy, it is possible to carry out immunofluorescent staining on thicker (e.g., 70 μm) sections. The very limited depth of focus results in excellent spatial resolution of the image, while the ability to scan through the depth (z-axis) of a thicker section means that one can obtain 3-dimensional information about the distribution of immunostaining with each antibody and the relationship between different immunolabelled structures.

---

* Color figures follow page 206.

10.2.5 PROBLEMS WITH IMMUNOSTAINING

Suboptimal results with immunocytochemistry can occur for a number of reasons. In some cases it appears that primary antibodies described as being suitable for immunocytochemistry simply will not work in the experimental conditions for which they are needed. If possible, it is best before purchasing a new antibody to see examples of immunostaining carried out under similar conditions to those in which it is to be used (e.g., in the appropriate species, and with a suitable fixative).

When testing a new antibody it is important to try a very wide range of dilutions; if the antibody is used too dilute, then obviously the staining will be weak or absent, but if the antibody is too concentrated, there may also be no specific immunostaining. We usually start by using a series of dilutions increasing in 4-fold steps (e.g., 1:250, 1:1000, 1:4000, 1:16,000, 1:64,000), and if necessary, refine this on the second occasion. Although manufacturers often suggest dilutions for immunocytochemistry, these should be taken only as a rough guide. The optimal dilution of primary antibody depends on the immunocytochemical technique. For example, avidin–biotin–peroxidase methods are far more sensitive than those involving immunofluorescence, and with immunofluorescence we often use a primary antibody at 10 times the concentration needed for immunoperoxidase. Secondary antibodies from a particular supplier are likely to behave consistently, and once the optimal concentration for each secondary antibody has been determined, it should not need to be varied.

Many published immunocytochemical protocols recommend addition of normal blocking serum (at concentrations of up to 10%) to all antibody solutions. This serum is obtained from the species in which the secondary antibody was raised, and is meant to reduce non-specific binding of the secondary antibody to the tissue, which would result in background staining. We have not found any improvement resulting from addition of blocking serum in light microscopic immunocytochemistry and do not normally use it; however, in cases of persistent background (non-specific) staining it may be worth trying. Another recommendation for reducing background staining is to increase the ionic strength of solutions because this reduces low-affinity binding of antibodies that can lead to background signal. We have found that with certain antibodies, dramatic improvement in the quality of immunostaining can be obtained by using PBS with 0.3M (instead of the usual 0.15M) NaCl, and we routinely use this in the antibody diluents and for rinsing.

It is sometimes necessary to ensure that immunostaining has penetrated fully through the thickness of the section, for example, when carrying out quantitative analysis of contacts between two neuronal populations. As stated above, treatment of sections with 50% ethanol immediately after sectioning leads to improved antibody penetration, and the addition of a detergent (0.3% Triton-X100, Sigma) to antibody solutions is also useful for this purpose.

10.2.6 IMMUNOCYTOCHEMICAL CONTROLS

Non-specific binding to tissue by immunoglobulins from the serum or from the secondary antibody may result in a significant amount of background staining. A common way to assess this is to process the tissue through all steps of the immunostaining procedure but omit the primary antibody. An essential but relatively simple way to

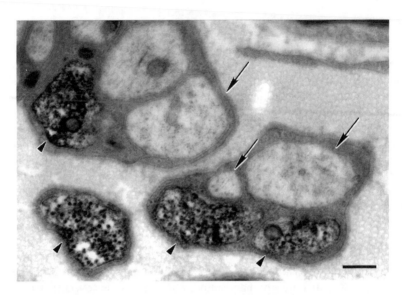

FIGURE 10.2 Immunocytochemical staining with an antibody directed against subunit 1 of the N-methyl-D-aspartate receptor (NMDAR1) at the EM level. A digital nerve from the hind toe of a rat was removed and stained *en bloc* with the pre-embedding ABC method described in Protocol 4. Dense immunoreaction product is observed in 4 unmyelinated axons (arrowheads); unlabelled axons are also observed (arrows). Bar = 0.25 μm.

assess primary antibody specificity is to perform an absorption control in which the primary antibody is incubated with an excess of its specific antigen (e.g., 100 μg of antigen/ml of antibody at the working dilution). These two components are incubated together for 16 to 24 h and then used as the primary antibody to immunostain tissue sections. Parallel staining should be performed using unabsorbed primary antibody at the working dilution. There should be no staining in the tissue exposed to the pre-absorbed antibody. If staining does occur in these sections, this staining must be due to some unknown antibody in the primary antiserum. At this point, the options are to find another source of antibody, or dilute the antibody to a working concentration where specific staining can be obtained but the absorption controls are successful.

10.2.7 COMBINING IMMUNOCYTOCHEMISTRY WITH OTHER TECHNIQUES

10.2.7.1 Anterograde or Retrograde Tract Tracing Techniques

There are various ways of combining neuroanatomical tract tracing techniques with immunocytochemistry. This approach is particularly straightforward if an unlabelled tracer substance (e.g., cholera toxin B subunit, *phaseolus vulgaris* leucoagglutinin) is used, and is detected with an antibody directed against it, as part of a multiple-immunolabelling procedure (Color Figure 1).[9] However, it is also possible to carry out immunocytochemistry after the use of other tracers, for example HRP,[10] fluorogold,[11] or biotinylated dextran (see 10.6 for a discussion of different tracers).[12]

Combining tract tracing and immunocytochemistry for brightfield microscopy can be used, for example, to reveal contacts between immunocytochemically identified axon terminals and cell bodies or dendrites of retrogradely labelled neurons.[10] The use of immunofluorescence and confocal microscopy in combined studies offers certain advantages: 3 different labels (e.g., a tracer substance and two endogenous antigens) can readily be detected, and in addition, co-localization of antigens in the same neuronal cell body or axon terminal can be examined. For example, we have used this approach to demonstrate contacts between substance P-containing axons and retrogradely labelled spinothalamic tract neurons that possess NK1 receptors (Color Figure 1).[9]

10.2.7.2 Intracellular Injection

Immunocytochemistry can also be combined with intracellular injection methods to examine neuronal circuits in which the injected neuron is involved. For brightfield microscopy, suitable intracellular labels include HRP or Neurobiotin (followed by avidin-peroxidase) that are revealed with DAB. The immunostain can be performed subsequently with an immunoperoxidase method, in which case metal-intensification of the DAB reaction product can be used to make it distinct from that in the injected neuron.[13] For epi-fluorescence or confocal microscopy, neurons can be injected directly with a fluorescent dye (e.g., Lucifer Yellow or rhodamine-dextran)[14] or with Neurobiotin, which is then revealed with avidin conjugated to a suitable fluorochrome. Lucifer Yellow is not ideal for combined studies because of its broad excitation and emission spectra, which result in bleed-through fluorescence.

10.2.7.3 Lectin Binding

Certain plant lectins, for example, *Bandeiraea simplicifolia* isolectin B4 (BSIB4) selectively bind to carbohydrate residues on the membranes of some small diameter primary afferents. BSIB4 has been injected into peripheral nerves and used as a transganglionic tracer, but it also can be applied to sections of perfusion-fixed dorsal root ganglion or spinal cord, in which case it binds to the cell bodies or central terminals of C-fibers (particularly those which do not contain neuropeptides). Biotinylated-BSIB4 applied to sections can be revealed with avidin conjugated to a fluorescent dye, and this can be combined with immunofluorescent detection of one or more antigens to provide information about the presence of receptors on particular populations of primary afferents.[15,16]

10.3 IMMUNOCYTOCHEMISTRY AT THE ELECTRON MICROSCOPIC LEVEL

10.3.1 General Principles

Many of the principles discussed in the previous sections are applicable to immunocytochemical staining at the EM level and these sections should be read carefully if the investigator is unfamiliar with general immunocytochemical procedures. Two basic approaches can be used to accomplish EM immunostaining: pre-embedding and post-embedding. For pre-embedding, the reaction can be carried out on either free floating Vibratome sections or small tissue blocks, and is frequently done with a peroxidase method. A wide variety of antigens can be detected with pre-embedding immunocytochemistry. For post-embedding, the reaction is carried out on ultrathin sections following plastic embedding and usually involves immunogold labelling. There are several advantages to post-embedding methods: (1) they provide superior spatial resolution; (2) they may allow access of antibodies to sites that are otherwise inaccessible (e.g., synaptic clefts); and (3) they offer the possibility of using serial sections for multiple labelling of profiles in different ultrathin sections. However, many antigens do not tolerate resin processing and the range of antigens that can be detected with post-embedding methods is, therefore, much more limited. The following sections will review specific details and caveats that are unique to immuno-staining at the EM level.

10.3.2 Tissue Preparation

10.3.2.1 Fixation

Fixatives for electron microscopic immunocytochemistry usually contain glutaral-dehyde (at various concentrations from 0.05 to 3.0%) as well as formaldehyde, since glutaraldehyde is needed to preserve ultrastructure (see Protocol 1). Although increasing the concentration of glutaraldehyde in the fixative improves the ultra-structure, it may cause loss of antigenicity and it is often necessary to use a con-centration of glutaraldehyde that represents a compromise between optimum ultra-structural preservation and satisfactory antigenicity. If a particular antibody will not recognize its antigen (no specific staining) in glutaraldehyde-fixed tissue, even when an extremely low concentration is added to the fixative, acrolein is another option for EM fixation.[17,18] Acrolein is a powerful irritant (it is an ingredient in tear gas) and so perfusions must be done in a very controlled environment under a fume hood.

10.3.2.2 Vibratome Sections for Pre-Embedding Immunocytochemistry

A Vibratome is particularly suitable for cutting sections for pre-embedding electron microscopic immunocytochemistry because it does not require freezing, or embed-ding of tissue in a supporting medium. It is best to avoid the use of detergents (which are otherwise used to enhance antibody penetration) as these damage the ultrastructure. Exposing the tissue to graded alcohols made in phosphate buffer (10%, 25%, 40%, 25%, 10%) for 5 min each, followed by thorough rinsing will enhance antibody penetration without disrupting the ultrastructure. For certain anti-bodies, the penetration of immunoperoxidase staining into Vibratome sections may be limited in the absence of detergents, and if this is the case it is advisable to work with relatively thin (25 μm) Vibratome sections. As described above, sections are incubated in 1% sodium borohydride,[4] which reduces the deleterious effects of glutaraldehyde on antigenicity.

10.3.2.3   Immunostaining Tissue *en Bloc*

We have had great success immunostaining various receptor populations in tissue blocks of human[19] and rat[2,20–22] skin (Fig. 10.2). In human tissue, 5-mm punch biopsies were fixed by immersion and the blocks were trimmed to 300 to 500 µm. For rat studies, tissue was fixed by aortic perfusion and glabrous skin was removed and cut into 300-µm blocks.

10.3.3   IMMUNOSTAINING METHODS

The immunoperoxidase methods described in the light microscopic section are routinely used for pre-embedding EM localization of antigens. (Protocol 4 details the EM immunostaining procedure, Protocol 7 gives details for mixing an embedding medium.) Single labelling at the EM level can be accomplished using an ABC kit (Vector Labs) or with avidin–HRP conjugate as described above.

Multiple immunolabelling at the EM level may be needed to verify that synaptic contacts between labelled terminals and identified neurons are present, or to demonstrate that a single dense core vesicle contains 2 different peptides. There are several ways to carry out multiple-immunolabelling:

1. It is possible to use two pre-embedding methods which give rise to distinctive labels (e.g., immunoperoxidase and immunogold followed by silver intensification).[12,23]
2. Pre-embedding immunoperoxidase labelling for one antigen can be followed by post-embedding immunogold detection of another.[24–27]
3. Two or more antigens can be detected on the same ultrathin section by using a post-embedding method with different sizes of gold particles for each antigen.
4. Different antigens can be detected on serial ultrathin sections with post-embedding immunocytochemistry.[23,28]

If pre- and post-embedding immunocytochemistry are to be combined, the tissue is prepared and immunostained as described in Protocol 4 for the first antigen. Following the embedding in plastic and thin sectioning, the plastic is etched off the sections, allowing the antigenic sites for the second label to be exposed for immunogold labelling (Protocol 6). Etching time becomes a major factor since too little etching of the plastic leaves the antigens hidden by the plastic; too much etching and antigenicity is destroyed.

10.4   THE TUNEL METHOD FOR STUDYING APOPTOSIS

Programmed cell death (PCD) is a key event in the development of the central nervous system (CNS).[29] Apoptosis is sometimes used as a synonym for PCD but the term is also used to describe a cytologic sequence of events that results in death of cells unrelated to developmental processes.[30] There are examples of apoptosis occurring in the adult CNS, with mechanical trauma cited as one cause.[31,32] For example, it has been shown that transection of the sciatic nerve in adult rats results in NMDA-receptor mediated degeneration of spinal cord dorsal horn cells as determined by the *in situ* TUNEL technique.[32] This technique was developed by Gavrieli et al.,[33] and is based on the observation that apoptosis and PCD are associated with DNA degradation. The appearance of a "ladder" of nucleolar DNA on agarose gels is a hallmark of apoptosis, although the demonstration of apoptosis with this technique necessitates grinding of the entire tissue, thus precluding identification of the cell types undergoing apoptosis. In contrast, the TUNEL method relies on *in situ* labelling of breaks in the DNA in nuclei of individual cells in tissue sections. As described in the original paper, TUNEL is the acronym for TdT-mediated dUTP-biotin nick end labelling. It relies on the binding of terminal deoxynucleotidyl transferase (TdT) to the 3'-OH end of DNA exposed by the break, followed by the synthesis of a labelled polydeoxynucleotide molecule. Thus, on histological sections, nuclear DNA is exposed by proteolytic treatment, and then TdT is used to incorporate biotinylated deoxyuridine into the site of the DNA breaks. Avidin-peroxidase amplifies the signal enabling visualization of apoptosis in the light microscope. A generalized protocol for the TUNEL method is described by Ben-Sasson et al.[34] This text is an excellent reference, providing a comprehensive collection of methods to study apoptosis.

10.5 *IN SITU* HYBRIDIZATION

In situ hybridization is a method that enables the precise anatomical localization and identification of specific nucleic acid sequences in DNA or mRNA. Initially, radio-isotopes (^{32}P, ^{35}S, ^{3}H) were the only labels available and autoradiography was the only means of detecting hybridized sequences. Currently, however, probes with non-radio-active labels are available eliminating the safety concerns and the extensive time needed for autoradiography and providing options for combining different labels in one experiment. The main probes in current use are complementary DNA (cDNA) probes, RNA probes (riboprobes), and oligonucleotide probes. A direct hybridization method is where the target of interest is bound directly to a non-fluorescent or a fluorochrome-labelled probe such that the probe–target hybrid can be immediately visualized with a microscope after the hybridization reaction. Indirect hybridization methods require the probe to have a reporter molecule such as digoxigenin that is detected by antibodies, or biotin that is detected by streptavidin. One must be careful in interpreting *in situ* hybridization reactions because the presence of mRNA for various enzymes, receptors, transmitters, structural proteins, etc. may not correlate with actual protein levels. There have been several occasions where high levels of message were found in the absence of detectable protein.[35] Detailed theory and methodological procedures can be found in several sources.[36,37] *In situ* hybridization techniques are applied to pain research to address a variety of questions including, but not limited to, localization of transmitters,[38,39] receptors,[40–42] channels,[43,44] and cytokines.[45]

10.6 NEURONAL TRACT TRACING

10.6.1 General Principles

When tracing pathways in the central and peripheral nervous system, it is first necessary to decide which is the most suitable tracer substance to use. Various factors need to be taken into consideration; for example, one may wish to examine the tissue with confocal microscopy, in which case a fluorescent marker will be needed. Alternatively, if the tissue is to be viewed with EM, then it will be necessary to use a tracer which can be rendered electron dense. If tract tracing is to be combined with other methods (e.g., immunocytochemistry), this will also need to be taken into consideration. Certain tracers are particularly suitable for anterograde tracing (i.e., they are taken up by cell bodies and dendrites and transported to axon terminals), while others are recommended for retrograde tracing (i.e., from axon terminal to cell body). Many tracers can travel equally well in both directions.

For *in vivo* studies, tracers are usually administered either by pressure injection or by ionophoresis through glass micropipettes, and injections into the CNS are targeted by means of stereotaxic coordinates. The procedure is carried out with suitable general anesthesia and surgery under aseptic conditions, and is followed by an appropriate post-operative recovery period. At the end of this period, the animal is re-anesthetized and fixed by perfusion (as described in 10.2.2.3 above). Tissue is removed, post-fixed and cut into sections (e.g., with a Vibratome), and if necessary, processed to allow detection of the tracer.

With CNS injections, it is important to examine the region containing the injection site to determine the spread of tracer, since it may have spread into structures other than the intended target, which may complicate the interpretation of the results. Most tracers can be taken up by non-terminal portions of axons, and uptake of tracer into axons that are merely passing through (or near) the injection site can also cause problems of interpretation.

10.6.2 NON-FLUORESCENT NEURONAL TRACERS

Many anterograde and retrograde tracers have been developed in the past. The advent of HRP[46] followed by its conjugation to wheat-germ agglutinin (WGA-HRP) and the B subunit of cholera toxin (CTb-HRP) revolutionized tract tracing. These markers provide a sensitive and relatively simple way to visualize pathways and have been used extensively, not only to trace central[47,48] and peripheral[49] pathways, but also to visualize sprouting of axonal fibers in the CNS.[50,51]

Several tracers have been shown to be particularly suited for anterograde transport specifically and these include *phaseolus vulgaris* leucoagglutinin (PHA-L), dextrans, and biocytin. PHA-L, a plant lectin, is transported at a rate of 2 to 6 mm/day.[52–55] Optimal labelling results when the PHA-L is iontophoresed and extended survival times are employed (10 d to 7 weeks).[56,57] Dextrans are polysaccharides that may be biotinylated or conjugated to fluorochromes (see 10.6.3.2). Biotinylated dextran amine requires a post-injection survival time of 2 to 21 d. In addition to being transported anterogradely, it may also be taken up by damaged axons and by axons of passage at the injection site.[58] Biocytin can be delivered by using ionotophoresis or pressure injection. It requires short post-injection survival times (24 to 48 h).[59,60] There are reports of biocytin being transported retrogradely but only after relatively large injections.[59,61] Both biotinylated dextran amine and biocytin can be revealed either by incubation of tissue sections in avidin–HRP followed by a peroxidase reaction, or with avidin conjugated to a fluorescent dye.[12]

10.6.3 FLUORESCENT DYES

The use of fluorescent dyes provides the opportunity for double and triple labelling with different fluorochromes, and thus allows simultaneous visualization of multiple pathways.[62–64] In addition, these dyes are particularly suitable for studies in which tract tracing is combined with immunocytochemistry (see 10.2.7.1). Some of the well-known fluorescent tracers are diamidino yellow, fast blue, propidium iodide, and fluorogold.[65] Recent advances in fluorescent markers include the development of lipophilic carbocyanine dyes, fluorescent dextrans, and fluorescent microspheres.

10.6.3.1 Carbocyanine Dyes

The lipophilic dialkylcarbocyanine dyes, such as DiI, DiO (Molecular Probes, Eugene, OR), and the dialkylaminostyryl dyes such as DiA are now widely used as retrograde and anterograde neuronal tracers in both living[66] and fixed[67] tissues. These dyes, including DiD and DiR, are intensely fluorescent and fade relatively slowly when exposed to photic excitation. In living tissue, these dyes do not affect cell viability, development, or physiological properties. It has been reported that DiI-labelled motoneurons remain viable in culture for up to 4 weeks and *in vivo* for up to 1 year. In living tissue the dyes travel at a rate of 6 mm/d,[66] diffusing in the plasma membrane. *In vivo*, however, they are soon internalized by endocytosis and thus membranous labelling is replaced by a granular cytoplasmic labelling. The dyes do not transfer from labelled to unlabelled cells as observed in culture where the dyes are retained by the retrogradely labelled cells and do not spread to other cells. When double labelling is called for, DiI and DiO are often paired. DiI is excited by green and fluoresces bright red-orange when viewed with rhodamine filters, while DiO is excited in blue and fluoresces green when viewed through fluorescein filters. DiI and DiO may be dissolved to permit microinjection for anterograde or retrograde labelling. DiI dissolves in ethanol and DiO in a mixture of ethanol and DMSO. DiO is more soluble in dimethylformamide (DMF). These solutions can be added directly to culture medium for labelling of cells. Crystals of dye, with or without pre-mixing with grease, may also be applied directly to the region of interest.[68]

In fixed tissue, the dyes label neurons via lateral diffusion in the plasma membrane at a rate of approximately 0.2 to 0.6 mm/d,[66,67] although slower rates have been reported and these rates are affected by several parameters including fixation time, temperature of storage, age of animal, and the dye used.[69] DiI has been shown to be an excellent anterograde and retrograde tracer in fixed tissue.[67] Again, there is little evidence of transfer of dye from cell to cell; however, care must be taken when interpreting results of labelled profiles after tissue sectioning, because some leakage may occur when the membrane is disrupted by the knife during sectioning.[66] A major advantage of using carbocyanine dyes in fixed tissue is that they allow inclusion of human autopsy material for anterograde and retrograde studies.[70] Also, the dyes are not internalized (endocytosed) by fixed cells as occurs in living cells, and thus, the fixed labelled cells retain the uniform membranous labelling indefinitely. Unfortunately, the very slow rates of transport of the dyes in cell membranes in fixed tissue mean that they are not suitable for studying long-distance neuronal connections. In fixed tissue, the dyes can be applied by placing crystals in appropriate regions to maximize incorporation by either cell bodies or terminals.

In addition to neuronal tracing, carbocyanine dyes have been used for many other applications such as detection of cell–cell fusion[71–73] and adhesion,[74] tracing cell migration during development[75,76] or after transplantation,[77,78] cytotoxicity assays,[79,80] and labelling of lipoproteins.[81,82]

10.6.3.2 Fluorescent Dextrans

Fluorescent dextrans, such as rhodamine-B dextran amine (RBD) and tetramethyl rhodamine-dextran amine (Fluoro-Ruby), are retrogradely and anterogradely transported.[83] Transport time varies between 6 to 14 d post-injection. These markers are taken up mainly by terminals and damaged fibers of passage.[83]

10.6.3.3 Fluorescent Microspheres

The use of microspheres or beads conjugated to fluorophores was introduced by Katz et al.[84] Currently, they can be purchased as polystyrene or latex beads and can be labelled with one of a variety (up to seven) dyes (Molecular Probes). In addition to tract tracing,[85] they can be used for labelling transplants[86–88] or cells in culture.[89] The microspheres are transported by damaged fibers of passage but not by intact fibers. Retrogradely labelled cells will remain labelled for a very extended period of time (at least 2 y post-injection).[90]

PROTOCOLS

PROTOCOL 1: PREPARATION OF 4% FORMALDEHYDE FIXATIVE

1. Phosphate buffer (PB), 0.2M stock: 0.2M NaH_2PO_4/Na_2HPO_4 mixed to pH 7.4.
2. *Heat 400 ml dH_2O to 60°C and add 40 g paraformaldehyde. Add conc NaOH dropwise until suspension clears. Add 500 ml 0.2M PB and make up to 1l with dH_2O.
3. Filter and use within 24 h.

* For EM immunocytochemistry, add 0.05 to 3% glutaraldehyde (EM grade) after solution has cooled to room temperature. With higher glutaraldehyde concentrations, the concentration of formaldehyde is often reduced to 1%.

PROTOCOL 2: LM IMMUNOPEROXIDASE METHOD FOR VIBRATOME SECTIONS

1. Rinse blocks of fixed tissue in PB.
2. Cut Vibratome sections (40 to 70 µm thick) into PB.
3. Place sections into 50% ethanol in dH_2O for 30 min immediately after cutting to enhance penetration of antibodies.
4. Rinse 3 × 5 mins in PBS.

5. Incubate sections 18 to 72 h in primary antibody diluted in PBS containing 0.3% Triton X-100 (PBST) at 4°C.
6. Rinse 3 × 5 mins in PBS.
7. Incubate 2 to 24 h in biotinylated secondary antibody (diluted in PBST) at 4°C. The secondary antibody should be directed against IgG of the species in which the primary antibody was raised.
8. Rinse 3 × 5 mins in PBS.
9. Incubate 2-24 h in avidin-peroxidase (Sigma, diluted 1:1000 in PBST) or ABC reagent (Vector Labs).
10. Rinse 3 × 5 mins in PBS.
11. Rinse 5 min in PB.
12. Incubate 5 to 15 min in DAB solution: the reaction should be monitored and stopped (step 13) when the staining intensity is adequate. The DAB solution consists of 0.05% 3,3′-diaminobenzidine in PB with 0.01% H_2O_2, and is made up just before use and filtered. DAB is a potential carcinogen and should be handled with extreme caution. To avoid repeated handling of the powder, we make up a stock solution of 1% DAB in water and freeze it in aliquots. Disposable gloves should be worn during all procedures involving DAB, and all solutions and glassware (and any spillages) should be treated with bleach immediately after use, to neutralize the DAB.
13. Rinse 3 × 5 min in PB.
14. Mount onto gelatinized slides and allow to dry overnight.
15. Dehydrate slide-mounted sections in ascending series of alcohols, clear in xylene (or equivalent) and coverslip with DPX.

Notes Bottles containing the sections are continuously agitated during all stages between 3 and 10.

This method can be adapted for slide-mounted (e.g., cryostat) sections as follows: A ring is drawn around the section(s) on the slide with a "PAP" pen so that the small droplets of reagents that subsequently will be applied to the section will not spread across the slide. All rinsing stages, together with the DAB incubation, are carried out in suitable slide jars. Antibody solutions and avidin–HRP conjugate are applied as droplets (20 to 50 μl, depending on section size), and the slides are kept in humid containers.

Nickel chloride (0.07%) can be added to the DAB reagent, in which case the brown DAB precipitate is replaced with a dark blue/black stain.

Protocol 3: Dual- or Triple-Labelling Immunofluorescence Method for Vibratome Sections

1. Rinse tissue blocks in PB.
2. Cut Vibratome sections (40 to 70 μm thick) into PB.
3. Place sections into 50% ethanol in dH_2O for 30 min.
4. Rinse 3 × 5 min in PBS.

5. Incubate 18 to 72 h in cocktail of primary antibodies (each raised in a different species) at appropriate dilutions in PBST at 4°C.

6. Rinse 3 × 5 min in PBS.

7. Incubate 2 to 24 h in cocktail of fluorescent species-specific secondary antibodies diluted in PBST at 4°C. It is vital that the secondary antibodies are raised in different species from those of the primary antibodies. For convenience we routinely use secondary antibodies raised in donkey (Jackson Immunoresearch, West Grove, PA). Donkey has become a universal donor for secondary antibodies in multiple-immunolabelling studies, since few (if any) primary antibodies are raised in this species. Obviously, the secondary antibodies chosen must be directed against the species in which the primary antibodies were raised, and each must be conjugated to a different fluorescent dye. The choice of fluorescent dyes depends on the wavelength of the laser lines. We use FITC, LRSC, and Cy5, since these are optimally excited by the lines of our Krypton-Argon laser (488 nm, 568 nm, 647 nm).

8. Rinse 3 × 5 min in PBS.

9. Mount sections on glass slides with a glycerol-based anti-fade mounting medium (e.g., Vectashield, Vector Labs). The use of anti-fade medium reduces the rate of photobleaching, which is otherwise a problem, particularly with FITC.

10. Apply a coverslip and seal around the edges with nail varnish.

11. Store slides in a –20° freezer until needed. Slides prepared in this way can be kept for extended periods (several years).

12. Examine with epi-fluorescence or confocal microscope.

PROTOCOL 4: EM IMMUNOPEROXIDASE METHOD FOR FREE FLOATING SECTIONS OR EN BLOC TISSUE

1. Rinse blocks of fixed tissue in PB.

2. Cut Vibratome sections (25 µm thick) into PB or trim tissue blocks to 300-µm slabs.

3. Rinse tissue in PB for 10 min.

4. Incubate tissue in 1% sodium borohydride in 0.1M PB for 1 h and then rinse 6 × 5 min in PBS.

5. To enhance penetration of antibodies, take sections through an ascending series of alcohols (10%, 25%, 40%, 25%, 10%) made in PB for 5 min each and then thoroughly rinse in PBS.

6. Block tissue in 10% normal goat serum (NGS) in PBS, pH7.4 for 1 h.

7. Incubate sections 18 to 72 h in primary antibody diluted in PBS.

8. Rinse 3 × 10 min in PBS.

9. Wash tissue in 3% normal goat serum (NGS) in PBS for 30 min.

10. Incubate tissue in biotinylated IgG in 1% NGS for 1 h.

11. Rinse tissue in 1% NGS and 3% NGS for 30 min each.

12. Incubate tissue in the ABC reagent (avidin–biotin–peroxidase complex, Vector Labs) or avidin–peroxidase (Sigma, diluted 1:1000 in PBS) for 2 h at room T°C.
13. Rinse 3 × 10 min in PB.
14. Incubate tissue DAB for 1 to 6 min. The DAB solution consists of 0.025% 3,3'-diaminobenzidine in PB with 0.01% H_2O_2, and is made up just before use and filtered.
15. Rinse 6 × 5 min in PBS.
16. Transfer tissue to glass vials for osmification and dehydration.

PROTOCOL 5: OSMIFICATION AND DEHYDRATION OF EM TISSUE

1. Place tissue in 1% osmium tetroxide in PB for 1 to 2 h depending on the size and thickness of the tissue.
2. After rinsing with PB, dehydrate with 50%, then 70% alcohol 10 min each.
3. Stain tissue *en bloc* with 1% uranyl acetate in 70% alcohol in the dark for 1 h.
4. Rinse with 70% alcohol.
5. Continue dehydration with 90% alcohol for 10 min; 100% alcohol 3 times, for 10 min each.
6. Rinse in propylene oxide (2 times for 5 min each).
7. Infiltrate tissue in a 1:1 mixture of propylene oxide and embedding medium (see Protocol 7) overnight at 4°C.
8. Transfer tissue to fresh embedding medium for 4 h.
9. Transfer tissue to molds filled with embedding medium and cure in the oven at 60°C overnight.

PROTOCOL 6: POST-EMBEDDING IMMUNOGOLD LABELLING

Ultrathin sections should be mounted on formvar-coated nickel slot grids. *Grids are etched or incubated in primary and immunogold solutions on drops placed on parafilm in a covered petri dish. +Grid washing is performed on a 24-well plate on a rotator. Each well is filled almost to the top with double-distilled H_2O (ddH_2O). One grid is placed in each well on the surface of the water. At the end of each wash, overfill the well with water to facilitate removal of the grid. Blotting excess fluid from a grid is accomplished by touching a small piece of filter paper to the edge of the grid. Filter all ddH_2O and Tris-buffered saline (TBS) before use.

1. *Etch sections with filtered 1% Periodic Acid (in ddH_2O) for 10 min. Blot.
2. +Wash grids 3 times for 5 min each with ddH_2O on slow rotator. Blot.
3. *Etch sections with filtered 1% sodium meta-periodate (in ddHd$_2O$) for 30 to 60 min. Blot.
4. +Wash grids 3 times for 5 min each with ddH_2O on a slow rotator. Blot.
5. *Incubate grids on drops of filtered 1% sodium borohydride (in ddH_2O) for 10 min. Blot.
6. +Wash grids 3 times for 5 min each with ddH_2O on a slow rotator. Blot.

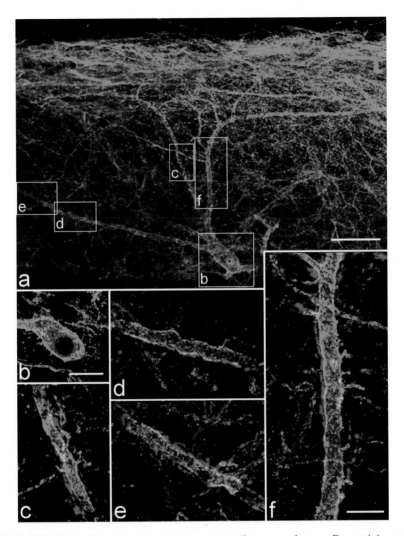

COLOR FIGURE 1 This figure demonstrates contacts between substance P-containing axons (red) and a neuron in lamina III of the rat spinal cord which expresses the NK1 receptor (green) and belongs to the spinothalamic tract. The images were obtained with a confocal microscope from a parasagittal Vibratome section which had been reacted by a triple-labelling immunofluorescence method. (a) This image shows NK1 receptor-immunostaining (FITC, green). The immunoreactive neuron has its cell body in the lower part of the picture, and dorsally directed dendrites which pass up into the superficial dorsal horn. Boxes show the regions illustrated at higher magnification in other parts of the figure. (b) The cell body has been scanned to reveal NK1 receptor (green) together with the tracer CTb (labelled with LRSC and shown as blue), which was injected into the thalamus 3 days previously. The presence of CTb in the soma indicates that the neuron belongs to the spinothalamic tract. (d–f) show contacts between substance P-immunoreactive axons (stained with Cy5 and shown as red) and various dendrites which belong to the neuron. Scale bars: a = 50 mm, b = 20 mm, c–f = 10 mm. (From Naim, M. et al., *J. Neurosci.*, 17, 5548, 1997. With permission.)

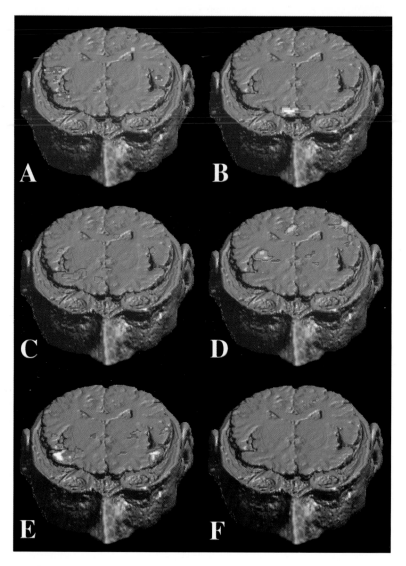

COLOR FIGURE 2 Differences in brain activity to pain in different clinical and experimental settings, taken from various studies ongoing in Apkarian's laboratory. (A) Shows cortical activity in a normal volunteer for thermal painful stimuli applied to the hand. (B) Cortical regions that are affected by changes in chronic pain are shown for patients suffering from reflex sympathetic dystrophy. (C) Cortical activity during spontaneous chronic back pain. (D) Cortical activity in the same subject as in (C), when the back pain is exacerbated by radicular pain. (E) Cortical activity in a normal subject, where the radicular pain is mimicked acutely. (F) Cortical activity in the same subject as in (E), when the leg is still manipulated but without causing pain.

7. *Blocking: place grids on drops of 1:30 NGS in 0.05M TBS, pH 7.2, containing 0.1% BSA for 30 min. Blot.
8. *Incubate grids on drops of primary antibody (diluted in filtered 0.05M TBS containing 0.01% BSA) for 1 to 3 d at 4°C in a moisturized chamber. Blot.
9. +Wash grids 3 times 5 min each in 0.05M TBS, pH7.2. Blot.
10. +Wash grids 5 min in 0.05M TBS containing 0.2% BSA, pH 7.2. Blot.
11. *Incubate grids on drops of filtered gold buffer, pH 8.2 for 10 min. Blot. (Gold buffer is made with 0.05M TBS, containing 0.25% BSA. pH 8.2 is very important!).
12. *Incubate grids on drops of gold-labelled secondary antibody (colloidal gold coated with anti-IgG, diluted 1:30 with gold buffer) for 90 min at room temperature. Blot.
13. +Wash grids 3 times 5 min each with 0.05M TBS, pH 7.2 on slow rotator. Blot.
14. +Wash grids 3 times 5 min each with ddH$_2$O on slow rotator. Blot.
15. Dry grids and stain with 3% uranyl acetate in 70% alcohol for 10 min followed by 0.4% lead citrate in ddH$_2$O for 5 min.

PROTOCOL 7: RECIPE FOR EPON-ARALDITE EMBEDDING MEDIUM

Measure by volume:

1. EPON 812 — 10 ml
2. Araldite 502 — 10 ml
3. Dodecanyl Succinic Anhydride (DDSA) — 24 ml

Mix first 3 ingredients well, then add

4. [2,4,6,(tri(dimethylaminomethyl)phenol)] (DMP-30) — 0.8ml

Source for all reagents: Electron Microscopy Sciences

REFERENCES

1. Côté, S.L., Ribeiro-Da-Silva, A., and Cuello, A., Current protocols for light microscopy immunocytochemistry, *Immunocytochemistry II*, Cuello, A., Ed., J. Wiley, Chichester, 1993.
2. Coggeshall, R.E. and Carlton, S.M., Ultrastructural analysis of NMDA, AMPA and kainate receptors on unmyelinated and myelinated axons in the periphery, *J. Comp. Neurol.*, 391, 78, 1998.
3. Llewellyn-Smith, I.J. and Minson, J.B., Complete penetration of antibodies into Vibratome sections after glutaraldehyde fixation and ethanol treatment: light and electron microscopy for neuropeptides, *J. Histochem. Cytochem.*, 40, 1741, 1992.
4. Kosaka, T., Nagatsu, I., Wu, J.-Y., and Hama, K., Use of high concentrations of glutaraldehyde for immunocytochemistry of transmitter-synthesizing enzymes in the central nervous system, *Neuroscience*, 18, 975, 1986.

5. Wolf, H.K., Buslei, R., Schmidtkastner, R., Schmidtkastner, P.K., Pietsch, T., Wiestler, O.D., and Blumcke, I., NeuN- a useful neuronal marker for diagnostic histopathology, *J. Histochem. Cytochem.*, 44, 1167, 1996.

6. Todd, A.J. and Sullivan, A.C., A light microscope study of the coexistence of GABA-like and glycine-like immunoreactivities in the spinal cord of the rat, *J. Comp. Neurol.*, 296, 496, 1990.

7. Somogyi, P., Hodgson, A.J., Chubb, I.W., Penke, B., and Erdei, A., Antisera to γ-aminobutyric acid. II. Immunocytochemical application to the central nervous system, *J. Histochem. Cytochem.*, 33, 248, 1985.

8. Ju, G., Hökfelt, T., Brodin, E., Fahrenkrug, J., Fischer, J.A., Frey, P., Elde, R.P., and Brown, J.C., Primary sensory neurons of the rat showing calcitonin gene-related peptide immunoreactivity and their relation to substance P-, somatostatin-, galanin-, vasoactive intestinal polypeptide- and cholecystokinin-immunoreactive ganglion cells, *Cell Tiss. Res.*, 247, 417, 1987.

9. Naim, M., Spike, R.C., Watt, C., Shehab, S.A.S., and Todd, A.J., Cells in laminae III and IV of the rat spinal cord which possess the neurokinin-1 receptor and have dorsally-directed dendrites receive a major synaptic input from tachykinin-containing primary afferents, *J. Neurosci.*, 17, 5548, 1997.

10. Doyle, C.A. and Maxwell, D.J., Direct catecholaminergic innervation of spinal dorsal horn neurons with axons ascending the dorsal columns in cat, *J. Comp. Neurol.*, 331, 434, 1993.

11. Pollock, R., Kerr, R., and Maxwell, D.J., An immunocytochemical investigation of the relationship between substance P and the neurokinin-1 receptor in the lateral horn of the rat thoracic spinal cord, *Brain Res.*, 777, 22, 1997.

12. Polgar, E., Shehab, S.A.S., Watt, C., and Todd, A.J., GABAergic neurons that contain neuropeptide Y selectively target cells with the neurokinin 1 receptor in laminae III and IV of the rat spinal cord, *J. Neurosci.*, 19, 2637, 1999.

13. Maxwell, D.J. and Jankowska, E., Synaptic relationships between serotonin-immuno-reactive axons and dorsal horn spinocerebellar tract cells in the cat spinal cord., *Neuroscience*, 70, 247, 1996.

14. Maxwell, D.J., Riddell, J.S., and Jankowska, E., Serotoninergic and noradrenergic contacts associated with premotor interneurons in spinal pathways from group II muscle afferents, *Eur. J. Neurosci.*, 12, 1271, 2000.

15. Averill, S., McMahon, S.B., Clary, D.O., Reichardt, L.F., and Priestley, J.V., Immu-nocytochemical localization of trkA receptors in chemically identified subgroups of adult rat sensory neurons, *Eur. J. Neurosci.*, 7, 1484, 1995.

16. Vulchanova, L., Riedl, M., Shuster, S.J., Stone, L.S., Hargreaves, K.M., Buell, G., Surprenant, A., North, R.A., and Elde, R., $P2X_3$ is expressed by DRG neurons that terminate in inner lamina II, *Eur. J. Neurosci.*, 10, 3470, 1998.

17. LaMotte, C.C., Vasoactive intestinal polypeptide cerebrospinal fluid-contacting neurons of the monkey and cat spinal central canal, *J. Comp. Neurol.*, 258, 527, 1987.

18. LaMotte, C.C. and deLanerolle, N.C., VIP terminals, axons, and neurons: distribution throughout the length of monkey and cat spinal cord, *J. Comp. Neurol.*, 249, 133, 1986.

19. Kinkelin, I., Brocker, E.-B., Koltzenburg, M., and Carlton, S.M., Localization of ionotropic glutamate receptors in peripheral axons of human skin, *Neurosci. Lett.*, 283, 149, 2000.

20. Carlton, S.M. and Coggeshall, R.E., Immunohistochemical localization of $5HT_{2A}$ receptors in peripheral sensory axons in rat glabrous skin, *Brain Res.*, 763, 271, 1997.

21. Carlton, S.M., Zhou, S., and Coggeshall, R.E., Evidence for the interaction of glutamate and NK1 receptors in the periphery, *Brain Res.*, 790, 160, 1998.

22. Coggeshall, R.E., Zhou, S., and Carlton, S.M., Opioid receptors on peripheral sensory axons, *Brain Res.*, 764, 126, 1997.

23. Todd, A.J., Watt, C., Spike, R.C., and Sieghart, W., Co-localization of GABA, glycine and their receptors at synapses in the rat spinal cord, *J. Neurosci.*, 16, 974, 1996.

24. Carlton, S.M. and Hayes, E.S., Dynorphin A(1-8) immunoreactive cell bodies, dendrites and terminals are postsynaptic to calcitonin gene-related peptide primary afferent terminals in the monkey dorsal horn, *Brain Res.*, 504, 124, 1989.

25. Carlton, S.M., Westlund, K.N., Zhang, D., Sorkin, L.S., and Willis,W.D., Calcitonin gene-related peptide containing primary afferent fibers synapse on primate spinothalamic tract cells, *Neurosci. Letts.*, 109, 76, 1990.

26. Hayes, E.S. and Carlton, S.M., Primary afferent interactions: analysis of calcitonin gene-related peptide-immunoreactive terminals in contact with unlabeled and GABA-immunoreactive profiles in the monkey dorsal horn, *Neuroscience*, 47, 873, 1992.

27. Lekan, H.A. and Carlton, S.M., Glutamateric and GABAergic input to rat spinothalamic tract cells in the superficial dorsal horn, *J. Comp. Neurol.*, 361, 417, 1995.

28. Todd, A.J., GABA and glycine in synaptic glomeruli of the rat spinal dorsal horn, *Eur. J. Neurosci.*, 8, 2492, 1996.

29. Glucksmann, A., Cell deaths in normal vertebrate ontogeny, *Biol. Rev.*, 26, 59, 1951.

30. Kerr, J.F.R., Wyllie, A.H., and Currie, A.R., Apoptosis: a basic biological phenomenon with wide-ranging implications in tissue kinetics, *Br. J. Cancer*, 26, 239, 1972.

31. Rink, A., Fung, K.-M., Trojanowski, J.Q., Lee, V.M.Y., Neugebauer, E., and McIntosh, T.K., Evidence of apoptotic cell death after experimental traumatic brain injury in the rat, *Am. J. Pathol.*, 147, 1575, 1995.

32. Azuke, J.J., Zimmermann, M., Hsieh, T.-F., and Herdegen, T., Peripheral nerve insult induces NMDA receptor-mediated, delayed degeneration in spinal neurons, *Eur. J. Neurosci.*, 10, 2204, 1998.

33. Gavrieli, Y., Sherman, Y., and Ben-Sasson, S.A., Identification of programmed cell death in situ via specific labeling of nuclear DNA fragmentation, *J. Cell Biol.*, 119, 493, 1992.

34. Ben-Sasson, S.A., Sherman, Y., and Gavrieli, Y., Identification of dying cells *in situ* staining, in *Cell Death*, 1st ed., Schwartz, L.M. and Osborne, B.A., Eds., Academic Press, New York, 1995.

35. Uhl, G.R., Zingg, H.H., and Habener, J.F., Vasopressin mRNA *in situ* hybridization: localization and regulation studied with oligonucleotide cDNA probes in normal and Brattleboro rat hypothalamus, *PNAS*, 82, 5555, 1985.

36. Eberwine, J.H., Valentino, K.L., and Barchas, J.D., *In Situ Hybridization in Neurobiology: Advances in Methodology*, Oxford University Press, New York, 1994.

37. Grunewald-Janho, S., Keesey, J., Leous, M., van Miltenburg, R., and Schoeder, C., *Nonradioactive in Situ Hybridization Application Manual*, 2nd ed., Boehringer Mannheim, Mannheim, Germany, 1996.

38. Ji, R.-R., Zhang, X., Wiesenfeld-Hallin, Z., and Hökfelt, T., Expression of neuropeptide Y and neuropeptide Y (Y1) receptor mRNA in rat spinal cord and dorsal root ganglia following peripheral tissue inflammation, *J. Neurosci.*, 14, 6423, 1994.

39. Ma, W. and Bisby, M.A., Ultrastructural localization of increased neuropeptide immunoreactivity in the axons and cells of the gracile nucleus following chronic constriction injury of the sciatic nerve, *Neuroscience*, 93, 335, 1999.

40. Nearl, C.R. Jr., Mansour, A., Reinscheid, R., Nothacker, H.P., Civelli, O., Akil, H., and Watson, S.J. Jr., Opioid receptor-like (ORL1) receptor distribution in the rat central nervous system: comparison of ORL1 receptor mRNA expression with (125)I-[(14)Tyr]-orphanin FG binding, *J. Comp. Neurol.*, 412, 563, 1999.

41. Li, H.S. and Zhao, Z.Q., Small sensory neurons in the rat dorsal root ganglia express functional NK-1 tachykinin receptor, *Eur. J. Neurosci.*, 10, 1292, 1998.
42. Kus, L., Sanderson, J.J., and Beitz, A.J., N-methyl-D-aspartate R1 messenger RNA and [125]MK-801 binding decrease in rat spinal cord after unilateral hind paw inflammation, *Neuroscience*, 68, 159, 1995.
43. Blackburn-Munro, G. and Fleetwood-Walker, S.M., The sodium channel auxiliary subunits beta1 and beta2 are differentially expressed in the spinal cord of neuropathic rats, *Neuroscience*, 90, 153, 1999.
44. Cummins, F.J., Fried, T.R., Black, J.A., and Waxman, S.G., *In vivo* NGF deprivation reduced SNS expression and TTX-R sodium currents in IB4-negative DRG neurons, *J. Neurophys.*, 81, 803, 1999.
45. Arruda, J.L., Colburn, R.W., Rickman, A.J., Rutkowski, M.D., and DeLeo, J.A., Increase of interleukin-6 mRNA in the spinal cord following peripheral nerve injury in the rat: potential role of 1L-6 in neuropathic pain, *Brain Res.*, 62, 228, 1998.
46. LaVail, J.H. and LaVail, M.M., Retrograde axonal transport in the central nervous system, *Science*, 176, 1416, 1972.
47. Carlton, S.M., Honda, C.N., Willcockson, W.S., Lacrampe, M., Zhang, D., Denoroy, L., Chung, J.M., and Willis, W.D., Descending adrenergic input to the primate spinal cord and its possible role in modulation of spinothalamic cells, *Brain Res.*, 543, 77, 1991.
48. Westlund, K.N., Carlton, S.M., Zhang, D., and Willis, W.D., Direct catecholaminergic innervation of primate spinothalamic tract neurons, *J. Comp. Neurol.*, 299, 178, 1990.
49. Swett, J.E. and Woolf, C.J., The somatotopic organization of primary afferent terminals in the superficial laminae of the dorsal horn of the rat spinal cord, *J. Comp. Neurol.*, 231, 66, 1985.
50. Woolf, C.J., Shortland, P., and Coggeshall, R.E., Peripheral nerve injury triggers central sprouting of myelinated afferents, *Nature*, 355, 75, 1992.
51. Lekan, H.A., Carlton, S.M., and Coggeshall, R.E., Sprouting of Aβ fibers into lamina II of the rat dorsal horn in peripheral neuropathy, *Neurosci. Lett.*, 208, 147, 1996.
52. Gerfen, C.R. and Sawchenko, P.E., An anterograde neuroanatomical tracing method that shows the detailed morphology of neurons, their axons and terminals: immuno-histochemical localization of an axonally transported plant lectin, *Phaseolus vulgaris* leucoagglutinin, *Brain Res.*, 290, 219, 1984.
53. Groenewegen, H.J. and Wouterlood, F.G., Light and electron microscopic tracing of neuronal connections with *Phaseolus vulgaris*-leucoagglutinin (PHA-L), and combina-tions with other neuroanatomical techniques, in *Analysis of Neuronal Microcircuits and Synaptic Interactions. Handbook of Chemical Neuroanatomy*, Vol. 8, Björklund, A., Hökfelt, T., Wouterlood, F.G., and van den Pol, A.N., Eds., Elsevier, Amsterdam, 1990.
54. Sawchenko, P.E. and Gerfen, C.R., Plant lectins and bacterial toxins as tools for tracing neuronal connections, *Trends Neurosci.*, 8, 378, 1985.
55. Ter Horst, G.J., Groenewegen, H.J., Karst, H., and Luiten, P.G.M., *Phaseolus vulgaris* leucoagglutinin immunohistochemistry. A comparison between autoradiographic and lectin tracing of neuronal efferents, *Brain Res.*, 307, 379, 1984.
56. Rouiller, E.M., Liang, F., Moret, V., and Wiesendanger, M., Trajectory of redirected corticospinal axons after unilateral lesion of the sensorimotor cortex in neonatal rat: a *phaseolus vulgaris*-leucoagglutinin (PHA-L) tracing study, *Exp. Neurol.*, 114, 53, 1991.
57. Wouterlood, F.G. and Jorritsma-Byham, B., The anterograde neuroanatomical tracer biotinylated dextran-amine: comparison with the tracer *phaseolus vulgaris*-leucoag-glutinin in preparations for electron microscopy, *J. Neurosci. Meth.*, 48, 75, 1993.

58. Pare, D. and Smith, Y., Thalamic collaterals of corticostriatal axons: their termination field and synaptic targets in cats, *J. Comp. Neurol.*, 372, 551, 1996.

59. King, M.A., Louis, P.M., Hunter, B.E., and Walker, D.W., Biocytin: a versatile anterograde neuroanatomical tract tracing alternative, *Brain Res.*, 497, 361, 1989.

60. Kita, H. and Armstrong, W., A biotin-containing compound N-(2-aminoethyl)biotin-amide for intracellular labeling and neuronal tracing studies: comparison with biocytin, *J. Neurosci. Meth.*, 37, 141, 1991.

61. McDonald, A.J., Neuroanatomical labeling with biocytin: a review, *NeuroReport*, 3, 821, 1992.

62. Dolleman-Van der Weel, M.J., Wouterlood, F.G., and Witter, M.P., Multiple antero-grade tracing, combining *Phaseolus vulgaris* leucoagglutinin with rhodamine- and biotin-conjugated dextran amine, *J. Neurosci. Meth.*, 51, 9, 1994.

63. Tseng, G.F., Parada, I., and Prince, D.A., Double-labelling with rhodamine beads and biocytin: a technique for studying corticospinal and other projection neurons in vitro, *J. Neurosci. Meth.*, 37, 121, 1991.

64. Fritzsch, B. and Sonntag, R., Sequential double labelling with different fluorescent dyes coupled to dextran amines as a tool to estimate the accuracy of tracer application and of regeneration, *J. Neurosci. Meth.*, 39, 9, 1991.

65. Bentivoglio, M. and Chen, S., Retrograde neuronal tracing combined with immuno-cytochemistry, *Immunohistochemistry II*, Cuello, A.C., Ed., Wiley, New York, 1993.

66. Honig, M.G. and Hume, R.I., Fluorescent carbocyanine dyes allow living neurons of identified origin to be studied in long-term cultures, *J. Cell Biol.*, 103, 171, 1986.

67. Kuffler, D.P., Long-term survival and sprouting in culture by motoneurons isolated from the spinal cord of adult frogs, *J. Comp. Neurol.*, 302, 729, 1990.

68. Vercelli, A., Assal, F., and Innocenti, G.M., Emergence of callosally-projecting neurons with stellate morphology in the visual cortex of the kitten, *Exp. Brain Res.*, 90, 346, 1992.

69. Vercelli, A., Repici, M., Garbossa, D., and Grimaldi, A., Recent techniques for tracing pathways in the central nervous system of developing and adult mammals, *Brain Res. Bull.*, 51, 11, 2000.

70. Burkhalter, A. and Bernado, K.L., Organization of corticocortical connections in human visual cortex, *PNAS*, 86, 1071, 1989.

71. Blumenthal, R., Sarkar, D.P., Durell, S., Howard, D.E., and Morris, S.J., Dilation of the influenza hemagglutinin fusion pore revealed by the kinetics of individual cell-cell fusion events, *J. Cell Biol.*, 135, 63, 1996.

72. Li, L.H., Hensen, M.L., Zhao, Y.L., and Hui, S.W., Electrofusion between hetero-geneous-sized mammalian cells in a pellet: potential applications in drug delivery and hybridoma formation, *Biophys. J.*, 71, 479, 1996.

73. Spotl, L., Sarti, A., Dierich, M.P., and Most, J., Cell membrane labeling with fluores-cent dyes for the demonstration of cytokine-induced fusion between monocytes and tumor cells, *Cytometry*, 21, 160, 1995.

74. Ramanathan, R., Wilkemeyer, M.F., Mittal, B., Perides, G., and Charness, M.E., Alcohol inhibits cell-cell adhesion mediated by human L1 [published erratum appears in *J. Cell. Biol.*, 133, 1139, 1996], *J. Cell Biol.*, 133, 381, 1996.

75. Fraser, S.E., Iontophoretic dye labeling of embryonic cells, *Meth. Cell Biol.*, 51, 147, 1996.

76. Ragnarson, B., Bengtsson, L., and Haegerstrand, A., Labeling with fluorescent carbo-cyanine dyes of cultured endothelial and smooth muscle cells by growth in dye-containing medium, *Histochemistry*, 97, 329, 1992.

77. Heredia, M., Santacana, M., and Valverde, F., A method using DiI to study the connectivity of cortical transplants, *J. Neurosci. Meth.*, 36, 17, 1991.

78. Zompa, E.A., Pizzo, D.P., and Hulsebosch, C.E., Migration and differentiation of PC12 cells transplanted into the rat spinal cord, *Int. J. Dev. Neurol.*, 11, 535, 1993.

79. Johann, S., Blumel, G., Lipp, M., and Forster, R., A versatile flow cytometry-based assay for the determination of short- and long-term natural killer cell activity, *J. Immunol. Meth.*, 185, 209, 1995.

80. Chang, L., Gusewitch, G.A., Chritton, D.B., Folz, J.C., Lebeck, L.K., and Nehlsen-Cannarella, S.L., Rapid flow cytometric assay for the assessment of natural killer cell activity, *J. Immunol. Meth.*, 166, 45, 1993.

81. Lohne, K., Urdal, P., Leren, T.P., Tonstad, S., and Ose, L., Standardization of a flow cytometric method for measurement of low-density lipoprotein receptor activity on blood mononuclear cells, *Cytometry*, 20, 290, 1995.

82. Barak, L.S. and Webb, W.W., Fluorescent low density lipoprotein for observation of dynamics of individual receptor complexes on cultured human fibroblasts, *J. Cell Biol.*, 90, 595, 1981.

83. Schmued, L., Kyriakidis, K., and Heimer, L., *In vivo* anterograde and retrograde axonal transport of the fluorescent rhodamine-dextranamine, Fluoro-Ruby, within the CNS, *Brain Res.*, 526, 127, 1990.

84. Katz, L.C., Burkhalter, A., and Dreyer, W.J., Fluorescent latex microspheres as a retrograde neuronal marker for *in vivo* and *in vitro* studies of visual cortex, *Nature*, 310, 498, 1984.

85. Cornwall, J. and Phillipson, O.T., Quantitative analysis of axonal branching using the retrograde transport of fluorescent latex microspheres, *J. Neurosci. Meth.*, 24, 1, 1988.

86. Jaeger, C.B. and Wolf, A.L., Neo-pathway formation of dissociated neural grafts demonstrated by immunocytochemistry, fluorescent microspheres, and retrograde transport, *Brain Res.*, 487, 225, 1989.

87. Sheen, V.L. and Macklis, J.D., Targeted neocortical cell death in adult mice guides migration and differentiation of transplanted embryonic neurons, *J. Neurosci.*, 15, 8378, 1995.

88. Stoppinni, L., Helm, G.A., Stringer, J.L., Lothman, E.W., and Bennett, Jr., J.P., *In vitro* and *in vivo* transplantation of fetal rat brain cells following incubation with various anatomic tracing substances, *J. Neurosci. Meth.*, 27, 121, 1989.

89. Hollander, H., Egensperger, R., and Dirlich, G., Size distribution of rhodamine-labelled microspheres retrogradely transported in cultured neurons, *J. Neurosci. Meth.*, 29, 1, 1989.

90. Divac, I. and Morgensen, J., Long-term retrograde labelling of neurons, *Brain Res.*, 524, 339, 1990.

11 Quantitative Morphology in Relation to Long-Term Pain States, Estimates of Neuron Numbers

Richard E. Coggeshall

CONTENTS

11.1 INTRODUCTION

Architectural changes in the nervous system have an important role in the development and maintenance of chronic pain; for example, the loss of primary afferent neurons following nerve lesions undoubtedly has profound functional consequences. To better define the role of morphologic changes in chronic pain, it will be necessary to correlate structural, functional, and molecular data. At present, among the quantitative morphologic data most frequently sought for such correlations are neuron numbers. Such data usually must be obtained from histologic sections because of the small size of the units to be counted. Estimating neuron numbers from histologic material is a non-trivial task, however, and many different methods have been proposed. Thus, we first discuss the assumptions underlying the various methods for these types of quantitative analysis. A particular task is to justify stereologic methods since these are at present the most efficient way to get unbiased numerical estimates, and also because these methods can be used for other quantitative estimates such as surface areas and volumes. A protocol for obtaining unbiased stereologic estimates of neuron numbers is then presented.

11.2 COUNTING METHODS

The major goal of studies on numbers of neurons in histologic sections is unbiased estimates because these estimates approach true numbers as sampling increases. By contrast, biased estimates approach incorrect numbers as sampling increases. The actual counting is only part of an unbiased sampling procedure, but it is the most controversial.[1,2] Thus, we particularly need to examine the assumptions that justify various counting methods, but guidelines toward all levels of unbiased sampling will be discussed.

Methods for counting neurons in histologic sections can be placed into one of four categories:

1. Profile counts
2. Serial reconstructions
3. Assumption-based, model-based, indirect, or 2-dimensional methods
4. Stereologic, direct, or 3-dimensional methods[3,4]

11.2.1 PROFILE COUNTS

Profiles are what is seen of a neuron in histologic sections. Counting profiles is the most common method for estimating neuron numbers.[4,5] The necessary assumption is that each profile uniquely represents a neuron. Since some neurons are usually cut into more than one profile by the knife, this assumption is almost always incorrect. Thus, estimates by this method are rarely unbiased. This can be seen in Fig. 11.1, which is a depiction of sectioned neurons in a reference space (nucleus, ganglion, etc.). It can be seen easily that the number of profiles far exceeds the number of objects that give rise to these profiles. Unfortunately, counting profiles is a very efficient method, which explains its popularity.

11.2.1.1 Profile Ratios

Investigators often use profile ratios (profile counts in control vs. experimental situations, e.g., numbers of mitotic figures per high power field in normal as compared to cancerous tissue) to see if neuron numbers change after a perturbation.[5] The assumption is that biases in the control and experimental cancel, and on this basis, it is further assumed that proportional changes in ratios indicate proportional changes in neuron numbers. The first problem with this reasoning is that profile ratios, just like profile counts, deal with profile numbers, not neuron numbers. Second, it is unlikely that biases cancel; for example, neither neurons nor the reference space (e.g., nucleus, lamina) can change size, shape, or orientation without changing profile ratios unrelated to changes in neuron numbers. But adjustments in both neurons and the reference space are likely after perturbations, so controls are necessary to show that the above changes do not take place. Thus, if controls are not obtained, and they usually are not, ratios of profile counts should not be used to show changes in neuron numbers.

11.2.2 SERIAL RECONSTRUCTIONS

One can reconstruct neurons in serial sections on the assumptions that the neurons can be followed from section to section and that each neuron can be seen in at least one section. These assumptions are not simplistic, but they are usually easily met and there is no sampling. When this is done, a completely accurate count is obtained (Fig. 11.1). Thus, this is a powerful method that yields the true number of neurons. The difficulty is that the method is too inefficient in any except the simplest cases.

11.2.3 ASSUMPTION-BASED, MODEL-BASED, INDIRECT, OR 2-DIMENSIONAL METHODS

The need for correcting profile counts because of sectioning and creating multiple profiles per neuron has been recognized for many years. This was first done by making geometric assumptions or models that led to formulae to convert profile counts into neuron counts.[6-10] The most important of these geometric assumptions is that the height (H or caliper height) of the objects being counted in the plane orthogonal to the plane of section can be determined.[2] The reason is that if H is known, then these methods will convert unbiased profile counts into unbiased estimates of neuronal numbers. The difficulty is that, in practice, H is determined by measuring diameters of neuronal nuclei from single sections; however, profile diameters are not H. For example, even if one is measuring a perfectly spherical object, where diameter equals H, most sections of neuronal nuclei do not pass through nuclear centers and thus do not measure nuclear diameters. The problem is intensified

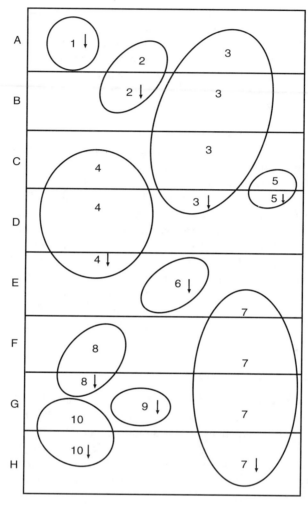

FIGURE 11.1 A schematic diagram of a reference space (nucleus, ganglion, lamina, etc.) containing 10 neurons (1 to 10) subdivided into 8 histological sections (A to H). It should be understood that an investigator would look down on each section in a microscope. Each part of a neuron seen in a section, even if it is the whole neuron (e.g., 1, 6, and 9), is a profile. Note that profile counts have no relation to neuron numbers, so neither profile counts nor their ratios should be used if an investigator desires unbiased estimates of neuron numbers. If one analyzed the tissue in serial sections, giving a number to each neuron where it is first seen, the result would be a completely accurate determination of neuron numbers. The difficulty is that this is an extremely inefficient method. Most indirect methods depend on an accurate determination of H, but they make this determination by measuring profile diameters, which by inspection can be seen to lead to an inaccurate estimate of H. Stereological techniques identify neurons when they appear in one section or focal plane but not the next, as indicated by arrows. Note that if one counts these arrows, a completely accurate estimate of neuron numbers is obtained.

if nuclei are not perfectly spherical. As an example, if one measures the diameters of any of the profiles of neurons 3 and 7 in Fig. 11.1, one would obtain a result very different from H. Thus, nuclear profile diameters are biased measures of H. It follows that estimates of neuron numbers that rely on such determinations of H will themselves be biased. To get unbiased estimates of H, one needs to measure object heights in the plane orthogonal to the plane of section in 3 dimensions, which requires serial reconstructions. Since this is almost never done, there is an unquantified bias in almost all estimates from these methods.

11.2.3.1 Terminology

The above methods are referred to as 2 dimensional because measurements are made on single histological sections (although sections have depth, the measurements are planar), assumption- or model-based because they depend on geometric assumptions or models, and indirect because they convert profile counts to neuron counts. This terminology is irritating, particularly the appellation "assumption-based," because all methods make assumptions.[1] In answer, it is true that any study that estimates neural numbers from histologic sections makes certain types of assumptions (see below), but indirect methods also depend on geometric assumptions that are almost never verified. The most accurate descriptor of these techniques, therefore, would be "geometric-assumption-based techniques," but this is cumbersome and in accord with previous suggestions, we will refer to these as indirect methods.[11]

11.2.3.2 Are Estimates from Indirect Methods Always Biased?

The answer is no. For example, if H is determined in 3 dimensions with unbiased sampling, unbiased profile estimates can be converted to unbiased estimates of cell numbers. On the other hand, H is almost never determined the correct way, and comparisons show considerable biases in many estimates using several indirect methods of this type.[12,13] The real difficulty is that one cannot tell without comparisons with unbiased estimates how biased estimates from indirect methods are. Unfortunately simple examination of the sections by an experienced histologist, as suggested by one group of workers,[1] will not provide the necessary data to show that the assumptions are met. Thus, comparisons with unbiased estimates are necessary, but it is usually more efficient simply to take the unbiased estimates rather than using them to calibrate other estimates from methods that depend on geometric assumptions.

11.2.4 STEREOLOGICAL METHODS

Stereology is a group of mathematical methods that relate object numbers (as well as other variables such as surface areas or volumes) to measurements from tissue sections. In this sense, all histologic numeric methods are stereologic. In practical terms, however, stereologic techniques for neuron numbers are those that identify each neuron in 3-dimensional space by a single unique point. The single unique point arises from the fact that any neuron, no matter its size, shape, or orientation, has a place where it will appear in one section and not the next (the physical disector)[14,15] or in one optical plane but not the next (the optical disector).[14,15] If other aspects of sampling are unbiased, this method will result in unbiased estimates of neural numbers in a reference space. As an example, if the neurons are counted by this method in Fig. 11.1 (arrows), an accurate count of neuron numbers results. It should be understood that the single unique point is key because one is actually counting neurons and not profiles. For this reason, there is no need to assume that each profile represents one neuron (profile counts) or that H can be measured on single sections (indirect methods).

11.2.4.1   Potential Biases in Histological Numeric Methods

The word bias probably causes more confusion than any other word pertaining to methods for estimating neural numbers. The first point is that methods are not intrinsically biased or unbiased, they simply provide biased or unbiased estimates to the extent that their assumptions are or are not met. Second, there are several types of bias. Among those potential biases common to all methods are observational, sampling, and methodological. Observational biases occur because repeated measurements of the same thing (e.g., section thickness) vary. For this type of bias, quality is important because the better the staining, sectioning, etc., the more repeatable the measurements and thus the less the observational biases. Sampling biases are, for example, when a sub-population of neurons is preferentially selected but the conclusions are applied to the whole population. Methodological biases are, for example, when machines do not perform properly (e.g., a microtome that does not cut constant thickness sections) or when not all neurons to be sampled are stained. These biases are potential in all histologic numeric methods.

11.2.4.2   Stereology — Absence of Specific Geometric Assumptions

That any method for estimating neuron numbers from histologic sections contains potential biases does not change the fact that indirect methods and profile counts depend on other assumptions specific to these methods (e.g., each profile represents a neuron uniquely for profile counts, or H can be measured from profiles for assumption-based methods). These specific assumptions are usually not met, and are almost never verified, thus estimates from methods that depend on these assumptions are usually biased. The advantage of stereologic methods is that these specific assumptions are not necessary, so we recommend stereology as the method of choice for unbiased estimates of neuronal numbers.

11.3   DISCUSSION OF THE STEPS IN A STEREOLOGIC PROTOCOL FOR ESTIMATING NEURON NUMBERS

11.3.1   INTRODUCTION

A stereologic numerical estimate is not restricted to neuron numbers but applies to any objects of biologic interest in any defined space in the body. For ease of understanding, we focus on the case where the region of interest is serially sectioned. More complete descriptions exist[11,16-18] of the theory of sampling and of sampling when complete sectioning of a reference space is impractical.

11.3.2 Serially Section the Tissue Containing the Reference Space

The reference space is serially sectioned and each section should be numbered sequentially. The only other point here is that although it is necessary to section through the entire reference space, it is not necessary to save all the sections. For example, if an investigator needs 10 sections for optical disectors, it would only be necessary to stain these 10 sections no matter how many were cut.

11.3.3 Select the Sections for Sampling

We use systematic random sampling. Thus, we make a preliminary assessment of approximately how many sections will be produced and how many are needed for an appropriate precision, and divide them to get section separation 'k'. Prior to sectioning, the investigator picks a random number (R) between section 1 and section k, and takes the Rth section, the R+kth section, etc. for sampling. This ensures that every neuron in the reference space has an equal chance of being sampled.

It is rare that the neurons in whole sections need be counted. Instead, one can choose relatively small regions in each section in which to count. As long as all these regions have themselves an equal chance of being chosen, the sampling remains unbiased. Many schemes are possible. We usually choose parts of sections by dropping grids randomly on montages of each section and then counting only in previously designated squares in this grid. When subdivisions are complete, inclusion and exclusion lines should be used to indicate which units to count and which to exclude when profiles touch the counting frames.[16,19]

Classic random sampling[20] would also provide unbiased estimates, but uniform random sampling is more efficient for histological work.[16,17,20] Furthermore, it is sometimes difficult to be truly random when picking sections by classic random sampling. In particular, sections from the center of a reference space or where the staining is of especially high quality are often preferentially selected.

It is not always necessary to serially section reference spaces because appropriate subdividing can reduce the amount of sectioning to a small part of the whole.[11,21] Serial sections facilitate fractionator analyses, however (see below).

11.3.4 Define the Reference Space

The next step is to define the reference space by outlining it in the sections chosen for sampling. This is necessary because if the reference space is not known, it is impossible to devise a sampling paradigm where all elements in the reference space have an equal chance of being sampled. Instead, all one can obtain is densities (numbers/volume). Unfortunately, densities are ambiguous, even if we ignore the sampling problem, because densities change with changes in the reference space (this is sometimes called the reference trap) as well as with changes in neural numbers.[22] For well-defined structures such as ganglia or prominent nuclei, outlining the reference space is simple. In other cases, it is more difficult, as for example when laminar boundaries are obscure. Nevertheless, it must be done if unbiased estimates of neuron numbers or proportional changes in these numbers are to be obtained.[11]

11.3.5 Count the Neurons

After sections and/or regions within the sections are selected, the next task is to count the neurons. The two recommended ways to do this are the *physical* disector and the *optical* disector.

11.3.5.1 The Physical Disector

The physical disector consists of reference and look-up section pairs. To choose these, we start with the Rth section as the reference section and use the next section as the look-up section. All neuronal profiles that appear in the reference section but not in the look-up section are counts, symbolized by Q- in the stereologic literature. To do this, we prepare montages of our chosen sections, draw outlines of the reference space with fiduciary marks and all labelled neuron profiles on transparent plastic and then overlay the drawings, and count those where a profile is found in one section and not the next (see below). Other matching schemes are used and may be better suited to particular situations. It is important to emphasize that neurons are being identified as unique points in 3-dimensional space, thus these are direct neuronal counts. Counts in all disectors from each animal are then summed.

It is not necessary that the look-up and reference sections be adjacent. It is, however, necessary to make certain that the distance between reference and look-up sections does not exceed the narrowest dimension of whatever one is counting or some objects will get lost between the sections and bias the estimate. It is best that the sections for the physical disector be relatively thin. A recommendation is that section thickness be no more than 1/3 to 1/4 of the caliper diameter of the objects being counted. One can increase efficiency by reversing reference and look-up sections and counting in the opposite direction. In this case, one must remember that two disector analyses are done for each chosen pair of sections.

11.3.5.2 The Optical Disector

Optical disectors are done in one relatively thick section. The procedure is to choose counting frames in an unbiased pattern (all regions of the chosen section have equal chances of being sampled), set upper and lower focal planes in the frame, and count all neurons or neuronal nuclei that come into focus as one focuses from the upper to the lower focal plane. Thus, each count is a neuron identified uniquely in 3-D space.

One must take special care with inclusion and exclusion boundaries since these are counts in volumes as opposed to frames in physical disectors, which are areal. An investigator does not count neurons or neuronal nuclei already in focus in the upper focal plane.

There is no hard and fast rule for how thick a section for an optical disector must be. Guidelines include guard zones on the top and bottom of the sections because of knife damage at each section surface and a reasonable distance between upper and lower focal planes to clearly allow neurons or their nuclei to come in and out of focus. We usually use 30-μm thick sections with an upper guard zone of 3 μm, a lower guard zone of 7 μm, and a disector height of 20 μm.

In our hands, if sections are dried onto slides before staining, there is shrinkage of 70 to 90% in the vertical axis, so it is difficult to get an ideal section thickness. It may also be difficult to get uniform staining throughout the thick section. There is no easy solution for these difficulties. Possibilities include staining the sections free-floating and using a water-soluble medium to have stain reagents penetrate from both sides of the section and obviate shrinkage, cutting thinner sections and doing physical disectors, and doing optical disectors on shrunken sections with correspondingly shrunken guard zones and focal plane separations and resultant uncertainties as to when objects are in focus. Trial and error will often be necessary.

The optical disector is a major time saver if such problems as shrinkage and stain penetration can be overcome. One simply focuses between focal planes at selected locations and counts the objects that come into focus. In addition, there is no difficulty in matching profiles from section to section as sometimes occurs with the physical disector. Thus, this procedure is highly recommended.

11.3.6 ESTIMATE NEURON NUMBERS IN THE REFERENCE SPACE

11.3.6.1 Fractionator

If the sampled fraction of a reference space is known, multiplying the counts by the reciprocal of the sampled fraction gives an unbiased estimate of neuron numbers in that space. For example, if one counts neurons in every 10th section, one multiplies the summed neuron counts by 10. If one samples only parts of each chosen section, one would multiply the disector counts by both section separation and the reciprocal of the proportion of each section that is sampled. This is a statistically powerful method that obviates the necessity of determining tissue shrinkage.

11.3.6.2 $N_V \times V_{ref}$

If a fractionator is not done, one estimates neuron numbers by multiplying numerical density (N_V) by the volume of the reference space (V_{ref}). N_V is the sum of neuron counts divided by the summed volume of the disectors, and V_{ref} is usually best determined by a Cavaliari estimate, done by summing areas of the reference space on 8 to 10 sections chosen in a systematic random pattern and multiplying by section separation and thickness.

11.4 A TYPICAL STEREOLOGIC PROTOCOL

There are many ways to do a stereologic analysis. Procedures will vary depending on whether one is doing a physical or optical disector analysis and a fractionator or $N_v \times V_{ref}$. Also, goals of different studies vary. The two basic principles, however, are that every neuron has an equal chance of being sampled and that the neurons are counted as single unique points in 3-dimensional space. We present below a protocol for a physical disector–fractionator analysis on a relatively small reference space, and will follow with brief comments on other analyses. For those with interest in pursuing these issues, detailed analyses with statistical evaluations of each step[11] and paradigmatic physical[11] and optical disector analyses[11] are available.[13,21,23,24]

11.4.1 Numbers of Primary Afferent Neurons in Rat Dorsal Root Ganglia

1. Determine k. In more detail, estimate the size of the reference space and choose a section thickness appropriate for the study (relatively thin for physical disectors, relatively thick for optical disectors). Cut some sections at this thickness to make certain that the staining and sectioning procedures are working well and to get a rough estimate of how many counts per unit area will be obtained. On this basis, select section separation (k).

   As an example, our reference space is rat lumbar dorsal root ganglia, which are approximately 2500 μm in length, and our section thickness is 4 μm. We find 5 disectors per ganglion gives an appropriate sample, and since we use physical disectors, we obtain 5 pairs of facing sections chosen in a uniform random pattern with k = 125 (625 total sections divided by 5, the chosen number of disectors).

2. Do uniform random sampling. Pick a random number R between the first and kth sections (in our case, between sections 1 and 125), stain the R and R + 1, the R + k and R + k + 1, the R + 2k and R + 2k + 1, ...sections until all sections have been chosen or excluded. This gives pairs of facing sections, designated reference and look-up sections, separated by section separation k. We discard the non-selected sections. It should be remembered that one must proceed all the way through the tissue or the sampling will be biased, so if the total number of sections is estimated incorrectly, one may end up with more or fewer disectors than planned.

3. Stain the chosen sections. In our case we stain with a standard buffered toluidine blue solution used for plastic embedded material.

4. Photograph each reference and look-up section.

5. Place transparent mylar sheets over each section and draw with a marking pen the outline of the reference space, fiduciary marks, and all neuron (or neuronal nuclear) profiles.

6. Overlay the sheets from each pair of sections, align the sheets by the fiduciary landmarks, and determine which profiles in the reference section are not found in the look-up section. These are counts, signified by Q- in the stereologic literature, and they represent neurons identified uniquely in 3-dimensional space.

7. Reverse reference and look-up sections and repeat the process.

8. Sum the counts from all disectors.

9. Multiply the summed counts by section separation divided by 2. The division by 2 is because 2 disectors were done at each section separation interval.

10. The result is an unbiased estimate of neuron numbers in the reference space as done by a physical disector–fractionator analysis.

11.4.2 Further Comments

11.4.2.1 Physical Disectors on Large Tissues

In most instances, it is neither desirable nor efficient to count the entire section as outlined above. For larger material, therefore, we place a grid that outlines squares of a known size over each montage. A proportion of these squares (e.g., every 10th square) is chosen in a uniform random pattern and the above counting procedure is done only in the chosen squares. In this case, the investigator would multiply the counts by section separation times the reciprocals of the chosen squares. If we did this in our material, we would multiply the summed counts by 125 (section separation) and 10 (the reciprocal of the proportion of each section counted). Thus, the basic procedures for doing physical disectors on large as opposed to small sections are the same; the sampling is just adjusted accordingly.

11.4.2.2 Optical Disectors

The overall procedure is essentially the same as above, except that one is doing single relatively thick sections rather than pairs of sections. Thus, one chooses k, picks the sections to be sampled in a uniform random pattern, stains the chosen sections, picks a fraction of each section to be examined by subdividing in a uniform random pattern, counts between focal planes in each subdivision with appropriate attention to exclusion and inclusion lines, sums the counts, and multiplies by section separation and the fraction of each section counted. This is a procedure suited to automation and is most efficient if such problems as staining all the way through the tissue and getting sections of appropriate thickness can be overcome.

REFERENCES

1. Guillery, R.W. and Herrup, K., Quantification without pontification: choosing a method for counting objects in sectioned tissues, *J. Comp. Neurol.*, 386, 2, 1997.
2. West, M.J., Stereological methods for estimating the total number of neurons and synapses: issues of precision and bias, *TINS*, 22, 51, 1999.
3. Coggeshall, R.E., A consideration of neural counting methods, *TINS*, 15, 9, 1992.
4. Coggeshall, R.E. and Lekan, H. A., Commentary on methods for determining numbers of cells and synapses. A case for more uniform standards of review, *J. Comp. Neurol.*, 364, 6, 1996.
5. Oorschot, D.E., Are you using neuronal densities, synaptic densities or neurochemical densities as your definitive data? There is a better way to go, *Prog. Neurobiol.*, 44, 233, 1994.
6. Abercrombie, M., Estimation of nuclear population from microtome sections, *Anat. Rec.*, 94, 239, 1946.
7. Konigsmark, B.W., Methods for counting of neurons, in *Contemporary Research Methods in Neuroanatomy*, Nauta, W.J.H. and Ebbesson, S.O.E., Eds., Springer Verlag, New York, 1970, 315.
8. Hendry, L.A., A method to correct adequately for the change in neuronal size when estimating neuronal numbers after nerve growth factor treatment, *J. Neurocytol.*, 5, 337, 1976.
9. Smolen, A.J., Wright, L.L., and Cunningham, T.J., Neuron number in the superior cervical sympathetic ganglion of the rat: a critical comparison of methods for cell counting, *J. Neurocytol.*, 12, 739, 1983.
10. Rose, R.D. and Rohrlich, D., Counting sectioned cells via mathematical reconstruction, *J. Comp. Neurol.*, 263, 365, 1987.
11. West, M.J. and Gundersen, H.J.G., Unbiased stereological estimation of the number of neurons in the human hippocampus, *J. Comp. Neurol.*, 296, 1, 1990.

12. Pover, C.M. and Coggeshall, R.E., Verification of the disector method for counting neurons, with comments on the empirical method, *Anat. Rec.*, 231, 573, 1991.

13. Pakkenberg, B., Moeller, A., and Gundersen, H.J.G., The absolute number of nerve cells in substantia nigra in normal subjects and in patients with Parkinson's disease estimated with an unbiased stereological method, *J. Neurol. Neurosurg. Psychiat.*, 54, 30, 1991.

14. Sterio, D.C., The unbiased estimation of number and sizes of arbitrary particles using the disector, *J. Microscopy*, 134, 127, 1984.

15. Gundersen, H.J., Bagger, P., Bendtsen, T.F., Evans, S.M., Korbo, L., Marcussen, N., Moeller, A., Nielsen, K., Nyengaard, J.R., Pakkenberg, B., Sorensen, F.B., Vesterby, A., and West, M.J., The new stereological tools: disector, fractionator, nucleator and point sampled intercepts and their use in pathological research and diagnosis, *APMIS*, 96, 857, 1988.

16. Gundersen, H.J., Bendtsen, T.F., Korbo, L., Marcussen, N., Moeller, A., Nielsen, K., Nyengaard, J.R., Pakkenberg, B., Sorensen, F.B., Vesterby, A., and West, M.J., Some new, simple and efficient stereological methods and their use in pathological research and diagnosis, *APMIS*, 96, 379, 1988.

17. Gundersen, H.J. and Jensen, E.B., The efficiency of systematic sampling in stereology and its predictions, *J. Microsc.*, 147, 229, 1987.

18. Howard, C.V. and Reed, M.G., *Unbiased Stereology: Three-Dimensional Measurement in Microscopy*, Bias Scientific, Oxford, U.K., 1998, 1.

19. Gundersen, H.J.G., Notes on the estimation of the numerical density of arbitrary profiles: the edge effect, *J. Microscopy*, 111, 219, 1977.

20. Weibel, E.R., *Stereological Methods*, Academic Press, New York, 1980.

21. Geinisman, Y., Gundersen, H.J., Van der Zee, E., and West, M.J., Unbiased stereological estimation of the total number of synapses in a brain region, *J. Neurocytol.*, 25, 805, 1996.

22. Braendgaard, H. and Gundersen, H.J.G., The impact of recent stereological advances on quantitative studies of the nervous system, *J. Neurosci. Meth.*, 18, 39, 1986.

23. Pakkenberg, B. and Gundersen, H.J.G., Total number of neurons and glial cells in human brain nuclei estimated by the disector and the fractionator, *J. Microscopy*, 150, 1, 1988.

24. Tandrup, T., A method for unbiased and efficient estimation of number and mean volume of specified neuron subtypes in rat dorsal root ganglion, *J. Comp. Neurol.*, 329, 269, 1993.

12 Functional Brain Imaging in Humans: Methodology and Issues

Karen D. Davis

CONTENTS

12.1 INTRODUCTION

A variety of techniques have been used to study pain and nociception at molecular, cellular, and system levels. However, the invasiveness of most of these techniques, such as single unit electrophysiology and anatomical tracing, limit their application to animal studies. Although a large body of information has already been derived by such animal studies, there still exists the dilemma of how to link human perception to brain activity in awake humans. The development of brain-imaging techniques has filled this gap to some extent and has begun to uncover brain mechanisms underlying many human conditions.

This chapter provides an overview of four imaging techniques that have been applied to the study of pain. Presented for each technique are a brief description of its basis, a summary of its advantages and disadvantages, and its application to the field of pain. These overviews are by no means exhaustive and the reader is directed to original technical publications and reviews for additional information. The aim of this format is to provide insight into each technique and its usage such that the reader can assess which approach to be most appropriate for a particular question.

The first three techniques are based on indirect measures of neuronal function. Indirect techniques measure vascular changes, such as blood flow or oxygenation, in the brain which arise in response to the metabolic needs of neural activity. These changes are thought to occur within a few seconds of increased neuronal activity in an area up to a few millimeters of the active neurons (see Kinahan and Noll, 1999) and can be detected at the brain surface and within deep structures. The fourth technique (MEG) provides a more direct measure of neuronal activity as it records magnetic field changes that more closely reflect the spatial location and temporal properties of neuronal activity, primarily at the surface of the brain. The basis of each of these techniques impacts the data emerging from that technique in terms of its ability to tell us something about neuronal activity and central pain mechanisms.

12.2 POSITRON EMISSION TOMOGRAPHY (PET)

12.2.1 HOW IT WORKS

Positron emission tomography (PET) technology for imaging human brain function has been improving continually since its early development in the 1970s. A PET image is created by the detection of positrons emitted from an intravenously injected radionuclide. The tracer will distribute itself throughout the body, including the brain, according to such factors as metabolic need, excretion, uptake, etc., and as such its concentration indirectly reflects neuronal activity. As the tracer decays, it will emit a positron which travels a few millimeters, collides with an electron, and releases two photons (gamma rays) in opposite directions. A series of PET detectors that surround the head picks up the signals which are used to create a tomographic image of the location of the radionuclides (Toga and Mazziotta, 1996; Orrison et al., 1995). To visualize the location of the PET data with respect to brain anatomy, data images can either be co-registered with a subject's own MRI or it can be transformed to fit onto a standardized atlas map.

12.2.2 RADIOCHEMICALS AND TYPES OF PET STUDIES

A variety of radiochemicals can be produced for PET studies. One important determinant in the choice of positron-emitting isotope is its half-life. On one end of the spectrum is ^{15}O which has a short half-life (~2 min) and is commonly used to measure cerebral blood flow in activation studies. On the other extreme is ^{18}F with a half-life of 110 min that is used as 18-fluorodeoxyglucose (FDG) to measure cerebral glucose metabolism for such applications as tumor detection. Other radionuclides with intermediate half-lives such as ^{11}C (20 min) have been used to study receptor binding of dopamine, benzodiazapine, and opiates (Kegeles and Mann, 1997; Toga and Mazziotta, 1996; Orrison et al., 1995; Scheffel, 1993; Phelps and Mazziotta, 1985).

There are three major types of PET studies. Receptor binding studies, typically with injection of a radioactive receptor antagonist or agonist, can identify the receptor density and binding properties of ligands in brain regions of interest. This type of PET experiment is particularly useful to study neurotransmitter abnormalities associated with neurological disorders such as epilepsy, Parkinson's disease, and Alzheimer's disease. A second type of PET study is the measurement of regional cerebral blood flow (rCBF) in the resting state. This application has particular importance for clinical studies of neurological abnormalities in disease or injury (Jonsson et al., 1998; Kegeles and Mann, 1997; Scheffel, 1993; Phelps and Mazziotta, 1985). The third and most popular type of scientific PET application is activation studies, typically with injection of $[^{15}O]H_2O$, which can identify task-related changes in blood flow. The basic premise is to assess statistically significant differences in the data obtained during control and test scans. The particulars of this analysis are beyond the scope of this review. There are several analysis packages currently used for a PET data set, one of the most popular being the statistical parametric mapping (SPM) program (Friston et al., 1995; Friston et al., 1994; Tamminga et al., 1994; Friston et al., 1991).

12.2.3 ADVANTAGES AND DISADVANTAGES

One of the greatest advantages of PET compared to fMRI (see below) is that the experiment is done in a relatively open, noise-free environment which can accommodate most experimental devices (e.g., to deliver stimuli, present tasks, record responses, etc.). Also, the subject pool is not restricted by the presence of any internal devices (e.g., pacemakers) as it is in fMRI studies. However, PET studies are relatively expensive and not widely available outside of special imaging centers that contain a PET scanner and a cyclotron to generate the radioactive tracers. The injection of a radioactive tracer is invasive and also impacts the total number of scans that can be obtained in a given timeframe in any one subject. This, in turn, impacts the ability to achieve high enough statistical power for single-subject analysis and repeat-subject studies. The most advanced 3D PET scanners have greater sensitivity for detection of lower radioactive doses and can allow for some single-subject studies (Halber et al., 1997), although the maximal allowable yearly radioactive dose is still restrictive. Another disadvantage of PET is the moderate to poor spatial resolution and the poor temporal resolution. The spatial resolution suffers not only because of the resolution of the original scans, but to the filtering that is performed on the data sets to allow for averaging within and across subjects and subsequent transformation into a standardized Talairach and Tournoux atlas format (Talairach and Tournoux, 1988). The temporal resolution of activation studies utilizing $[^{15}O]H_2O$ is of the order of 1 min, the time required to obtain each scan.

12.2.4 PET IMAGING OF PAIN

In the first brain-imaging studies of acute pain, Talbot et al. (1991) identified four cortical regions of activation to noxious heat stimuli: primary and secondary somatosensory cortex (SI, SII), anterior insula, and the anterior cingulate cortex (ACC). Jones et al. (1991a) reported heat pain-evoked activation in the ACC, lentiform nucleus, and thalamus. Subsequent studies have confirmed the involvement of these areas in thermal-, mechanical-, and laser-evoked pain and have also identified additional sites of pain-related activation such as the prefrontal cortex, supplemental motor cortex, basal ganglia, and cerebellum (Aziz et al., 1997; Derbyshire et al., 1997; Svensson et al., 1997; Xu et al., 1997; Casey et al., 1996; Casey et al., 1994; Coghill et al., 1994), and the hypothalamus and periaqueductal gray (Hsieh et al., 1996). Across these studies there is some variability in the location, laterality, and statistical significance of activations. Although the thalamus was identified in many studies, the spatial resolution of PET, data averaging, and filtering restrict the interpretation of the precise thalamic nucleus involved in the pain response. The different stimulation methods (site, modality, intensity, etc.), statistical methods (data averaging, statistical cutoffs, etc.), attentional context, and individual variability likely contribute to the inconsistencies in cortical and thalamic activations. For further descriptions and discussion of these variabilities see Peyron et al., 1999; Derbyshire et al., 1997; and Hsieh et al., 1996.

The widespread cortical activations identified in pain studies have been implicated in affective, cognitive, and reflexive responses to a painful stimulus. These findings demonstrate that there is a distributed network of many brain areas that are recruited by a painful stimulus that contributes to the multidimensional experience evoked by the stimulus (Coghill et al., 1999; Coghill et al., 1994). Since many dimensions of pain can be graded, it was not surprising to find pain intensity-related activations in many regions including the SI, SII, anterior insula, ACC, and motor areas (Coghill et al., 1999). However, a study that used hypnosis to manipulate pain unpleasantness independently of pain intensity uncovered a relationship between unpleasantness and ACC activation (Rainville et al., 1997). Interestingly, Craig et al. (1996) reported that the ACC was uniquely activated in both real pain and during an illusion of pain evoked by simultaneous warm and cool stimuli.

A small number of PET studies looked at changes associated with chronic pain. Iadorola et al. (1995) found that the resting blood flow was significantly decreased in the thalamus contralateral to the side of neuropathic pain and allodynia. Hsieh et al. (1995) confirmed this finding and also found increased rCBF in the anterior insula, prefrontal, posterior parietal cortex, and ACC in neuropathic pain patients. A study of patients with lateral-medullary infarcts (Peyron et al., 1998) showed allodynia-evoked activation in most pain areas implicated in normal pain (SI, SII, anterior insula, etc.) but not the ACC. The authors suggested that the abnormal ACC response was related to the unpleasant nature of the allodynia. This is in contrast to the ACC activation evoked in an experimental models of allodynia (Iadarola et al., 1998).

A few studies have used PET to study the effects of thalamic or cortical stimulation in patients with chronic pain. One study (Duncan et al., 1998) found activation of the anterior insula during thalamic stimulation that incidentally also evoked thermal sensations. Another study (Davis et al., 2000) identified an activation in anterior ACC throughout thalamic stimulation and a delayed more posterior ACC activation. The relevance of these findings to chronic pain is unclear since the activations were not related to the amount of pain relief at the time of the PET scanning. PET scans obtained during motor cortex stimulation for chronic pain revealed activation of the thalamus, ACC, anterior insula, and frontal cortex (Garcia-Larrea et al., 1999; Peyron et al., 1995).

Further information about the human pain system has been derived from PET opiate receptor-binding studies, typically with [^{11}C]diprenorphine to map the distribution of μ, δ, and κ receptors (see Sadzot et al., 1991). This technique can identify receptor-rich areas such as the thalamus, ACC, prefrontal and frontal cortex (Vogt et al., 1995; Jones et al., 1991b; Sadzot et al., 1991). Future developments in opiate receptor-binding studies may provide important insights into chronic pain treatment.

12.3 SINGLE PHOTON EMISSION COMPUTERIZED TOMOGRAPHY (SPECT)

12.3.1 HOW IT WORKS

Since only a few SPECT studies of pain have been published, the technique will be mentioned only briefly. The technique of single-photon emission computerized tomography (SPECT) is similar to PET but not as popular for scientific study. However, because SPECT is readily available in many clinical centers, it has been used successfully by some investigators. Like PET, SPECT scanners have a gamma camera that detects emissions from decaying isotopes. Typical activation studies utilizing intravenous injection of isotopes with short half-lives are not possible with SPECT because the isotopes must emit only single photons (e.g., ^{123}I and ^{99}Tc). These isotopes have very long half-lives (often in the order of hours) and are used to measure perfusion. Cerebral blood flow (CBF) can also be measured using the inhalation of Xenon-133. The long retention of SPECT tracers allows for flexibility in the timing of data acquisition after injection of the overall effect of an injury, disease, stimulus, treatment, or other manipulation (Orrison et al., 1995; Prichard and Brass, 1992). However, the poor spatial and temporal resolution of SPECT cannot detect transient or finely localized cerebral changes.

12.3.2 SPECT IMAGING OF PAIN

The relatively few SPECT studies of pain examined different types of pain with variable findings. In normal volunteers, Apkarian et al. (1992) identified a decrease in cortical rCBF in the SI region associated with a long (3 min) sustained contralateral heat pain stimulus, whereas Di Piero et al. (1997; 1994) reported an increased rCBF in SI contralateral to a tonic cold-pain stimulus. It is not clear why these two studies produced opposite findings. Di Piero et al. (1997) studied cluster headache patients who were in a pain-free state. When a cold-pain stimulus was applied to the side ipsilateral to the headache side, there were smaller CBF increases in the contralateral SI and thalamus compared to normals and the patient's other (non-headache) side (Di Piero et al., 1997). This study suggests an underlying abnormality in pain processing in cluster headache patients, even in their pain-free periods. In the few other studies of chronic pain, variable results were found. There are case reports of painful restless legs syndrome and spinal cord injury pain showing an increased blood flow in the SI, ACC, and thalamus contralateral to chronic pain condition (Ness et al., 1998; San Pedro et al., 1998). A small study of central pain (Canavero et al., 1993), found some hypoperfusion of the frontoparietal region during maneuvers that increased the patient's pain. However, this was not a consistent finding among patients. Hypoperfusion was also noted in several other case reports; in the caudate of a patient with spinal cord injury pain (Ness et al., 1998), and in the thalamus of patients with central pain (Tanaka et al., 1997; Pagni and Canavero, 1995).

12.4 FUNCTIONAL MAGNETIC RESONANCE IMAGING (fMRI)

12.4.1 HOW IT WORKS

Magnetic resonance imaging (MRI) is a well-developed technology commonly used in medicine to examine structure. In the early 1990s, revolutionary developments led to the application of MRI to the study of brain function. The first functional MRI (fMRI) studies in humans used intravenous gadolinium as a contrast agent to enhance visualization of the small changes in MR signals in the visual cortex evoked by flashing lights (Belliveau et al., 1991). However, it was soon realized that visualization of these signals did not require injection of a contrast agent because the body has its own natural contrast agent, deoxygenated hemoglobin (deoxyHb) (Ogawa et al., 1992; Ogawa et al., 1990a; Ogawa et al., 1990b). Instead of injecting a tracer or contrast agent, most fMRI studies now rely upon the blood oxygenation level dependent (BOLD) effect. The basis of the BOLD effect is that the increased neuronal firing during a cognitive event or execution of a task sets in motion a series of hemodynamic changes that ultimately modifies the magnetic field and hence increases the MRI signal. Although the exact details of these events are still under investigation, it is generally thought that an increased metabolic demand (due to increased neuronal activity) results in an increase in blood flow beyond metabolic needs such that the final ratio of deoxyHb/oxyHb actually is reduced. It is the reduction in deoxyHb that alters the magnetic field properties and produces the increased MRI signal (DeYoe et al., 1994).

12.4.2 TYPES OF fMRI EXPERIMENTS

The most common form of an fMRI experiment is a blocked-design activation study. This entails alternating relatively long periods, typically 30 to 60 s, of a task with a control. A statistical comparison is made between the MR signal at each pixel in each of the brain slices acquired during the task and control. The change in signal evoked by a task is typically small, of the order of 1 to 5%. Therefore, large numbers of slice acquisitions are necessary to increase detection of the signal and statistical power. The block design was necessary in the studies conducted in the early to mid 1990s because of the relatively long acquisition times (up to 6 s) required to image each brain slice and having to collect data from the same slice repeatedly to enhance the signal-to-noise ratio. However, with the advent of fast, echoplanar speed imaging (EPI) this restriction has been lifted. With EPI, it is possible to image a single slice of the brain in under 100 ms and the entire brain in < 2 s. This enhanced imaging speed allows for repetitive imaging of multiple slices within a block, and for event-related non-block design imaging. Event-related imaging is conceptually similar to the post-stimulus histograms that represent an effect in electrophysiological studies. That is, an analysis is performed to identify a time-locked response to a brief event or stimulus. The MR signals that are detected in fMRI are related to the hemodynamic response that follows neuronal activity. This vascular response takes the form of a gamma variate function ~10 to 12 s in duration that peaks at 4 to 6 s (Buckner, 1998; Dale and Buckner, 1997; Buckner et al., 1996). The detection of task-related responses in block design vs. an event-related paradigm are somewhat different. In a block design of say 60 s per block, the search for task-related responses can typically employ a "boxcar" function that approximates the duration of the task or stimulus, shifted by ~4 s for a hemodynamic delay. However, the effect of using a precise hemodynamic response is not crucial because of the long block. In contrast, in an event-related design whereby the stimulus is short, say 1 s, the detection of short task-related responses must take the predicted hemodynamic response into consideration. Many fMRI software packages can perform event-related analyses.

12.4.3 ADVANTAGES AND DISADVANTAGES

One advantage of fMRI is the wide availability of 1.5T clinical scanners. In the mid 1990s, much of the development of fMRI was greatly facilitated by the presence of MRIs in most major hospitals. Although it is now becoming more common for sophisticated imaging centers to have dedicated research MRIs, it is still quite feasible to create research and clinical fMRI groups within a clinical hospital setting. The noninvasive nature of the technique facilitates recruitment of volunteers and eliminates restricting the number of scans in an individual subject which, in turn, allows for repeated sessions and more data acquisition within each experimental session. Therefore, the large volumes of data that can be collected result in enhanced statistical power for single-subject analysis. The good spatial (1 to 2 mm is possible) and temporal (100s of ms is possible) resolution of fMRI allows for a choice of single-subject or group studies that employ either event-related or block designs. The resolution also allows for visualization of regions of activation in the cortex and subcortical regions. However, functional imaging in the brainstem and spinal cord is hampered by the size of the structures and the artifacts associated with movement of fluids and blood. The statistical analysis and display of fMRI data are now facilitated by several well-developed and widely available software packages.

The disadvantages of fMRI include its expense, limited availability if time-sharing on a clinical scanner, and restriction to subjects who do not have internal magnetic or metallic devices such as a pacemaker. Subjects must also be able to tolerate the loud imaging noise and the somewhat claustrophobic MR tube. The technical limitations include poor access to the face and upper body and the requirement for specially constructed non-ferromagnetic stimulation devices.

12.4.4 fMRI Studies of Pain

Although fMRI of pain is still in its infancy, great strides have been made to develop ways to investigate brain mechanisms underlying both chronic and acute pain. Although there are some differences between studies, most surveys of stimulus-related responses to noxious electrical stimulation of the skin or peripheral nerve (Davis et al., 2000a; Bucher et al., 1998; Oshiro et al., 1998; Davis et al., 1997; Davis et al., 1995), or noxious heat/cold (Apkarian et al., 1999; Becerra et al., 1999; Gelnar et al., 1999; Ploghaus et al., 1999; Davis et al., 1998a; Davis et al., 1998b), mechanical (Disbrow et al., 1998), chemical (Porro et al., 1998) or visceral (Binkof-ski et al., 1998) stimulation have revealed activation in the areas previously identified by PET to be involved in pain; that is, the SI, SII, anterior insula, ACC, and thalamus. However, many fMRI studies have examined pain-related activations in individual subjects with much finer spatial and temporal resolution possible with PET, which has allowed for more precise localization of activations and examination of the intersubject variability in anatomical structures (e.g., various sulci patterns) and functional activations (Kwan et al., 2000; Bushnell et al., 1999; Gelnar et al., 1999; Ploghaus et al., 1999; Davis et al., 1998a; Porro et al., 1998).

Most recently, psychophysical assessment and fMRI have been combined to examine the pain-related activations related to some particular aspect of the pain experience. This approach has been used to separate those activations due to the mere presence of a stimulus (e.g., due to attention and other non-specific cognitive factors) vs. those related to the subjects' actual sensory experiences. Toward this goal, pain ratings have been obtained in parallel psychophysical sessions (Apkarian et al., 1999) or during the imaging session (Davis et al., 2000b,c; Davis et al., 1998b; Porro et al., 1998; Davis et al., 1997). The latter approach has the added benefit of enabling the individual variability in psychophysical response and brain response to be examined. A word of caution about interpretation of these results should be mentioned because ratings of only one component of pain can be obtained at a time. Since the magnitude and time course of many components of the pain experience (e.g., intensity and affect) are similar (see Davis et al., 2000b), brain activations correlated to the obtained ratings may be related to other sensations that are not being monitored.

fMRI has also examined the relationship and interaction of pain, attention, and anticipation. Pain and attention tasks on their own activate nearby but separate regions in the posterior and anterior ACC, respectively (Davis et al., 1997). Similarly, both pain and its anticipation activate the thalamus, SI, SII, and basal ganglia, but activate slightly different regions within the ACC, insula, and cerebellum (Ploghaus et al., 1999). For instance, anticipation of pain activated the anterior ACC whereas the pain itself activated the posterior ACC.

A number of factors have impeded the study of chronic pain with fMRI. The technique is not naturally suited to detect baseline changes (unlike PET which can measure altered basal blood flow) but rather is typically used to identify task-related activations. Furthermore, it is difficult to repetitively alter the level of chronic pain on a time scale suitable for fMRI; however, a few studies have manipulated chronic pain in groups of patients. A study of amputees with phantom limb pain modifiable by use of a myoelectric prosthesis reported cortical reorganization correlated to the use of the prosthesis (Lotze et al., 1999). In a preliminary study of back-pain patients, exacerbation of the pain evoked by leg raising was used to manipulate pain levels (Krauss et al., 1999). fMRI of the effect of spinal and brain stimulation for pain control is also being developed as a means to study analgesic mechanisms (Kiriakopoulos et al., 1997). Further study of chronic pain with fMRI will likely benefit from development of chronic pain models. An fMRI study of one such model (Baron et al., 1999) (capsaicin-induced mechanical hyperalgesia) found prefrontal and motor activations associated with mechanical hyperalgesia. The locations of these activations suggest an involvement of cognitive processes, orientation, and motor and behavioral planning in response to pain.

12.5 MAGNETOENCEPHALOGRAPHY (MEG)

12.5.1 HOW IT WORKS

Magnetoencephalography (MEG) is a technique that detects weak magnetic fields within the human brain. The basis of MEG is that neuronally generated electrical currents induce a perpendicularly oriented magnetic field which can be detected directly outside the head by MEG detectors. Therefore, MEG is most sensitive for cortical neuronal activity parallel to the skull. The ultrasensitive MEG detectors are superconducting quantum interference device (SQUID) magnetometers. A modern MEG device contains up to 300 channels which nearly surround the head. Since the MEG signals are small, the recordings must be obtained in a magnetically shielded room. The current sources responsible for the MEG signals, assumed to represent the site of synchronous neuronal activity of cortical pyramidal cells, can be derived from the equivalent current dipole model. Within a selected temporal frame, solving for this inverse problem (i.e., the cortical source of the magnetic signals recorded outside the brain) forms the basis for MEG data analysis software (Hari and Forss, 1999; Näätänen et al., 1994; Gallen and Bloom, 1993)

12.5.2 ADVANTAGES AND DISADVANTAGES

The major advantages of MEG are that it is noninvasive and has excellent temporal resolution. The noninvasive nature of the technique allows for repeat studies on single subjects. The fine temporal resolution (milliseconds) allows for inferences about neuronal activity not possible with indirect techniques that measure vascular changes (i.e., PET, SPECT, and fMRI). For example, the time course of task-evoked activations could provide clues about cortical connectivity (Hari and Forss, 1999). MEG data can also be superimposed onto a high-resolution MRI to provide good spatial location of source signals. However, there are also some disadvantages that have impeded the proliferation of MEG as a common imaging technique. Most importantly, the high cost of the device and the necessity for a magnetically shielded room have restricted the number of MEG facilities to a few centers in the world. Although the analysis of MEG data and the inverse problem are being developed to assess subcortical activations, most studies are restricted to the cortex. Also, the

MEG signals detected are dominated by those produced by synchronous activity in large numbers of neurons. Therefore, activity in small numbers of neurons will likely go undetected in a typical MEG experiment. Finally, the spatial extent of a region of activation cannot be determined with MEG.

12.5.3 MEG STUDIES OF PAIN

There are numerous MEG studies of sensory systems and plasticity but a paucity of pain studies. One of the earliest pain studies (Howland et al., 1995) identified responses to electrical stimulation of the digit at short latencies in the contralateral SI (at 40 to 60 ms) followed by longer latencies in the ipsilateral and bilateral SII/insula (100 to 250 ms). Interpretation of these findings should consider that no controls were obtained of responses evoked by non-painful levels of stimulation. Indeed, subsequent studies demonstrated early (< 100 ms) contralateral SI dipoles for both non-painful and painful levels of electrical stimulation and longer latency (> 100 ms) pain-specific dipoles bilaterally in SII and ACC (Kitamura et al., 1997; Kitamura et al., 1995). MEG studies of pain evoked by painful CO_2 applied to the nasal mucosa also found long latency responses in SII (Hari et al., 1997). A study of painful laser-evoked responses reported that both the contralateral SI and SII dipoles occurred at ~130 ms, suggestive of parallel processing of thalamocortical inputs to these two cortical regions (Ploner et al., 1999). MEG also has been used to follow cortical plasticity in patients with phantom limb pain. Flor and colleagues have reported correlations between the extent of cortical plasticity and phantom limb pain in traumatic and congenital amputees (Flor et al., 1998; Flor et al., 1995).

12.6 COMPARISON OF TECHNIQUES: WHEN TO USE WHAT TECHNIQUE

The advantages and disadvantages of each imaging technology discussed above must be considered when deciding which method is most appropriate for a particular study. The techniques vary considerably in terms of their spatial and temporal resolution, cost, availability, subject and equipment limitations, and types of possible experimental designs. Each technique has been used successfully to some degree to study particular aspects of pain, as shown by the examples in Fig. 12.1. The choice of technique should be considered carefully when developing hypotheses and interpreting clinical and experimental data. For example, pre-surgical mapping requires a high degree of spatial accuracy, such as that possible with fMRI, but may not be particularly concerned with temporal resolution. On the other hand, the fine temporal data provided by MEG can provide important information for studies of connectivity. The ability to study individual subjects is also crucial to some clinical studies and those of individual characteristics. PET may be a good choice to assess moderate to large-group responses whereas fMRI may be better suited to assess the variability among small numbers of individual responses.

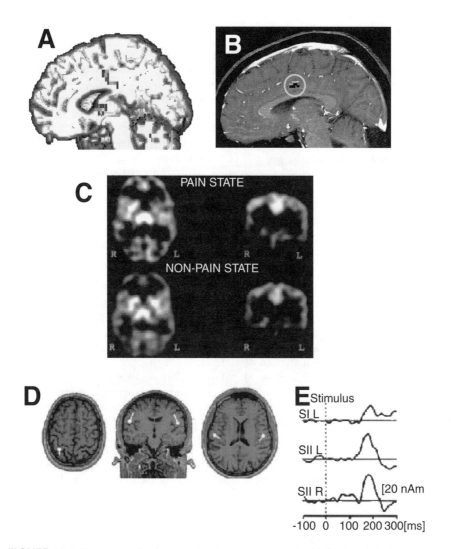

FIGURE 12.1 Examples of pain-related activations obtained with PET, fMRI, SPECT and MEG. (A) PET image of activation in the posterior ACC associated with thalamic stimulation in a group of chronic pain patients. (From Davis et al., 2000c. With permission.) (B) fMRI image of a heat pain-related response in the posterior ACC of an individual subject. (From Kwan et al., 85, 359, *Pain*, 2000. With permission.) (C) SPECT images of the regional cerebral blood flow increases in the thalamus, cingulate and SI during a pain compared to a non-pain state in a patient with spinal cord injury pain. (From Ness et al., *Pain*, 78, 139, 1998. With permission.) (D, E). MEG images and source strengths and times of the painful laser-evoked responses (dipoles) in SI and SII of an individual subject. (From Ploner et al., *J. Neurophysiol.*, 8, 3100, 1999. With permission.) ACC, anterior cingulate cortex; SI, primary somatosensory cortex; SII, secondary somatosensory cortex.

REFERENCES

Apkarian, A.V., Darbar, A., Krauss, B.R., Gelnar, P.A., and Szeverenyi, N.M., Differentiating cortical areas related to pain perception from stimulus identification: temporal analysis of fMRI activity, *J. Neurophysiol.*, 81, 2956, 1999.

Apkarian, A.V., Stea, R.A., Manglos, S.H., Szeverenyi, N.M., King, R.B., and Thomas, F.D., Persistent pain inhibits contralateral somatosensory cortical activity in humans, *Neurosci. Lett.*, 140, 141, 1992.

Aziz, Q., Andersson, J.L.R., Valind, S., Sundin, A., Hamdy, S., Jones, A.K.P., Foster, E.R., Langstrom, B., and Thompson, D.G., Identification of human brain loci processing esophageal sensation using positron emission tomography, *Gastroenterology*, 113, 50, 1997.

Baron, R., Baron, Y., Disbrow, E., and Roberts, T.P., Brain processing of capsaicin-induced secondary hyperalgesia: a functional MRI study, *Neurology*, 53, 548, 1999.

Becerra, L.R., Breiter, H.C., Stojanovic, M., Fishman, S., Edwards, A., Comite, A.R., Gonzalez, R.G., and Borsook, D., Human brain activation under controlled thermal stimulation and habituation to noxious heat: an fMRI study, *Magn. Reson. Med.*, 41, 1044, 1999.

Belliveau, J.W., Kennedy, D.N.J., McKinstry, R.C., Buchbinder, B.R., Weisskoff, R.M., Cohen, M.S., Vevea, J.M., Brady, T.J., and Rosen, B.R., Functional mapping of the human visual cortex by magnetic resonance imaging, *Science*, 254, 716, 1991.

Binkofski, F., Schnitzler, A., Enck, P., Frieling, T., Posse, S., Seitz, R.J., and Freund, H.J., Somatic and limbic cortex activation in esophageal distention: a functional magnetic resonance imaging study, *Ann. Neurol.*, 44, 811, 1998.

Bucher, S.F., Dieterich, M., Wiesmann, M., Weiss, A., Zink, R., Yousry, T.A., and Brandt, T., Cerebral functional magnetic resonance imaging of vestibular, auditory, and nociceptive areas during galvanic stimulation, *Ann. Neurol.*, 44, 120, 1998.

Buckner, R.L., Event-related fMRI and the hemodynamic response, *Hum. Brain Mapping*, 6, 373, 1998.

Buckner, R.L., Bandettini, P.A., O'Craven, K.M., Savoy, R.L., Petersen, S.E., Raichle, M.E., and Rosen, B.R., Detection of cortical activation during averaged single trials of a cognitive task using functional magnetic resonance imaging, *Proc. Natl. Acad. Sci. USA*, 93, 14878, 1996.

Bushnell, M.C., Duncan, G.H., Hofbauer, R.K., Ha, B., Chen, J.I., and Carrier, B., Pain perception: is there a role for primary somatosensory cortex? *Proc. Natl. Acad. Sci. USA*, 96, 7705, 1999.

Canavero, S., Pagni, C.A., Castellano, G., Bonicalzi, V., Bello', M., Duca, S., and Podio, V., The role of cortex in central pain syndromes: preliminary results of a long-term technetium-99 hexamethylpropyleneamineoxime single photon emission computed tomography study, *Neurosurgery*, 32, 185, 1993.

Casey, K.L., Minoshima, S., Berger, K.L., Koeppe, R.A., Morrow, T.J., and Frey, K.A., Positron emission tomographic analysis of cerebral structures activated specifically by repetitive noxious heat stimuli, *J. Neurophysiol.*, 71, 802, 1994.

Casey, K.L., Minoshima, S., Morrow, T.J., and Koeppe, R.A., Comparison of human cerebral activation patterns during cutaneous warmth, heat pain and deep cold pain, *J. Neurophysiol.*, 76, 571, 1996.

Coghill, R.C., Sang, C.N., Maisog, J.M., and Iadarola, M.J., Pain intensity processing within the human brain: a bilateral, distributed mechanism, *J. Neurophysiol.*, 82, 1934, 1999.

Coghill, R.C., Talbot, J.D., Evans, A.C., Meyer, E., Gjedde, A., and Bushnell, M.C. Distributed processing of pain and vibration by the human brain, *J. Neurosci.*, 14, 4095, 1994.

Craig, A.D., Reiman, E.M., Evans, A., and Bushnell, M.C., Functional imaging of an illusion of pain, *Nature*, 384, 258, 1996.

Dale, A.M. and Buckner, R.L., Selective averaging of rapidly presented individual trials using fMRI, *Hum. Brain Mapping*, 5, 329, 1997.

Davis, K.D., Kwan, C.L., Crawley, A.P., and Mikulis, D.J., Functional MRI study of thalamic and cortical activations evoked by cutaneous heat, cold and tactile stimuli, *J. Neurophysiology*, 80, 1533, 1998a.

Davis, K.D., Kwan, C.L., Crawley, A.P., and Mikulis, D.J., Event-related fMRI of pain: entering a new era in imaging pain, *Neuroreport*, 9, 3019, 1998b.

Davis, K.D., Kwan, C.L., Crawley, A.P., and Mikulis, D.J., Electrical nerve stimulation can be used as a tool in fMRI studies of pain- and tingling-evoked activations, *Pain Res. Manag.*, 5, 81, 2000a.

Davis, K.D., Kwan, C.L., Crawley, A.P., and Mikulis, D.J. fMRI of cortical and thalamic activations correlated to the magnitude of pain, in *Proc. 9th World Congr. Pain*, Devor, M., Rowbotham, M.C., and Wiesenfeld-Hallin, Z., Eds., IASP Press, Seattle, 9, 479, 2000b.

Davis, K.D., Taub, E., Duffner, F., Lozano, A.M., Tasker, R.R., Houle, S., and Dostrovsky, J.O., Activation of the anterior cingulate cortex by thalamic stimulation in patients with chronic pain: a positron emission tomography study, *J. Neurosurg.*, 92, 64, 2000c.

Davis, K.D., Taylor, S.J., Crawley, A.P., Wood, M.L., and Mikulis, D.J., Functional MRI of pain- and attention-related activations in the human cingulate cortex, *J. Neurophysiol.*, 77, 3370, 1997.

Davis, K.D., Wood, M.L., Crawley, A.P., and Mikulis, D.J., fMRI of human somatosensory and cingulate cortex during painful electrical nerve stimulation, *Neuroreport*, 7, 321, 1995.

Derbyshire, S.W.G., Jones, A.K.P., Gyulai, F., Clark, S., Townsend, D., and Firestone, L.L., Pain processing during three levels of noxious stimulation produces differential patterns of central activity, *Pain*, 73, 431, 1997.

DeYoe, E.A., Bandettini, P., Neitz, J., Miller, D., and Winans, P., Functional magnetic resonance imaging (FMRI) of the human brain, *J. Neurosci. Meth.*, 54, 171, 1994.

Di Piero, V., Ferracuti, S., Sabatini, U., Pantano, P., Cruccu, G., and Lenzi, G.L., A cerebral blood flow study on tonic pain activation in man, *Pain*, 56, 167, 1994.

Di Piero, V., Fiacco, F., Tombari, D., and Pantano, P., Tonic pain: a SPET study in normal subjects and cluster headache patients, *Pain*, 70, 185, 1997.

Disbrow, E., Buonocore, M., Antognini, J., Carstens, E., and Rowley, H.A., Somatosensory cortex: a comparison of the response to noxious thermal, mechanical, and electrical stimuli using functional magnetic resonance imaging, *Hum. Brain Mapping*, 6, 150, 1998.

Duncan, G.H., Kupers, R.C., Marchand, S., Villemure, J.-G., Gybels, J.M., and Bushnell, M.C., Stimulation of human thalamus for pain relief: possible modulatory circuits revealed by positron emission tomography, *J. Neurophysiol.*, 80, 3326, 1998.

Flor, H., Elbert, T., Knecht, S., Wienbruch, C., Pantev, C., Birbaumer, N., Larbig, W., and Taub, E., Phantom-limb pain as a perceptual correlate of cortical reorganization following arm amputation, *Nature*, 375, 482, 1995.

Flor, H., Elbert, T., Mühlnickel, W., Pantev, C., Wienbruch, C., and Taub, E., Cortical reorganization and phantom phenomena in congenital and traumatic upper-extremity amputees, *Exp. Brain Res.*, 119, 205, 1998.

Friston, K.J., Frith, C.D., Liddle, P.F., and Frackowiak, R.S.J. Comparing functional (PET) images: the assessment of significant change, *J. Cereb. Blood Flow Metab.*, 11, 690, 1991.

Friston, K.J., Holmes, A.P., Worsley, K.J., Poline, J.-P., Frith, C.D., and Frackowiak, R.S.J., Statistical parametric maps in functional imaging: a general linear approach, *Hum. Brain Mapping*, 2, 189, 1995.

Friston, K.J., Worsley, K.J., Frackowiak, R.S.J., Mazziotta, J.C., and Evans, A.C., Assessing the significance of focal activations using their spatial extent, *Hum. Brain Mapping*, 1, 210, 1994.

Gallen, C.C. and Bloom, F.E., Brain imaging: mapping the brain with MSI, *Curr. Biol.*, 3, 522, 1993.

Garcia-Larrea, L., Peyron, R., Mertens, P., Gregoire, M.C., Lavenne, F., Le Bars, D., Convers, P., Maugui, Sindou, M., and Laurent, B., Electrical stimulation of motor cortex for pain control: a combined PET-scan and electrophysiological study, *Pain*, 83, 259, 1999.

Gelnar, P.A., Krauss, B.R., Sheehe, P.R., Szeverenyi, N.M., and Apkarian, A.V., A comparative fMRI study of cortical representations for thermal painful, vibrotactile, and motor performance tasks, *Neuroimage*, 10, 460, 1999.

Halber, M., Herholz, K., Wienhard, K., Pawlik, G., and Heiss, W.D., Performance of a randomization test for single-subject ^{15}O-water PET activation studies, *J. Cereb. Blood Flow Metab.*, 17, 1033, 1997.

Hari, R. and Forss, N., Magnetoencephalography in the study of human somatosensory cortical processing, *Philos. Trans. R. Soc. Lond. [Biol.]*, 354, 1145, 1999.

Hari, R., Portin, K., Kettenmann, B., Jousmäki, V., and Kobal, G., Right-hemisphere preponderance of responses to painful CO_2 stimulation of the human nasal mucosa, *Pain*, 72, 145, 1997.

Howland, E.W., Wakai, R.T., Mjaanes, B.A., Balog, J.P., and Cleeland, C.S., Whole head mapping of magnetic fields following painful electric finger shock, *Cognit. Brain Res.*, 2, 165, 1995.

Hsieh, J.C., Belfrage, M., Stone-Elander, S., Hansson, P., and Ingvar, M., Central representation of chronic ongoing neuropathic pain studied by positron emission tomography, *Pain*, 63, 225, 1995.

Hsieh, J.C., Ståhle-Bäckdahl, M., Hägermark, Ö., Stone-Elander, S., Rosenquist, G., and Ingvar, M., Traumatic nociceptive pain activates the hypothalamus and the periaqueductal gray: a positron emission tomography study, *Pain*, 64, 303, 1996.

Iadarola, M.J., Berman, K.F., Zeffiro, T.A., Byas-Smith, M.G., Gracely, R.H., Max, M.B., and Bennett, G.J., Neural activation during acute capsaicin-evoked pain and allodynia assessed with PET, *Brain*, 121, 931, 1998.

Iadarola, M.J., Max, M.B., Berman, K.F., Byas-Smith, M.G., Coghill, R.C., Gracely, R.H., and Bennett, G.J., Unilateral decrease in thalamic activity observed with positron emission tomography in patients with chronic neuropathic pain, *Pain*, 63, 55, 1995.

Jones, A.K.P., Brown, W.D., Friston, K.J., Qi, L.Y., and Frackowiak, R.S.J., Cortical and subcortical localization of response to pain in man using positron emission tomography, *Proc. R. Soc. Lond. [Biol.]*, 244, 39, 1991a.

Jones, A.K.P., Qi, L.Y., Fujirawa, T., Luthra, S.K., Ashburner, J., Bloomfield, P., Cunningham, V.J., Itoh, M., Fukuda, H., and Jones, T., *In vivo* distribution of opioid receptors in man in relation to the cortical projections of the medial and lateral pain systems measured with positron emission tomography, *Neurosci. Lett.*, 126, 25, 1991b.

Jonsson, C., Pagani, M., Ingvar, M., Thurfjell, L., Kimiaei, S., Jacobsson, H., and Larsson, S.A., Resting state rCBF mapping with single-photon emission tomography and positron emission tomography: magnitude and origin of differences, *Eur. J. Nucl. Med.*, 25, 157, 1998.

Kegeles, L.S. and Mann, J.J., *In vivo* imaging of neurotransmitter systems using radiolabeled receptor ligands, *Neuropsychopharmacology*, 17, 293, 1997.

Kinahan, P.E. and Noll, D.C., A direct comparison between whole-brain PET and BOLD fMRI measurements of single-subject activation response, *Neuroimage*, 9, 430, 1999.

Kiriakopoulos, E.T., Tasker, R.R., Nicosia, S., Wood, M.L., and Mikulis, D.J., Functional magnetic resonance imaging: a potential tool for the evaluation of spinal cord stimulation: technical case report, *Neurosurgery*, 41, 501, 1997.

Kitamura, Y., Kakigi, R., Hoshiyama, M., Koyama, S., Shimojo, M., and Watanabe, S., Pain-related somatosensory evoked magnetic fields, *Electroencephalogr. Clin. Neurophysiol.*, 95, 463, 1995.

Kitamura, Y., Kakigi, R., Hoshiyama, M., Koyama, S., Watanabe, S., and Shimojo, M., Pain-related somatosensory evoked magnetic fields following lower limb stimulation, *J. Neurol. Sci.*, 145, 187, 1997.

Krauss, B.R., Grachev, I., Szeverenyi, N.M., and Apkarian, A.V., Imaging the pain of back pain, *Soc. Neurosci. Abstr.*, 25, 141, 1999.

Kwan, C.L., Crawley, A.P., Mikulis, D.J., and Davis, K.D., An fMRI study of the anterior cingulate cortex and surrounding medial wall activations evoked by noxious cutaneous heat and cold stimuli, *Pain*, 85, 359, 2000.

Lotze, M., Grodd, W., Birbaumer, N., Erb, M., Huse, E., and Flor, H., Does use of a myoelectric prosthesis prevent cortical reorganization and phantom limb pain?, *Nat. Neurosci.*, 2, 501, 1999.

Näätänen, R., Ilmoniemi, R.J., and Alho, K., Magnetoencephalography in studies of human cognitive brain function, *Trends Neurosci.*, 17, 389, 1994.

Ness, T.J., San Pedro, E.C., Richards, J.S., Kezar, L., Liu, H.G., and Mountz, J.M., A case of spinal cord injury-related pain with baseline rCBF brain SPECT imaging and beneficial response to gabapentin, *Pain*, 78, 139, 1998.

Ogawa, S., Lee, T.M., Kay, A.R., and Tank, D.W., Brain magnetic resonance imaging with contrast dependent on blood oxygenation, *Proc. Natl. Acad. Sci. USA*, 87, 9868, 1990a.

Ogawa, S., Lee, T.-M., Nayak, A.S., and Glynn, P., Oxygenation-sensitive contrast in magnetic resonance image of rodent brain at high magnetic fields, *Magn. Reson. Med.*, 14, 68, 1990b.

Ogawa, S., Tank, D.W., Menon, R., Ellermann, J.M., Kim, S.-G., Merkle, H., and Ugurbil, K., Intrinsic signal changes accompanying sensory stimulation: functional brain mapping with magnetic resonance imaging, *Proc. Natl. Acad. Sci. USA*, 89, 5951, 1992.

Orrison, W.W. Jr., Lewine, J.D., Sanders, J.A., and Hartshorne, M.F., *Functional Brain Imaging*, Mosby-Year Book, St. Louis, 1995.

Oshiro, Y., Fujita, N., Tanaka, H., Hirabuki, N., Nakamura, H., and Yoshiya, I., Functional mapping of pain-related activation with echo-planar MRI: significance of the SII-insular region, *Neuroreport*, 9, 2285, 1998.

Pagni, C.A. and Canavero, S., Functional thalamic depression in a case of reversible central pain due to a spinal intramedullary cyst. Case report, *J. Neurosurg.*, 83, 163, 1995.

Peyron, R., Garcia-Larrea, L., Deiber, M.P., Cinotti, L., Convers, P., Sindou, M., Mauguière, F., and Laurent, B., Electrical stimulation of precentral cortical area in the treatment of central pain: electrophysiological and PET study, *Pain*, 62, 275, 1995.

Peyron, R., Garcia-Larrea, L., Gr, g.M., Costes, N., Convers, P., Lavenne, F., Mauguiere, F., Michel, D., and Laurent, B., Haemodynamic brain responses to acute pain in humans: sensory and attentional networks [In Process Citation], *Brain*, 122, 1765, 1999.

Peyron, R., García-Larrea, L., Grégoire, M.C., Convers, P., Lavenne, F., Veyre, L., Froment, J.C., Mauguière, F., Michel, D., and Laurent, B., Allodynia after lateral-medullary (Wallenberg) infarct — A PET study, *Brain*, 121, 345, 1998.

Phelps, M.E. and Mazziotta, J.C., Positron emission tomography: human brain function and biochemistry, *Science*, 228, 799, 1985.

Ploghaus, A., Tracey, I., Gati, J.S., Clare, S., Menon, R.S., Matthews, P.M., and Rawlins, J.N., Dissociating pain from its anticipation in the human brain, *Science*, 284, 1979, 1999.

Ploner, T., Schmitz, F., Freund, H.J., and Schnitzler, A., Parallel activation of primary and secondary somatosensory cortices in human pain processing, *J. Neurophysiol.*, 81, 3100, 1999.

Porro, C.A., Cettolo, V., Francescato, M.P., and Baraldi, P., Temporal and intensity coding of pain in human cortex, *J. Neurophysiol.*, 80, 3312, 1998.

Prichard, J.W. and Brass, L.M., New anatomical and functional imaging methods, *Ann. Neurol.*, 32, 395, 1992.

Rainville, P., Duncan, G.H., Price, D.D., Carrier, B., and Bushnell, M.C., Pain affect encoded in human anterior cingulate but not somatosensory cortex, *Science*, 277, 968, 1997.

Sadzot, B., Price, J.C., Mayberg, H.S., Douglass, K.H., Dannals, R.F., Lever, J.R., Ravert, H.T., Wilson, A.A., Wagner, Jr., H.N., Feldman, M.A., and Frost, J.J., Quantification of human opiate receptor concentration and affinity using high and low specific activity [11C]diprenorphine and positron emission tomography, *J. Cereb. Blood Flow Metab.*, 11, 204, 1991.

San Pedro, E.C., Mountz, J.M., Mountz, J.D., Liu, H.G., Katholi, C.R., and Deutsch, G., Familial painful restless legs syndrome correlates with pain dependent variation of blood flow to the caudate, thalamus, and anterior cingulate gyrus, *J. Rheumatol.*, 25, 2270, 1998.

Scheffel, U., *In vivo* brain receptor imaging, *J. Clin. Immunoassay*, 16, 300, 1993.

Svensson, P., Minoshima, S., Beydoun, A., Morrow, T.J., and Casey, K.L., Cerebral processing of acute skin and muscle pain in humans, *J. Neurophysiol.*, 78, 450, 1997.

Talairach, J. and Tournoux, P., *Co-Planar Stereotaxic Atlas of the Human Brain*, Thieme Medical Publ., New York, 1988.

Talbot, J.D., Marrett, S., Evans, A.C., Meyer, E., Bushnell, M.C., and Duncan, G.H., Multiple representation of pain in human cerebral cortex, *Science*, 251, 1355, 1991.

Tamminga, C.A., Conley, R.R., and Wong, D.F., Human brain receptors, IV: Human in vivo receptor imaging, *Am. J. Psychiatry*, 151, 639, 1994.

Tanaka, S., Osari, S., Ozawa, M., Yamanouchi, H., Goto, Y., Matsuda, H., and Nonaka, I., Recurrent pain attacks in a 3-year-old patient with myoclonus epilepsy associated with ragged-red fibers (MERRF): a single-photon emission computed tomographic (SPECT) and electrophysiological study, *Brain Dev.*, 19, 205, 1997.

Toga, A.W. and Mazziotta, J.C., *Brain Mapping. The Methods*, Academic Press, San Diego, 1996.

Vogt, B. A., Watanabe, H., Grootoonk, S., and Jones, A. K. P., Topography of diprenorphine binding in human cingulate gyrus and adjacent cortex derived from coregistered PET and MR images. *Human Brain Mapping*, 3, 1, 1995.

Xu, X.P., Fukuyama, H., Yazawa, S., Mima, T., Hanakawa, T., Magata, Y., Kanda, M., Fujiwara, N., Shindo, K., Nagamine, T., and Shibasaki, H., Functional localization of pain perception in the human brain studied by PET, *Neuroreport*, 8, 555, 1997.

13 Methods for Imaging Human Brain Pathophysiology of Chronic Pain

A. Vania Apkarian, Igor D. Grachev,
Beth R. Krauss, and Nikolaus M. Szeverenyi

CONTENTS

ABSTRACT

Magnetic resonance (MR) imaging techniques are rapidly expanding to provide
anatomical, physiological, and biochemical information that can be used to
noninvasively study the human brain across these dimensions. In this chapter we
concentrate on the application of functional MR imaging and MR spectroscopy to
understand the brain pathophysiology of chronic pain states. We demonstrate the
novel hypotheses that can be tested by these methods in specific patient populations,
and argue that the approaches developed are generic enough to be applicable to a
large list of clinical pain conditions. We emphasize functional magnetic resonance
imaging (fMRI) adapted specifically to understand the neuronal activity of the brain
that underlies clinical pain states, and studies of brain biochemistry of chronic pain
patients in comparison to normal subjects using *in vivo* hydrogen magnetic resonance
spectroscopy (^{1}H-MRS). The results we have obtained by both approaches indicate
that the brain pathophysiology of chronic pain is different from normal subjects,
and both methods show abnormalities in the same brain regions: patients with chronic
sympathetically mediated pain and patients with chronic back pain show that the
prefrontal cortex is critically involved in such conditions.

    We also briefly describe new MR techniques, including morphological and
diffusion-limited MR imaging which provides information regarding axonal pathways. Both methods remain to be used in pain research. We provide some speculative
pointers regarding how these methods may be used to further advance our understanding of brain mechanisms of clinical pain states.

13.1 INTRODUCTION

Pain can be broadly subdivided into two categories: acute and chronic. Acute pain is
the pain experienced in everyday behavior; it is experienced with environmental
stimuli that cause injury or have the potential for causing injury. Temporally it is
closely related to the external stimulus and persists only for the duration of healing.
It serves to protect the organism from injury and its physiological mechanisms,
anatomical pathways as well as peripheral and central receptors, and neurotransmitters
mediating and modulating such pains are becoming well understood. Drugs acting
on peripheral or central receptors, non-steroidal antiinflammatory drugs (NSAIDs),
and morphine derivatives, properly treat acute pain. Noninvasive brain-imaging methods have been used by a number of labs to study acute and/or experimental pain. The
application of functional MR imaging to study brain networks underlying acute pain
states is reviewed in a separate chapter in this book. Here we only want to add that
the data accumulated from many labs over the last 10 years provide a relatively
consistent pattern of brain activity underlying acute pain perception.

In contrast, pain that persists past the normal time of healing is chronic pain. This time period may be as short as 1 month, but more conservatively, pains persisting for longer than 6 months after the end of healing are classified as chronic. Chronic pain results in an enormous medical cost. The National Chronic Pain Outreach Association estimates that chronic pain affects one in every three Americans and costs the U.S. economy $90 billion every year. It should be emphasized, however, that the precise estimate of the incidence of chronic pains remains unknown, and it may be even higher than the numbers quoted here. Only a small proportion of people precipitate into a chronic pain state (10 to 20%, the estimate varies by type). Currently there is no scientific knowledge as to the differences between those individuals who progress into a chronic pain state vs. those who do not. The primary focus of the research in this laboratory over the last 5 years has been the development of specific methodologies that can be used to study the brain pathophysiology and chemistry of chronic pain patients. Our bias is that chronic pain states are abnormal brain states that have not been systematically studied, primarily because of a lack of adequate technologies.

13.2 OVERVIEW OF FUNCTIONAL BRAIN IMAGING USING fMRI

Functional magnetic resonance imaging is a completely noninvasive method with which scientists can indirectly monitor brain activity in awake humans while the participants are subjected to some stimulus or are instructed to perform some task. Standard MRI is used to examine the anatomy of the brain by documenting differences in magnetization of the tissue, where gray to white matter differences in fat-to-water ratios are used to examine anatomy. In contrast, functional MRI detects small fluctuations in tissue magnetization properties secondary to neuronal activity. Because fMRI is noninvasive and, as far as is known, totally harmless, there is no limit to the number of studies performed by fMRI, enabling the study of individual subjects and/or patients as many times as necessary. Changes in neuronal activity are accompanied by changes in cerebral blood flow, blood volume, level of oxygenation, and tissue metabolic rate. Transient changes in these parameters produce differences in the balance between paramagnetic deoxyhemoglobin and diamagnetic oxyhemoglobin in red cells. Changes in this balance result in transient local magnetization changes, monitored directly by fMRI (Kwong et al., 1992). This method has been used to visualize brain activations in a very large set of stimuli and task conditions, including for thermal painful stimuli (e.g., Davis et al., 1995, 1998; Gelnar et al., 1999; see Chapter 8, this volume). In such studies, the participant is subjected to a stimulus-control condition, repeated many times. By assuming that the stimulus and control situations are reproducible and are different brain states, one can determine the brain regions where local magnetization is different between the two states and infer that these regions participate in the perceptual change accompanying stimulus-control sequences. This type of analysis is also used in Positron Emission Tomography (PET) where radionucleotide-tagged metabolic markers are injected in the participants and the differential distribution of these radioactive markers between the two states indicates the brain regions that distinguish between the states. Such comparisons between two stimulus conditions are called state-dependent design and the analysis of such fMRI data called state-dependent analysis.

13.2.1 WHAT DOES fMRI MEASURE?

The advantages of the fMRI technique over other brain imaging modalities should be quite evident. It is a non-invasive technique, allowing repeated measurement in the same individuals. It has a high spatial resolution and relatively high temporal resolution. The equipment required to perform such measurements is quite widely available with new higher field MR scanners becoming fMRI capable. The methodology, however, has sometimes been criticized because the exact metabolic processes underlying the signal changes still remain unknown and cannot be interpreted in quantitative units, e.g., cc of blood flow. Some progress is being made in this area. A number of studies have recently been able to identify and quantify the specific cellular molecular mechanisms of neuronal activity that are coupled to energy metabolism we observe in fMRI measures (Magistretti et al., 1999; Van Zijl et al., 1998; Kennan et al., 1998). The study by Zijl et al. (1998) formulates a fundamental theory for MRI signal changes in perfused tissue and then validates the model *in vivo* in cat brain. The theory correctly predicts the magnitude of fMRI (blood oxygen level dependent, BOLD) signal intensity changes on brain activation, thereby providing a sound physiological basis for these studies. Kennan et al. (1998) measured the contribution of blood volume changes in relation to blood oxygenation-induced susceptibility changes in hemoglobin. These types of measurements begin to provide a quantitative measure of blood-volume response to stimulation.

13.2.2 STATISTICAL AND TECHNICAL ISSUES

The details of fMRI data analysis approaches are outlined in Chapter 12. Here we want to highlight some specific statistical issues. In subsequent sections we emphasize the methodologies we have advanced to study chronic pain states.

We have developed an extensive set of tools to help generate fMRI data as well as analyze the resulting images. We rely on computer-controlled stimulators, which are synchronized with MR image collection. In this manner we can accurately parcel images as belonging to a stimulus or a control state. One of the technical problems we have overcome was determining what kind of electronic circuitry could be attached to the patient without producing artifacts on the images, since small magnetic field disturbances cause large artifacts in MR images. Using RF filters and our own penetration panels in the MR room, we found that we could successfully monitor finger position using low-impedance potentiometer-based devices. We have also successfully measured and controlled skin surface temperature in small areas, applied direct electrical stimulation to the median nerve, and presented auditory stimuli with an MR compatible music system. The rating and logging of the patient's responses to the stimulus are an important technical achievement. It enables a correlation analysis of the fMRI data with the patient's perception of the pain.

We use a Pentium computer on a mobile cart that is wheeled into the operator's area of the scanner. The computer runs LabView software and has a National Instruments LabPC+ interface card to communicate with the outside world. Amplifiers and associated buffer circuitry are also integrated into this mobile cart. It has a TTL interface to the scanner to identify the time of image data collection. After a functional MR scan, raw data are transmitted over a LAN computer network to any of a number of computers, either PCs running the Linux operating system (Red Hat) or SUN SPARC workstations. The images are reconstructed and analyzed using either t-test or correlation analysis. We have written in-house software in C with a TCl/Tk interface to streamline the data analysis. Our software is often accessed from a PC in the scanner operating area using Exceed software. This software maps the X-display of a Unix workstation to a PC and allows us to see the result of our fMRI scan in as little as 10 min after data collection. Alternatively, we use MEDx software (Sensor Systems) for transforming activation maps into a standard brain atlas coordinate system or when we want to do multi-subject population averaging (see Krauss and Apkarian, 1998).

If the fMRI signal was strong enough, comparing two images of the brain in two distinct states would suffice to delineate the brain regions engaged in differentiating between the states. Unfortunately, the fMRI signal is very small (0.5 to 3%, a function of many variables, including scanner strength, voxel size, relaxation time, magnet homogeneity, etc.) and is embedded in a large amount of noise. This necessitates signal averaging. In addition to machine noise, fMRI signal is also contaminated with physiological noise: heart rate and breathing, as well as low frequency fluctuations whose source remains uncertain. Signal averaging is also needed from a statistical point of view — if brain activity changes are to be asserted within a confidence range, then they need to be repeated enough to establish the consistency of the observations. Depending on the voxel size used and the number of slices examined, a given fMRI study routinely examines anywhere from 50,000 to 100,000 brain voxels. This complicates the statistical analysis with the need for adequate corrections for repeated measures. A number of labs have proposed different approaches to overcome these complications. We should point out that there is no clear consensus regarding the best approach for fMRI signal analysis, and it remains a subject that is hotly debated and changing fast (see Friston et al., 2000; Everitt and Bullmore, 1999; Brammer et al., 1997; Bandettini and Cox, 2000; Kiebel et al., 1999). It is important to keep in mind that brain activations determined by fMRI are always a result of averaging and filtering. Different labs and software packages use variously sized spatial and temporal filters. This clearly complicates direct comparisons of results between groups. Despite the statistical differences, when two states are compared, by generating a t-map, where the average difference between them is compared relative to the variability of the signal, or by a correlation map, or by using a non-parametric comparison between MR signal intensity fluctuations, such experimental designs are very similar to those used in PET studies. This approach, state-related design and analysis, assumes that the difference between the two cognitive/perceptual states is unique and consistently sustained for the duration of each state. Therefore, brain regions that statistically significantly differ between the two conditions are due to the unique change in the cognitive state.

A number of labs have now determined the temporal properties of brain hemo-dynamics (Bandettini and Cox, 2000; Cohen et al., 1997; Toni et al., 1999). These studies indicate that the brain hemodynamic response has a characteristic time course and can be viewed as a typical response to a transient input, which in linear systems theory is equivalent to the impulse response function. This characterization in turn has led to new experimental designs called event-related fMRI design and analysis in which very brief stimuli are presented and the related hemodynamic responses searched for in the MR images. A main advantage of the event-related design is the ability to present short stimuli with randomly varying intervals between events, which reduces predictability and expectation by the subjects. This approach is just beginning to be used in pain research.

13.3 APPLICATION OF fMRI TO ACUTE/EXPERIMENTAL PAIN TASKS

Both fMRI and PET have been used to identify brain regions involved in acute, or experimental, pain states. The results generally indicate that a number of cortical regions participate in coding acute pain. These usually include multiple somatosensory regions, some motor regions, insula, cingulate, and prefrontal cortex (see Fig. 13.4). The specifics of these patterns depend on stimulus intensity, duration, size, as well as attention and anticipation (see Chapter 12 this volume; Apkarian, 1995; Apkarian et al., 2000; Bushnell et al., 1999; Treede et al., 1999). It should be noted that there are important differences between the labs regarding the brain areas activated in these state-related pain tasks; at least some of these must be a consequence of assuming that the perceived pain is constant for the stimulus duration, as well as the amounts of contamination due to factors closely related to a painful state such as anxiety, attentional shifts, and expectational differences.

13.4 APPLICATION OF fMRI TO CLINICAL PAIN STATES

In this section two examples of fMRI studies are presented in two different chronic-pain patient populations. In the first case we use a state-related design in chronic sympathetically mediated pain patients (SMP of CRPS type I), in which fMRI scans are done while a thermal painful stimulus is applied to the hand that suffers from chronic pain. The task is repeated after a peripheral sympathetic block decreases the chronic pain. Differences in the brain activation network before and after the block indicate the brain regions that are involved in the chronic pain. The second patient group is chronic back-pain sufferers. In this group we use a novel technique where the ongoing chronic pain is rated during fMRI scans. These ratings are used to directly identify the brain regions mediating the chronic back pain.

13.4.1 State-Related Example — Studies in Chronic Sympathetically Mediated Pain Patients

The experimental setup is outlined in Fig. 13.1. Normal subjects or patients are scanned when they move their hands from a warm surface to a painful hot surface,

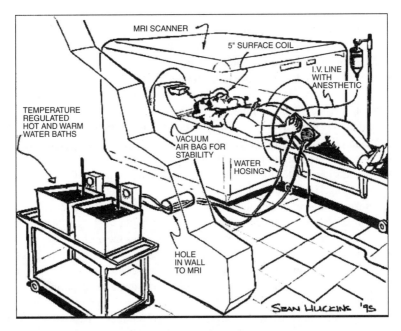

FIGURE 13.1 Experimental setup for studying brain network underlying chronic sympathetically mediated pain. The subject's head is immobilized in the fMRI scanner. The axillary space of the hand associated with chronic pain is cannulated. During an fMRI scan the patient or volunteer moves the hand between two surfaces heated by circulating water connected to water baths located in the adjoining room. After completion of the initial fMRI scan, a local anesthetic is injected through the cannula to block sympathetic outflow and reduce the chronic pain. A few minutes later the fMRI painful thermal task is repeated.

repeated 6 times in a given fMRI scan where the switches are done by a verbal cue. Following the completion of the first set of fMRI scans, Lidocaine (local anesthetic) is injected through a cannula implanted into the axillary space of the arm in chronic pain. Within 20 min after this injection the chronic pain subsides significantly, and the fMRI scans are repeated for the same thermal painful task.

Four different sets of studies were conducted in such patients (Apkarian et al., 2001). The early studies targeted the parietal cortex; however, the results showed that the responses in this region of the cortex, before and after sympathetic blockade, did not correlate with the large decreases in both the chronic pain and the stimulus-induced pain. In a second group of SMP patients we examined activity in the frontal cortex, contralateral to the chronic SMP suffering hand. These studies did show cortical activity changes correlated with the changes in pain following sympathetic blocks. Therefore, in a third group of SMP subjects both anterior and parietal regions were studied. The three sets of studies were limited in the amount of brain we could examine, necessitating separate functional scans to examine cortical activity in the contralateral frontal and parietal cortices. More recently, we were able to scan the whole brain. Therefore, in a fourth group of patients we have examined whole-brain activity changes prior to and immediately after sympathetic blocks. We have also studied cortical activity in response to thermal stimuli in three groups of normal subjects. In the first group we focused on parietal cortical responses. In the second group frontal cortical responses were studied. The third group of normal subjects also underwent sympathetic blocks identical to those used in the SMP patients, and their responses were studied in both the parietal and frontal cortices.

The mean pain ratings for thermal stimuli were similar between SMP patients and control subjects. Also, sympathetic blocks highly significantly decreased the SMP spontaneous pain, as well as the stimulus ratings in patients and in the volunteers. Pre-block, frontal cortical activity in the SMP subjects in response to the thermal stimulus was significantly higher than the frontal cortical activity in the normal subjects (see Color Figure 2),* and this activity significantly decreased postblock. In contrast, frontal cortical activity in normal subjects was not changed postblock. These results indicate that the prefrontal cortex is hyperactive to thermal painful stimuli in SMP patients, and that a sympathetic block that transiently decreases the magnitude of the chronic pain also reverses the prefrontal hyperactivity. It is important to note that the sympathetic blocks were performed at a level that blocked sympathetic outflow but minimally affected sensory fibers. We do not wish to dwell on the details of the results but to simply illustrate the experimental design. The study shows that a painful stimulus in chronic pain patients activates brain regions very differently from that seen in normal subjects. The weakness of the design, however, is that it is not clear whether the resultant differences in activation are a simple reflection of the brain network underlying chronic pain or rather a reflection of some complex interaction between the chronic network and the thermal pain perception network. This complexity has prompted us to seek simpler and more direct methods for studying the brain network underlying chronic pain states.

13.4.2 PAIN CONSCIOUSNESS AND SUBJECTIVITY — ITS TEMPORAL CHARACTERISTICS

The exact definition of pain consciousness is a matter of debate and intense interest in philosophical circles. Here pain consciousness is defined functionally by the subjective report of pain and is used as a parameter that can be distinguished from the painful stimulus. We take the simple position that if a subject is in pain, then he/she can indicate how much and for how long the pain persists. Documenting this temporal variation of the experience of pain gives us a specific tool with which we can examine brain networks involved in pain consciousness.

Unlike touch, vision, or audition, neural systems underlying pain perception are slow. As a result, pain perception can often be dissociated from the stimulus. There is a complex relationship between the intensity of pain experienced and its temporal properties. Hardy et al. (1968) showed that when a thermal painful stimulus is mild the reported pain may be delayed by seconds from the start of the stimulus and be transient. As the thermal painful stimulus intensity is increased, its perception becomes more constant and proceeds to outlast the stimulus; however, this temporal pattern seems unique for different individuals. Such temporal dissociations between stimulus

---

* Color figures follow page 206.

and perception may become far more dramatic in pain abnormalities. This temporal variability suggests an alternative explanation for positron emission tomography (PET) studies (Derbyshire et al., 1997) designed to identify brain regions involved in coding pain intensity. When different intensity painful stimuli are applied for 60 to 90 s and pain is evaluated by a single numeral over the whole duration, the brain regions correlated to the pain ratings may simply reflect the extent of their temporal involvement with the stimulus, rather than their involvement in coding pain intensity.

13.4.3 DIFFERENTIATING THE BRAIN PAIN NETWORK ALONG STIMULUS AND PERCEPTION DIMENSIONS

The need for studying the subjective experience of pain was eloquently stated years ago by Donald D. Price. In the first chapter of his book (1988), he states:

> The definition of pain as "an unpleasant sensory and emotional *experience* associated with actual or potential tissue damage…" (Merskey 1986) leaves us in a very interesting philosophical position with regard to the study of pain. If pain is defined subjectively as an experience, then the scientific study of pain ultimately has to study and even measure that experience. [Underline is our emphasis.]

This is exactly what we are attempting to achieve. In an early fMRI study of cortical regions activated by a thermal painful stimulus, we simply compared the brain areas significantly activated during the painful state as compared to the nonpainful warm control state (Gelnar et al., 1999). The results of the study showed that multiple brain regions are activated in an experimental painful task.

Subsequently, we analyzed the fMRI thermal painful stimulus results in relation to the time course of the stimulus and pain perception (Apkarian et al., 1999). The time course of the pain perception was determined in another group of subjects who were required to continuously rate the intensity of the subjectively perceived pain when they were presented with the same thermal painful stimulus as used in the fMRI study. The average ratings indicated that although the painful stimulus was constant, the perceived pain monotonically increased throughout the stimulus duration. By taking advantage of this temporal dissociation between stimulus characteristics and pain perception, we interrogated the brain activity with both time curves and identified cortical regions more closely related to either the stimulus or the perception (Apkarian et al., 1999). This analysis showed a gradual transition of information processing anteroposteriorly in the parietal cortex. Within this region, activity in the anterior areas more closely reflected thermal stimulus parameters, while activity more posterior was better related to the temporal properties of pain perception. The study indicates that pain subjectivity can be used to identify the functional differences of the brain network involved in pain perception.

13.4.4 USING PAIN SUBJECTIVITY TO EXAMINE THE BRAIN NETWORK UNDERLYING CHRONIC BACK PAIN

Up to 80% of all adults will eventually experience back pain, and it is a leading reason for physician office visits, hospitalization and surgery, and work disability.

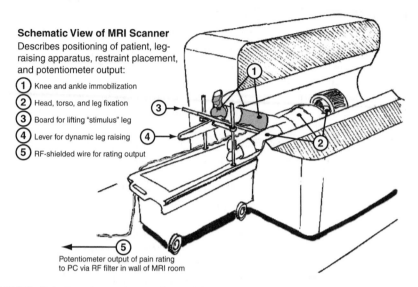

Schematic View of MRI Scanner
Describes positioning of patient, leg-raising apparatus, restraint placement, and potentiometer output:

1 Knee and ankle immobilization
2 Head, torso, and leg fixation
3 Board for lifting "stimulus" leg
4 Lever for dynamic leg raising
5 RF-shielded wire for rating output

Potentiometer output of pain rating to PC via RF filter in wall of MRI room

FIGURE 13.2 Experimental setup for studying the brain network underlying chronic back pain. The subject's head and body are strapped to the gantry to reduce head motion artifacts. The subject's thumb and forefinger are instrumented to enable monitoring their positions during fMRI scans. The finger span during fMRI indicates the magnitude of pain experienced. The time curves generated from these finger movements are correlated with the brain images to identify activation related to the pain perception.

Clearly, back pain is one of society's most significant nonlethal medical conditions. And yet the prevalence of back pain is perhaps matched in degree only by the lingering mystery accompanying it.

In the United States back problems are estimated to constitute 25% of all disabling occupational injuries (Cavanaugh and Weinstein, 1994), and it is the 5th most common reason for visits to the clinic (Deyo, 1998); for example, 40% of all visits to neurosurgeons and orthopedists is for back pain complaints (Cavanaugh and Weinstein, 1994). There are no brain imaging studies of back pain most likely because, until the advent of our approach, there were no obvious ways of manipulating the pain. Deyo (1998) goes on to state that about 85% of the back pain population does not have a definitive diagnosis. For this reason we examine low-back-pain patients where the diagnosis is more clearly defined, i.e., low back pain with radicular involvement. Moreover, this population is the one most prone to be treated surgically. Thus, this group affords the opportunity of relating brain pathophysiology to surgical outcomes. The clinical diagnosis for radicular back pain includes straight leg raising. Our functional studies use this same procedure in the scanner to examine the associated brain activity.

Figure 13.2 illustrates the experimental design that we have recently implemented for fMRI studies of chronic back pain patients. The subjects continuously rate their pain intensity through a logging device (potentiometer attached to the fingers of the hand opposite to the manipulated leg, connected to a computer running LabView); the finger span between thumb and pointer finger indicates intensity of pain. Subjects

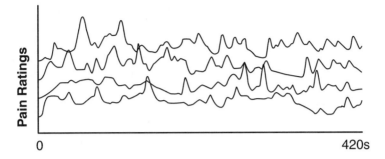

FIGURE 13.3 Time variations of pain ratings in a chronic back pain patient. The four tracings were collected during four consecutive fMRI scans. The y-axis corresponds to a visual analog scale; it is arbitrarily shifted to show the four curves. The subject rates the intensity of the pain by using a device connected to the thumb and forefinger; the larger the finger span, the higher the intensity of pain.

are strapped to the scanner gantry at multiple locations to minimize head movement artifacts, the knee is locked in an extended position, and in normal subjects the heel is dorsiflexed. The leg is moved up and down between two positions, one of which may cause pain. In normal subjects this procedure causes extreme stretching of multiple tendons at the heel and the knee, resulting in a deep pain involving joints, muscles, and tendons. In back pain patients, the leg raising exacerbates the sciatic and/or back pain. We have discovered that when back pain patients just lie on the couch of the scanner they often complain of significant exacerbation of their pain. Figure 13.3 plots spontaneous pain ratings in a chronic back pain patient.

The pain ratings are low-pass filtered by convolving the raw data with the brain hemodynamic function. The resultant time functions are then correlated with the fMRI scan images to identify the brain areas involved in these perceptions. The associated cortical activations in such back pain patients indicates that the prefrontal cortex is the primary brain region involved in the ongoing spontaneous pain. This pattern shifts to include the supplementary somatosensory region when the back pain is exacerbated by leg raising (see Color Figure 2). The design for this group of patients is an extension of the method we used to differentiate between stimulus coding and perception of pain. The briefly outlined results indicate similarity to the results in SMP patients and differences from the pain network identified in normal subjects. In both chronic pain patient groups the frontal cortex activity seems to differentiate them from the control subjects' pain network; however, it should be noted that the details of the prefrontal activations seems different between the two chronic pain states. Importantly, the approach is generic enough that it should be easily adaptable to study a large variety of clinical pain states.

13.5 OVERVIEW OF MEASURING BRAIN BIOCHEMISTRY NONINVASIVELY

Nuclear magnetic resonance spectroscopy or MR spectroscopy (MRS) is a method used in chemistry and physics laboratories for the analysis of molecular interactions and for the identification of chemical compounds. Recently this approach has been adapted for *in vivo* spectroscopic analysis of the brain, which has direct implications on the noninvasive clinical assessment of brain biochemistry. The key to this method is to localize MR signals to a specific volume, an approach commonly used in anatomic MRI. Since in all biological tissues, including the brain, water is the predominant chemical, observing weak signals from metabolites with concentrations thousands of times smaller than water requires methods for suppressing the water signal. Advances in MR technology in automating voxel positioning and suppressing the water signal have made *in vivo* MRS a relatively simple approach that can be used in studying brain chemistry (see Salibi and Brown, 1998). MRS can be viewed as a noninvasive method analogous to performing brain biopsies. MRS can be used to study the spectra of compounds in the brain that contain odd-numbered nuclei, such as ^{1}H, ^{17}Li, ^{13}C, ^{17}O, ^{19}F, and ^{31}P. This methodology has the potential for revolutionizing our understanding of human pathological brain states. In this sense it has been adapted and most extensively used in neurology and psychobiology research (e.g., Salibi and Brown, 1998). Spectra obtained with ^{31}P enable the measurement of high-energy phosphates such as adenosine tri-, di-, and monophosphates and creatinine phosphate. ^{31}P measurements have been obtained in patients suffering from stroke, various types of brain tumors, multiple sclerosis, Alzheimer's disease, and epilepsy. Proton spectra (^{1}H MRS) enable measurement of concentrations of a number of metabolites and excitatory and inhibitory neurotransmitters. The latter method also has been used to examine the brain biochemistry of various patient populations.

There are two major approaches in proton spectra: (1) Single-voxel method PRESS (point-resolved spectroscopy) or STEAM (simulated echo-acquisition mode). The single-voxel method provides multichemical spectra in a single selected region. (2) Multi-voxel method is used to obtain spectra in multiple regions simultaneously (CSI). Using this method, imaging can be done for each or multiple chemicals in the spectrum obtained from multiple voxels. While the multi-voxel method is advantageous for overall analysis of the distribution of multiple chemicals, it suffers from contamination from non-brain tissue. The single-voxel method is a slower method of collecting data, but provides better regional selectivity and less signal from surrounding tissues. The latter method is especially advantageous in cases where chemical changes are expected to be localized to specific brain regions.

^{1}H MRS can be used for measuring the chemical concentrations (i.e., peaks or intensities) of up to 9 different compounds in the living brain: N-acetyl aspartate (NAA), choline (Cho), glutamate (Glu), glutamine (Gln), γ-aminobutyric acid (GABA) [Glu + Gln + GABA is very often measured as Glx due to overlap between the peaks of these compounds], myo– and scyllo–inositol complex (Ins), glucose (Glc), lactate (Lac), and creatine (Cr). These measurements are performed as relative ratios among peaks (most commonly relative to Cr) or as absolute concentrations. Absolute concentration measurement requires an external or internal standard, which remains difficult to perform routinely especially because the external standard method is sensitive to magnetic field inhomogeneities due to separation of the standard and the measured volume (Salibi and Brown, 1998).

The ^{1}H-MRS spectra are usually characterized by three major peaks: NAA at 2.02 parts per million (ppm), Cr at 3.0 ppm, and Cho at 3.2 ppm. NAA is the dominant peak in normal adult brain spectra. The Cr spectrum is a combination of creatine and phosphocreatine (Michaelis et al., 1993). The proton Cho signal is a combination of Cho and Cho-containing compounds: Cho plasmogen, glycerophos-phorylcholine, phosphorylcholine, cytedine-diphosphate-choline, acetylcholine, and phosphatidylcholine (Michaelis et al., 1993).

A growing literature shows depletion of brain NAA in several neurodegenerative diseases (reviewed in Salibi and Brown, 1998), suggesting that NAA is a neuronal marker (this chemical is present only in living, mature neurons and not glia). Subsequent breakdown of NAA leads to aspartate which is an excitatory amino acid neurotransmitter. Recent reports suggest that:

NAA is also required for brain lipid biosynthesis and fatty acid α-hydroxy-lation (growth and development of the brain);

NAA is a precursor of a putative neurotransmitter N-acetyl-aspartyl-glutamate;

NAA biosynthesis may be related to neuronal protein synthesis;

NAA may be a storage form of aspartate; and

NAA is also a marker for mitochondrial dysfunction (reviewed in Faull et al., 1999).

Cr resonance is considered to be more stable than the NAA peak and is commonly used as a reference. However, this peak is abnormal with hypoxia, trauma, stroke, and tumor (Salibi and Brown, 1998). The Cr level is involved in energy metabolism in the brain (ATP production).

Choline is a precursor of phosphatidylcholine, as well as the metabolite of acetylcholine. Cho-containing compounds are involved in membrane phospholipid metabolism. This peak was found to be increased in fast growing tumors such as glioma, suggesting its role in increased cell proliferation and membrane turnover.

The other observable chemicals are Glu, 2.35 ppm; Gln, 2.15 ppm; GABA, 2.25; Ins, 3.60 ppm; Glc, 3.43 ppm; and Lac, 1.3 ppm (see Fig. 13.4). These smaller peaks are contaminated by signals from other chemicals and proteins, although the prominent signals (chemical shifts) are localized as indicated (Salibi and Brown, 1998). Glu and GABA are the major excitatory and inhibitory neurotransmitters of the brain which are released by approximately 90% of the cortical neurons and synapses (Shulman and Rothman, 1998; Magistretti et al., 1999). Glucose utilization is commonly considered a marker of general metabolic rate, and involved in the metabolic pathway of the tricarboxylic acid cycle of neurons and astrocytes (Gruetter et al., 1996; Shulman and Rothman, 1998; Magistretti et al., 1999). Within the brain, glucose is processed glycolytically, resulting in the release of lactate as an energy substrate for neurons (Magistretti et al., 1999). The energy demands of glutamatergic neurons account for 80 to 90% of total cortical glucose usage (Sibson et al., 1998). Functions of Ins are less clear, but it is considered to be a second messenger that liberates Ca^{2+} from the endoplasmic reticulum via intracellular receptors, linking to Ca^{2+} homeostasis, protein phosphorylation, and membrane trafficking.

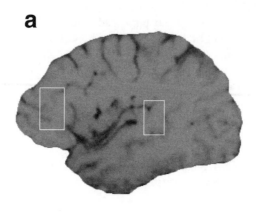

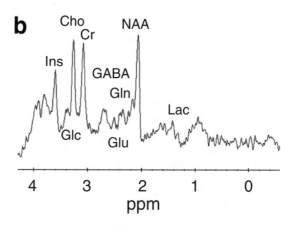

FIGURE 13.4 Localization of MR spectroscopy and the corresponding spectrum. (a) The voxel located in the prefrontal cortex is shown. (b) The spectrum obtained from this region in one subject is shown. The chemicals located at different shifts are indicated. (Adapted from Grachev et al., *Pain*, 89, 7, 2000. With permission.)

13.5.1 BRAIN REGIONAL DIFFERENCES IN BIOCHEMISTRY BETWEEN CHRONIC BACK PAIN PATIENTS AND NORMAL SUBJECTS

We have used proton spectroscopy to study the brain chemistry in chronic back pain patients and compare them to age- and sex-matched normal subjects. Six separate single-voxel measurements were done in 6 different left-brain regions (Grachev et al., 2000; see also, Grachev and Apkarian, 2000). The main hypothesis tested in this study was that in chronic pain patients brain areas showing hyperactivity in fMRI studies should also show abnormal brain chemistry. This was tested across three brain regions: thalamus, cingulate cortex, and dorsolateral prefrontal cortex (DLFPC), where we quantified the concentrations of N-acetyl aspartate (NAA), choline (Cho), glutamate (Glu), glutamine (Gln), γ-aminobutyric acid (GABA), myo– and scyllo–inositol complex (Ins), glucose (Glc), and lactate (Lac) relative to the concentration for creatine/phosphocreatine complex (Cr), which is commonly used as an internal standard. All chronic back pain subjects underwent clinical evaluation and perceptual measures of pain and anxiety.

13.5.1.1   Localized *in Vivo* 3D Single-Voxel ¹H-MRS Brain Examination

During MRS imaging the subject is placed on the scanner bed, and the whole-head gradient coil is positioned over the head. The subject's head is immobilized using a vacuum beanbag. Automated global shimming is performed to optimize the magnetic field homogeneity over the entire brain volume, as well as for each specific regional volume.

13.5.1.2   Acquisition of Magnetic Resonance Images for Localizing MRS Volumes

All MRS experiments are performed on a 1.5 Tesla General Electric clinical imaging instrument. High-resolution sagittal and axial views were used for the selection of volumes of interest. T1-weighted multislice spin echo scout images (TR = 500 ms; TE = 12 ms; 2NEX; 256×256 matrix; FOV = 24×24 cm) of the entire brain were obtained with 6.0-mm slice thickness and a 0.5-mm gap between slices, imaging 20 slice locations.

13.5.1.3   Selection of Volumes of Interest

Localized 3D single-voxel ¹H MRS is then performed in sagittal (cingulate and DLPFC) and in axial (thalamus, insula, orbitofrontal, and sensorimotor cortex) locations in the left hemisphere of right-handed normal volunteers and back pain patients. Positioning of each 8-cm³ voxel was performed by an experienced neuroanatomist and adjusted to the individual brain's sulcal topography. We used 8-cm³ voxels for each analyzed volume, although the shape of these voxels was different for each brain region. Fig. 13.4 shows the location of an example volume.

13.5.1.4   MRS Data Collection and Processing

Proton localized spectra were collected using STEAM sequence (probe-s PSD, TR = 1500 ms, TE = 30 ms). All spectra were transformed into a standardized scale using the Scion Image analysis package (1998, web site http://www.scioncorp.com). Proton spectra were analyzed by measuring heights at specified peaks, with the investigator blinded to both the location and the subject. The concentrations of biochemicals were measured relative to an internal standard peak for creatine/phosphocreatine complex (Cr). The peak heights ratio method is currently used in most clinical ¹H MRS because it is simple and requires no technical expertise or software besides that supplied with the imager. This approach is precise and accurate if the peak height is directly proportional to the peak area. Fig. 13.4 shows a typical proton MRS spectrum of the normal human brain obtained from DLPFC.

13.5.1.5 Perceptual Measures of Pain and Anxiety

All chronic back pain subjects underwent perceptual measures of pain and anxiety minutes before brain imaging. These tests included assessments of pain (short form of the McGill Pain Questionnaire; Melzack, 1987) and anxiety levels (the State-Trait Anxiety Inventory; Spielberger et al., 1983).

13.5.1.6 Statistical Analyses

Multi-way analysis of variance (ANOVA) was used to differentiate between back pain patients and volunteers, across brain regions and chemicals. The outcome variable was chemical concentration relative to Cr peak. Post-hoc comparisons identified specific regional chemical differences. Chemical network differences between patients and volunteers were examined using correlation analysis. Pearson's correlation, calculated for pairs of regional chemicals for each subject grouping, was used as the outcome variable in multi-way ANOVA, where the results are regarded as descriptive metrics. Univariate comparisons were then employed to test specific hypotheses derived from this analysis. In the back pain patients, the relationship between pain perception and brain regional chemistry was also analyzed using multi-way ANOVA.

Our results (Grachev et al., 2000) show that chronic back pain alters the human brain chemistry. Reductions of N-acetyl aspartate and glucose were demonstrated in the dorsolateral prefrontal cortex. Cingulate, sensorimotor, and other brain regions showed no chemical concentration differences. These findings provide direct evidence of abnormal brain chemistry in chronic back pain. Decreases in N-acetyl aspartate are documented in various conditions involving neuronal cell damage and loss, including stroke, multiple sclerosis, Alzheimer's disease, epilepsy, and several neurodegenerative disorders (Salibi and Brown, 1998), suggesting that our results provide evidence for a link between chronic pain and neuronal loss and degeneration, specifically in the prefrontal cortex.

13.5.2 BRAIN BIOCHEMISTRY REFLECTS SUBJECTIVE PROPERTIES OF CHRONIC PAIN

In chronic back pain, the interrelationship between chemicals within and across brain regions was abnormal, and there was a specific relationship between regional chemicals and perceptual measures of pain and anxiety.

To ascertain whether chemical changes reflect part of a larger-scale pattern for neurochemical organization, correlation analysis was used to determine the putative relationships between chemicals within and across brain regions. These chemical interrelationships, which we identify as chemical connectivity, were assessed for normal subjects and chronic back pain patients. If the observed chemical changes in the prefrontal cortex in chronic back pain are a reflection of chemical connectivity with other brain regions, then the pattern of chemical interrelationships within and across brain regions should be different between the two groups. Correlation analysis for 8 metabolites for 3 brain regions detected significantly high positive and negative correlations across 24 dimensions. This analysis revealed regional brain segregation,

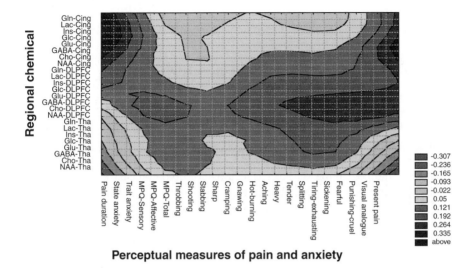

Perceptual measures of pain and anxiety

FIGURE 13.5 Contour plot of correlations between relative chemical concentrations in three brain regions and behavioral measurements of chronic back pain. The brain regions included were cingulate, Cing; and dorsolateral prefrontal, DLPFC; and cortices and thalamus, Tha. Eight chemicals (Gln, Lac, Ins, Glc, Glu, GABA, Cho, and NAA) are shown for each brain region. Twenty-three behavioral descriptors for pain and anxiety are shown on the x-axis. Correlation strengths for each brain regional chemical across the behavioral measure are shown as indicated on the grayscale bar. (Adapted from Grachev et al. *Pain*, 89, 7, 2000.)

across all chemical connectivity for the two groups of subjects. For both groups, the dominant positive correlations occur within brain regions and negative correlations are seen across brain regions; however, the pattern of chemical connectivity is different in chronic back pain subjects as compared to the normal subjects. In chronic back pain the positive chemical connectivity is weaker in the DLPFC, and the negative chemical connectivity is stronger between the thalamus and cingulate. We consider these changes in chemical connectivity an abnormal pattern.

If the observed abnormal chemical connectivity pattern in chronic back pain is specific for this diagnostic group, then the regional chemical variations should be related to the perceptual measures of pain and anxiety. The correlations of these regional chemicals with 23 perceptual descriptors are shown in Fig. 13.5. We used a multivariate approach to test whether this chemical–perceptual network in chronic back pain is different across brain regions and pain/anxiety perception. The mean strength of correlations between DLPFC and perceptual measures of pain and anxiety is 9 times stronger than the strength of correlations for cingulate, and 6 times stronger than for thalamus. Post-hoc comparisons revealed specific functional relationships between each perceptual descriptor and regional metabolites. The state and anxiety trait was found to be related to prefrontal and cingulate chemicals; the duration of pain was related to cingulate chemicals; and the sensory, affective, and total components of the McGill questionnaire were related to prefrontal metabolites. These findings indicate that the abnormal pattern of chemical connectivity in chronic back pain, which we described above, is related to pain perception. Therefore, this study indicates not only that brain chemistry is different in chronic pain patients but also that the abnormal changes reflect specific perceptual parameters.

13.6 CONCLUSIONS AND NEW DIRECTIONS

The methodologies presented above illustrate two fMRI examples for studying the brain networks of chronic pain states, and the application of MRS to comparing brain biochemistry between control subjects and chronic pain patients. The results, although presented rather briefly, repeatedly indicate the same brain structures as important in chronic pain. All three studies point to the prefrontal cortex as either involved in the perception of chronic pain and/or compromised as a result of its involvement in this perception. The differences observed in the brain network for pain perception between normal subjects and chronic pain patients imply reorganization of the pain network, consistent with the studies by Flor et al. (1995, 1997) and Ramachandran et al. (1995). The abnormal levels of NAA found in the prefrontal brain of patients with chronic back pain indicate that chronic pain may be associated with neural degeneration. Recent advances in MR technology and in processing of MR data present the possibility of testing both hypotheses: anatomical reorganization and neuronal degeneration, directly in the human brain.

Using MR morphometric techniques, where gray matter volumes may be compared directly in patients and normal volunteers, one can directly test the presence of neuronal degeneration in chronic pain states. Several approaches have been developed recently to study morphometry of the whole brain, subcortical regions (i.e., brain segmentation), or multiple cortical areas (i.e., cortical parcellation). The most sophisticated method, based on conceptual models of the cortex, was developed to divide the entire neocortex into multiple topographically defined brain areas per hemisphere (Rademacher et al., 2000; Caviness et al., 1999). This method has been applied to normal subjects (Rademacher, 2000; Caviness et al., 1999) and to different brain disorders (e.g., Grachev et al., 1999). Although the method is very time consuming (approximately 40 h for one brain) and requires certain skills in neuro-anatomy, the technique allows estimating relative differences in volumes in specific areas of the entire cortex. Another alternative morphometric approach involves voxel-wise comparisons of the local concentration of gray matter across different patient populations using parametric statistics (Ashburner and Friston, 2000, SPM99 package). This approach, voxel-based morphometry, is simpler, not biased to one particular structure, and gives a comprehensive assessment of anatomical differences throughout the brain. It has been tested among different patient populations (reviewed in Ashburner and Friston, 2000), including idiopathic headache syndrome (May et al., 1999).

If one hypothesizes that cortical reorganization with chronic pain is accompanied by rewiring of anatomical connectivity of the cortex, then this, too, can be tested directly by the new diffusion-limited MR imaging techniques (see Poupon et al., 2000). These diffusion-based MR images coupled with sophisticated analysis methods can delineate axonal pathways in the human brain. Therefore, it is now possible to examine the axonal pathways of specific brain regions and compare these pathways in different populations of pain patients. Yet another approach that may add significant new information regarding the properties of fMRI activation is path analysis. This approach (e.g., Bullmore et al., 2000) can identify the connectivity weights between activated brain regions. As a result one can test for subtle differences in connectivity between brain regions, especially when similar structures are activated in different pain states.

These anatomical, morphometric, and path analysis approaches together with the functional and biochemical methods outlined above provide an opportunity for unraveling the fundamental functional processes that accompany chronic pain. Such understanding should revolutionize our understanding, and consequently the proper treatment of chronic pain states.

ACKNOWLEDGMENTS

We would like to thank all the collaborators who made these studies possible. The contributions of P.A. Gelnar, P.S. Thomas, B. Frederickson, G. Tillapaugh-Fay, S. Huckins, A. Darbar, and M. Fonte have been critical to the success of these studies. This research was funded by NIH/NINDS grant NS 35115.

REFERENCES

Apkarian, A.V., Functional imaging of pain: new insights regarding the role of the cerebral cortex in human pain perception, *Sem. Neurosci.*, 7, 279, 1995.

Apkarian, A.V., Darbar, A., Krauss, B.R., Gelnar, P.A., and Szeverenyi, N.M., Differentiating cortical areas related to pain perception from stimulus identification: temporal analysis of fMRI activations in a painful thermal task, *J. Neurophysiol.*, 8, 2956, 1999.

Apkarian, A.V., Gelnar, P.A., Krauss, B.R., and Szeverenyi, N.M., Cortical response patterns to thermal pain depend upon stimulus size: a functional MRI study, *J. Neurophysiol.*, 8, 3113, 2000.

Apkarian, A.V., Thomas, P.S., Krauss, B.R., and Åver Szeverenyi, N.M., Prefrontal cortical hyperactivity in sympathetically mediated chronic pain, *Neurosci. Letts.*, 2001, in press.

Ashburner, J. and Friston, K.J., Voxel-based morphometry — the methods, *NeuroImage*, 11, 805, 2000.

Bandettini, P.A. and Cox, R.W., Event-related fMRI contrast when using constant interstimulus interval: theory and experiment, *Magnet. Reson. Med.*, 43, 540, 2000.

Brammer, M.J., Bullmore, E.T., Simmons, A., Williams, S.C., Grasby, P.M., Howard, R.J., Woodruff, P.W., and Rabe-Hesketh, S., Generic brain activation mapping in functional magnetic resonance imaging: a nonparametric approach, *Magnet. Reson. Imaging*, 15, 763, 1997.

Bullmore, E., Horwitz, B., Honey, G., Brammer, M., Williams, S., and Sharma T., How good is good enough in path analysis of fMRI data? *Neuroimage*, 11, 289, 2000.

Bushnell, M.C., Duncan, G.H., Hofbauer, R.K., Ha, B., Chen, J.I., and Carrier, B., Pain perception: is there a role for primary somatosensory cortex? *Proc. Natl. Acad. Sci. USA*, 96, 7705, 1999.

Cavanaugh, J.M. and Weinstein, J.N., Low back pain: epidemiology, anatomy and neurophysiology, in *Textbook of Pain*, third ed., Wall, P.D. and Melzack, R., Eds., Churchill Livingstone, New York, 1994, 441–455.

Caviness, Jr., V.S., Lange, N.T., Makris, N., Herbert, M.R., and Kennedy, D.N., MRI-based brain volumetrics: emergence of a developmental brain science, *Brain Dev.*, 21, 289, 1999.

Cohen, J.D., Peristein, W.M., Braver, T.S., Nystrom, L.E., Noll, D.C., Jonides, J., and Smith, E.E., Temporal dynamics of brain activation during a working memory task, *Nature*, 386, 604, 1997.

Davis, K.D., Kwan, C.L., Crawley, A.P., and Mikulis, D.J., Functional MRI study of thalamic and cortical activations evoked by cutaneous heat, cold, and tactile stimuli, *J. Neurophysiol.*, 80, 1533, 1998.

Davis, K.D., Wood, M.L., Crawley, A.P., and Mikulis, D.J., fMRI of human somatosensory and cingulate cortex during painful electrical nerve stimulation, *NeuroReport*, 7, 321, 1995.

Derbyshire, S.W., Jones, A.K., Gyulai, F., Clark, S., Townsend, D., and Firestone, L.L., Pain processing during three levels of noxious stimulation produces differential patterns of central activity, *Pain*, 73, 431, 1997.

Deyo, R.A., Low back pain, *Scientific American*, August, 49, 1998.

Everitt, B.S. and Bullmore, E.T., Mixture model mapping of the brain activation in functional magnetic resonance images, *Human Brain Mapping*, 7, 1, 1999.

Faull, K.F., Rafie, R., Pascoe, N., Marsh, L., and Pfefferbaum, A., N-acetylaspartic acid (NAA) and N-acetylaspartylglutamic acid (NAAG) in human ventricular, subarachnoid, and lumbar cerebrospinal fluid, *Neurochem. Res.*, 24, 1249, 1999.

Flor, H., Elbert, T., Knecht, S., Wienbruch, C., Pantev, C., Birbaumer, N., Larbig, W., and Taub, E., Phantom-limb pain as a perceptual correlate of cortical reorganization following arm amputation, *Nature*, 375, 482, 1995.

Flor, H., Braun, C., Elbert, T., and Birbaumer, N., Extensive reorganization of primary somatosensory cortex in chronic back pain patients, *Neurosci. Lett.*, 224, 5, 1997.

Friston, K., Phillips, J., Chawla, D., and Buchel, C., Nonlinear PCA: characterizing interactions between modes of brain activity, *Philos. Trans. R. Soc. (Lond.)*, 355, 135, 2000.

Gelnar, P.A., Krauss, B.R., Sheehe, P.R., Szeverenyi, N.M., and Apkarian, A.V., A comparative fMRI study of cortical representations for thermal painful, vibrotactile, and motor performance tasks, *NeuroImage*, 10, 460, 1999.

Grachev, I.D. and Apkarian, A.V., Chemical heterogeneity of the living human brain: a proton MR spectroscopy study on the effects of sex, age and brain region, *NeuroImage*, 11, 554, 2000.

Grachev, I.D., Fredrickson, B.E., and Apkarian, A.V., Abnormal brain chemistry in chronic back pain: an *in vivo* proton magnetic resonance spectroscopy study, *Pain*, 89, 7, 2000.

Grachev, I.D., Berdichevsky, D., Rauch, S.L., Heckers, S., Kennedy, D.N., Caviness, V.S., and Alpert, N.M., A method for assessing the accuracy of intersubject registration of the human brain using anatomic landmarks, *NeuroImage*, 9, 250, 1999.

Gruetter, R., Novotny, E.J., Boulware, S.D., Rothman, D.L., and Shulman, R.G., 1H NMR studies of glucose transport in the human brain, *J. Cereb. Blood Flow Metab.*, 16, 427, 1996.

Hardy, J.D., Stolwijk, A.J., and Hoffman, D., Pain following step increase in skin temperature, in *The Skin Senses*, D.R. Kenshalo, Ed., Charles C Thomas, Springfield, 1968, 444.

Kennan, R.P., Scanley, B.E., Innis, R.B., and Gore, J.C., Physiological basis for BOLD MR signal changes due to neuronal stimulation: separation of blood volume and magnetic susceptibility effects, *Mag. Res. Med.*, 40, 840, 1998.

Kiebel, S.J., Poline, J.B., Friston, K.J., Holmes, A.P., and Worsley, K.J., Robust smoothness estimation in statistical parametric maps using standardized residuals from the general linear model, *NeuroImage*, 10, 756, 1999.

Krauss, B. and Apkarian, A.V., Group average activation maps of functional MRI: methodology of identifying group brain areas activated during painful thermal stimuli, motor and vibrotactile tasks in humans, *Riv. Neuroradiol.*, 11, 135, 1998.

Kwong, K.K., Belliveau, J.W., Chesler, D.A., Goldberg, I.E., Weisskoff, R.M., Poncelet, B.P., Kennedy, D.N., Hoppel, B.E., Cohen, M.S., Turner, R., Cheng, H.-M., Brady, T.J., and Rosen, B.R., Dynamic magnetic resonance imaging of human brain activity during primary sensory stimulation, *Proc. Natl. Acad. Sci.*, 89, 5675, 1992.

Magistretti, P.J., Pellerin, L., Rothman, D.L., and Shulman, R.G., Energy on demand, *Science*, 283, 496, 1999.

May, A., Ashburner, J., Buchel, C., McGonigle, D.J., Friston, K.J., Frackowiak, R.S., and Goadsby, P.J., Correlation between structural and functional changes in brain in an idiopathic headache syndrome, *Nature Med.*, 5, 836, 1999.

Merskey, H., Classification of chronic pain by the International Association for the Study of Pain subcommittee on taxonomy, *Pain*, 3, 217, 1986.

Melzack, R., The short-form McGill Pain Questionnaire, *Pain*, 30, 191, 1987.

Michaelis, T., Merboldt, K.D., Bruhn, H., Hanicke, W., and Frahm, J., Absolute concentrations of metabolites in the adult human brain in vivo: quantification of localized proton MR spectra, *Radiology*, 187, 219, 1993.

Poupon, C., Clark, C.A., Frouin, V., Regis, J., Bloch, I., Le Bihan, D., and Mangin, J., Regularization of diffusion-based direction maps for the tracking of brain white matter fascicles, *NeuroImage*, 12, 184, 2000.

Price, D.D., Physiological mechanisms of pain inhibition, in *Psychological and Neural Mechanisms of Pain*, Raven Press, New York, 1988, chap. 1.

Rademacher, J., Aulich, A., Reifenberger, G., Kiwit, J.C., Langen, K.J., Schmidt D., and Seitz, R.J., Focal cortical dysplasia of the temporal lobe with late-onset partial epilepsy: serial quantitative RMI, *Neuroradiology*, 46, 4305, 2000.

Ramachandran, V.S., Rogers-Ramachandran, D., and Cobb, S., Touching the phantom limb, *Nature*, 377, 489, 1995.

Salibi, N. and Brown, M.A., *Clinical MR Spectroscopy: First Principles*, Wiley Liss, New York, 1998.

Shulman, R.G., and Rothman, D.L., Interpreting functional imaging studies in terms of neurotransmitter cycling, *Proc. Natl. Acad. Sci. USA*, 95, 11993, 1998.

Sibson, N.R., Dhankhar, A., Mason, G.F., Rothman, D.L., Behar, K.L., and Shulman, R.G., Stoichiometric coupling of brain glucose metabolism and glutamatergic neuronal activity, *Proc. Natl. Acad. Sci. USA*, 95, 316, 1998.

Spielberger, C.D., Gorsuch, R.L., Lushene, R., Vagg, P.R., and Jacobs, G.A., *Manual for the State-Trait Anxiety Inventory*, Consulting Psychologists Press, Palo Alto, CA, 1983.

Toni, I., Schluter, N.D., Josephs, O., Friston, K., and Passingham, R.E., Signal-, set- and movement-related activity in the human brain: an event-related fMRI study, *Cerebral Cortex*, 9, 35, 1999.

Treede, R.D., Kenshalo, D.R., Gracely, R.H., and Jones, A.K.P., The cortical representation of pain, *Pain*, 79, 105, 1999.

Van Zijl, P.C.M., Eleff, S.M., Ulatowski, J.A., Oja, J.M., Ulug, A.M., Traystman, R.J., and Kauppinen, R.A., Quantitative assessment of blood flow, blood volume and blood oxygenation effects in functional magnetic resonance imaging, *Nature Med.*, 4, 159, 1998.

14 Methods for Induction and Assessment of Pain in Humans with Clinical and Pharmacological Examples

*Thomas Graven-Nielsen, Märta Sergerdahl,
Peter Svensson, and Lars Arendt-Nielsen*

CONTENTS

14.1 INTRODUCTION

Pain, in general, and musculoskeletal pain, in particular, are common and have great personal, socio-economic, and psychological impact. Pain is a highly relevant clinical problem, and a better understanding of the basic neurophysiological mechanisms involved in pain is needed to obtain better prevention and treatment strategies.

The mechanisms involved in pain are often difficult to resolve from clinical studies due to the great variability among patients, which may be caused by different pain intensities and the duration of their pain condition. Furthermore, the origin of the pain is often difficult or impossible to locate. Human experimental pain models applied to healthy volunteers are suggested as a suitable strategy to investigate aspects of the neurobiological mechanisms involved in pain. Experimental pain research involves two separate topics: (1) standardized activation of the nociceptive system, and (2) measurements of the evoked responses. One important advantage with experimental pain studies is that the cause–effect relationship is known. In this situation healthy volunteers transiently become patients with a well-defined pain. In addition, experimental techniques may be used in clinical studies to quantify the sensitivity of the nociceptive system in pain patients and in pharmacological studies.

Pain is a multidimensional perception, and the reaction to a single standardized stimulus of a given modality can only represent a limited fraction of the entire pain experience. Combinations of different stimulation and assessment techniques can be used to gain advanced differentiated information about the nociceptive system under normal and pathophysiological conditions. Obvious psychological differences are present in the experience of a well defined, short-lasting, controlled, experimental pain stimulus and a chronic, often intractable pain condition. Experimental techniques are, however, the only way to bridge the gap between basic science and clinical application. Thus, the use of experimental pain can elucidate basic neurobiological mechanisms that are affected by pain.

Several techniques have been used to induce and assess pain in humans.[1-9] Most of these publications have concentrated on induction and assessment of single nociceptive stimuli applied to the skin. This chapter describes the battery of methods available for induction of skin and muscle pain. Techniques for visceral pain will be briefly discussed as this topic is covered in Chapter 5.[10] Other models such as experimental pain from joints, ligaments, oral mucosa, nasal mucosa, and teeth are beyond the scope of the present work.

Understanding of the basic aspects of muscle pain in humans is limited as most knowledge originates from experiments using anaesthetized animals and because experimental pain research has been based mainly on skin pain. The lack of understanding of the neural mechanisms involved in muscle pain has led to much philosophical speculation about mechanisms related to etiology, pathogenesis, diagnosis, and treatment. Thus, muscle pain must be considered as a major challenge in experimental pain research and will, therefore, be emphasized. This review is divided into three sections:

1. Experimental methods available for induction of pain from skin, muscles, and viscera in humans
2. Available techniques for assessment of human experimental pain
3. Examples of how experimental pain models can provide quantitative methods in clinical and pharmacological research

14.2 INDUCTION OF EXPERIMENTAL PAIN IN HUMANS

Beecher[11] and later Gracely[5] have defined several important characteristics which should be true for the ideal experimental pain stimulus. The main characteristics for an ideal pain stimulus are (1) to evoke a distinct pain sensation with minimal tissue damage, (2) to have a relationship between stimulus intensity and pain intensity, (3) to have reliable inter- and intra-session reproducibility, and (4) to excite nociceptors exclusively.

14.2.1 CUTANEOUS PAIN

The pioneering studies by von Frey[12] and Goldscheider[13] showed that pain could be elicited if specific stimulus modalities and configurations were applied to the skin. Quantitative studies of the human responses to painful stimuli followed.[1,14] Different techniques for the induction of cutaneous pain are outlined in Table 14.1.

TABLE 14.1
The Various Stimulus Modalities Available for the Induction of Controlled Experimental Cutaneous Pain in Humans

Modality	Stimulus
Electrical	Transcutaneous
	Intracutaneous
	Intraneural
Thermal	
Heat	Radiant (light, infrared)
	Contact thermode
Cold	Ice water
	Contact thermode
	Cold air
Mechanical	Pinprick
	Pressure
	Impact stimuli
	Pinch
Chemical	Capsaicin
	Mustard oil
	Hydrogen ions
	Inflammatory-related substances (serotonin, bradykinin, histamine, potassium ions, substance P)

14.2.1.1 Electrical

In 1851 von Helmholtz[15] introduced the method of electrical stimulation of the skin. Electrical stimulation is a non-physiological technique bypassing receptor transduction and depolarizing the afferent fibers directly. As the thick myelinated mechanosensitive afferents are activated at lower stimulus intensities than unmyelinated fibers, pain cannot be elicited without concurrent activation of a tactile sensation. For cutaneous electrical stimulation, the challenge is to minimize the spread of current to adjacent structures. The use of intracutaneous electrodes can minimize the current and, hence, spread. For surface stimulation, the Tursky electrode[16] is adequate as the current is restricted between an inner disc and a surrounding concentric annulus. Although constant current is normally used, the problem is still to maintain the skin impedance at a constant level. Bipolar, saline-soaked felt electrodes are widely used (especially for trancutaneous nerve stimulation of, e.g., the sensory sural nerve) as the impedance can be maintained at a fairly constant level if the felt tips are kept wet. Intraneural electrical stimulation of cutaneous afferents has been used to compare projected areas of pain with receptive fields determined with microneurography.[17–19] In recent years the electrical stimulation devices have been computer controlled, providing the possibility of delivering a preprogrammed series of stimuli with different intensities, durations, and frequencies.

14.2.1.2 Thermal

Goldscheider[13] introduced heat to evoke experimental pain. One of the first attempts to measure pain thresholds with the application of a hot object with graded temperatures was carried out by Elo and Nikula.[20] The major problem with this technique was evocation of touch and pressure sensations together with heat pain. Alrutz[21] suggested that radiant heat could solve this problem. The Hardy-Wolff-Goodell[1,22] non-contact heat stimulator was based on focused white light. The major disadvantage of photic stimulation is wavelength-dependent reflections from the skin.

In recent years, heat stimulation of high intensity by argon, CO_2, Nd-YAG, copper vapour lasers,[23-25] thulium YAG crystal laser,[26] GaInAs/GaA1As laser diodes,[27] and Xenon light[28] has been developed for cutaneous and mucosal[29,30] stimulation. Normally, the laser pulses used are of short duration (20 to 200 ms) and elicit a distinct pricking pain, which can be followed by a burning pain at the highest stimulus intensities. The nociceptors in the skin are divided into three main classes: C-mechano-heat, type I Aδ-mechano-heat, and type II Aδ-mechano-heat. Type I mechano-heat-nociceptors exhibit a slowly increasing response with a latency of several seconds to heat stimuli of high intensity and long duration.[31] If the skin is activated by rapid heat stimuli from, e.g., a laser, the type II Aδ-mechano-heat-nociceptors are activated[32] together with some activation of warmth receptors (C-fibers). When laser stimulation is used, it is very important to monitor skin parameters, as the thresholds decrease with increasing skin temperature and with decreasing skin thickness.[24]

For clinical purposes, thermodes based on the Peltier principle are widely used. The advantage of this method is that the same device can easily assess cold pain, cold, warm, and heat pain.[33] The disadvantages are (1) the temperature rise-times are slow (1 to 2°C/s), which mainly excites C-fibers and not Aδ-fibers, which are excited at higher rise-times,[34] and (2) the device touches the skin during examination, which can be a problem when testing allodynic areas in patients. The different pathways excited depend on the rise-time, which illustrates the importance of using a multi-modal stimulation approach to gain detailed information about the nociceptive system. Recently, a device based on heating foils has been developed. With this device very rapid temperature increases (30°C/s) can be obtained depending on the electrical current delivered to the foil.[35,36] The foil has a low thermal capacity and is glued (thermal conducting glue) to a Peltier element by which active cooling (15°C/s) rapidly returns the skin temperature to its baseline value.

Cold pain stimulation by immersion of the hand into ice water (called cold-pressor test) was initially developed to assess vasomotor reactions.[37] This method has been widely used, especially for testing drug efficacy.[38] The technique is based on water (0 to 2°C) with crunched ice, but it is important to have circulation in the water to avoid local heating around the immersed hand. This type of stimulus is, however, not specific to the skin; subcutaneous tissues might be stimulated. Recently, stimulators based and on the Peltier (contact probe) or Ranque-Hilsch vortex tube (air) principles have been used for more specific and localized cold stimulation.[39-42]

14.2.1.3 Mechanical

Painful mechanical stimulation can be achieved with pressure algometers or by solenoid stimulators with a blunt needle attached. The von Frey hair was originally made from horse hairs of different diameter, but is now replaced by graded nylon filaments often attached to a vibrating solenoid.[43–46] Standardized application of the von Frey hairs and room temperature and humidity are important parameters for obtaining a stable stimulus.[47] The tactile perception is probably Aβ-fiber mediated and the tactile pain sensation is Aδ-fiber mediated, which again illustrates the significance of using different stimulation modalities to obtain advanced knowledge on the nociceptive system. More recently, a high-speed impact stimulation[48–50] has been used with the velocity of the projectile determining the painfulness of the impact. Subcutaneous tissue can be stimulated with pressure applied to a skinfold by a pincher mounted on a pressure algometer.[51]

14.2.1.4 Chemical

Topical capsaicin (1%), the pungent ingredient of chili pepper, applied via moisturizing cream, induces primary and secondary hyperalgesia (see 14.2.1.6) with burning pain of moderate intensity.[52] In contrast, an acute and intense pain is induced by intradermal injection of capsaicin.[53–55] Topical application of mustard oil for a few minutes gives rise to burning pain and inflammatory reactions.[52,56–58] Capsaicin- and mustard oil-induced pain is mainly mediated by C-fibers.[58–61] Other chemicals, related to the inflammatory process, have been reported to produce pain when given intradermally.[62–65] Hydrogen-ions produce pain at a pH of approximately 5.2 when continuously infused intradermally.[66]

14.2.1.5 Summation

Nociceptive activity from the skin can summate either temporally or spatially. Central summation of pain was described a century ago,[12] and alteration in temporal and spatial summation has been suggested more recently as an important factor for various pain conditions, e.g., hyperalgesia.

The phenomenon of a repetitive nociceptive stimulus causing exaggerated perceptions of human pain is called temporal summation, which is assumed to be related to the wind-up detected in animal dorsal horn neurons.[67] In animals, an N-methyl-D-aspartate (NMDA) receptor antagonist[68] decreases wind-up, which is the increased neuronal firing of dorsal horn neurons to a train of stimuli.[68] Temporal summation is probably the initial part of wind-up because repeated stimulation for at least 20 s causes central sensitization in animal preparations,[69] which is not seen after a short duration of temporal summation in humans.[70] The relationship between temporal summation and wind-up is also indicated because both phenomena are inhibited by the blocking of the NMDA-receptor.[68,71–74] Recordings from nociceptive neurons in the spinothalamic tract[75] and dorsal horn[76] have provided neurophysiological evidence for central summation of nociceptive activity. Temporal summation is more pronounced for C-fiber-mediated second pain than for Aδ-fiber-mediated first pain.[77] Price et al.[77] used repetitive heat pulses for stimulation and found that summation could occur down to 0.3 Hz. Later the summation frequency was found to be intensity dependent.[78]

Spatial summation of warmth exists.[79,80] Whereas Stevens and Marks[80] estimated that no spatial summation existed for pain, this has now been contradicted.[24,29,81–84] Spatial summation seems not to be pronounced close to the pain threshold compared with supra-threshold intensities.[85] In the early studies in which no spatial summation was found,[1,85] the stimulation areas were larger than those used more recently. This could reflect inhibitory mechanisms triggered when a critical area of stimulation is reached.[86] Spatial summation of nociceptive activity has been observed both within[83] and between dermatomes.[84,87]

Another aspect of spatial summation is the central summation of activity in different nociceptive afferents. Andersen et al.[28] used the nociceptive reflex and psychophysical ratings and found a significant summation between pricking (Aδ-fibers) and burning (C-fibers) pain. Under a pathological condition with cutaneous hyperalgesia, the activation of the Aβ-afferents causes pain (allodynia). The question is whether summation under such conditions exists between non-nociceptive mechanosensitive pathways and nociceptive pathways. Under normal conditions, no such summation exists. In an experimental study, Andersen et al.[88] induced secondary hyperalgesia by topical application of capsaicin. Pain was induced by electrical pricking pain stimuli well outside the secondary hyperalgesic area with and without concurrent activation of the secondary area by tactile stimulation, which was perceived as pain due to allodynia. When simultaneous application of the two stimuli was made, significant facilitation of the nociceptive withdrawal reflex and psychophysical ratings was obtained. This indicated that the central sensitization associated with cutaneous hyperalgesia results in summation between non-nociceptive and nociceptive pathways.

For all stimulus modalities, the activated area increases with increased stimulus intensity; for electrical stimulation the current field is increased, for mechanical stimulation the deformed area is increased, and for thermal stimulation the heat (light) spreads to a large area. This spatial recruitment of afferent fibers is important, as spatial summation will increase for increasing stimulus intensity. For thermal stimulation the total duration at which the temperature is above nociceptive level increases at increasing stimulus intensity. Receptor activation time often outlasts the stimulus due to the thermal capacity of the tissue. This results in increased temporal summation for increased stimulus intensity. In studies of stimulus-response functions, central integrative mechanisms play an increased role for increased stimulus intensity.

14.2.1.6 Induction of Primary and Secondary Hyperalgesia

As a model for pathological changes, experimental hyperalgesia is important for basic and pharmacological studies. The hyperalgesic area developing around a painful skin stimulation or injury is termed secondary hyperalgesia and is related to central sensitization.[89] Desensitization or sensitization (primary hyperalgesia) of nociceptors in the stimulated area is dependent on peripheral mechanisms.[89]

Intradermal injection of capsaicin is the most commonly used method in humans to induce and assess central sensitization seen as secondary hyperalgesia.[53,60,89–95]

A small amount of capsaicin (i.e., 50 to 300 μg) produces a pronounced nociceptor discharge and evokes an intense burning spontaneous pain, allodynia to brush and pinprick, and facilitated temporal summation to mechanical stimulation in the secondary hyperalgesic area. Hyperalgesia to heat and pain on suprathreshold heat stimulation has been demonstrated close to the injection site (primary hyperalgesia),[53,60] whereas the heat pain threshold to radiant heat stimulation was unchanged in the secondary hyperalgesic area.[96] The sensory changes induced by topical administration of capsaicin[52–54,59,61,73,88] are very similar to those of intradermal injection, although it has a longer onset time of allodynia to brush and pinprick due to the slow penetration of capsaicin. Hyperalgesia to heat and pinprick in the application area (primary hyperalgesia) is probably due to peripheral sensitization, but central sensitization cannot be excluded.[49,89] Hyperalgesia to tonic pressure is also found in the primary area.[49] Mechanical hyperalgesia and allodynia to brush and pinprick stimulation are prominent in the secondary area. The persistence of this phenomenon for more than 30 min seems to be common to the capsaicin models, irrespective of intradermal or topical administration.

Topical mustard-oil application in humans is shown to produce secondary hyperalgesia to tactile stimulation.[52,57,97] The sensory disturbances also include reduced heat-pain threshold in the application area, increased warmth-detection threshold in the secondary hyperalgesic area, and a decreased cold-detection threshold in both the primary and secondary hyperalgesic areas. In addition, pressure pain thresholds are transiently reduced in the primary area, but not in the secondary hyperalgesic area.[52] The inter-session variability in the areas of brush and punctate allodynia is approximately 25 and 50%, respectively, and the duration of a stable punctate allodynia has been documented for 1h.[97]

The skin burn model produces a neurogenic inflammation resulting in a primary hyperalgesic zone at the site of the heat stimulation surrounded by a secondary hyperalgesic zone. Experimental skin burn has been demonstrated to reduce mechanical (brush and pinprick), thermal, and electrical pain thresholds and to facilitate suprathreshold pain in the secondary area.[57,98–101] Also, it induces facilitation of mechanically induced temporal summation, both in the primary and the secondary hyperalgesic areas,[102] but not to the electrical stimulation.[103] The duration of secondary mechanical hyperalgesia and reduced heat-pain threshold in the injured area extends for 6 h, and the reduction of tactile pain threshold in the burn area persists for 24 h after burn. The inter-session variability is approximately 30%.[101] The skin burn model has lately been combined with topical capsaicin to create a model that is more stable and has long-lasting sensory disturbance in order to be able to study analgesic drugs.[104] This model maintains an area of secondary hyperalgesia to punctuate stimuli for 4 h. Pathophysiological changes are similar to the heat burn model. In general, the different skin models on primary and secondary hyperalgesia are suitable for studies of phenomena involved in central sensitization.

14.2.2 MUSCLE PAIN

Sensory manifestations of muscle pain are seen as a cramp-like, diffuse aching pain in the muscle, pain referred to distant somatic structures, and modifications in the

TABLE 14.2
The Various Stimulus Modalities Available for the Induction of Controlled Experimental Muscle Pain in Humans

Type	Modality	Stimulus
Endogenous	Ischemia	Tourniquet + contractions
	Exercise	Concentric contractions
		Eccentric contractions
Exogenous	Electrical	Intramuscular
		Intraneural
	Mechanical	Pressure
	Chemical	Hypertonic saline, potassium, levo-ascorbic acid, capsaicin bradykinin, serotonin, calcitonin gene-related peptide, neurokinin A, substance P, hydrogen ions

superficial and deep sensitivity in the painful areas.[105–107] These manifestations are different from cutaneous pain, which is normally superficial and localized around the injury with a burning and sharp quality.[107] Various procedures can induce muscle pain, and they can be divided into endogenous and exogenous techniques.[108] The endogenous techniques are methods that provoke muscle pain by natural stimuli, for example, by ischemia or by exercise (Table 14.2). The exogenous techniques are external interventions, e.g., electrical stimulation of muscle afferents or injection of algogenic substances.

14.2.2.1 Ischemia

Lewis[109] proposed induction of muscle pain by ischemia. A tourniquet is applied, and after a period of voluntary muscle contractions a very unpleasant tonic pain sensation develops. The number of contractions, the level of force, and the duration are important determinants for the resulting pain.[110] The method is found to be reliable[111] and has been used for human analgesic assay.[112] This method induces a widespread pain in the entire occluded limb (skin, periosteum, muscle, etc.), and, therefore, it is not possible to induce muscle pain in a specific muscle. Accumulation of algogenic substances accompanies ischemic contractions, which are likely to cause pain.[113,114] The tourniquet model excites mainly C-fibers.[115,116]

14.2.2.2 Exercise

Exercise-induced muscle pain by concentric muscle work is normally short lasting and a result of impaired blood flow during work. It may, therefore, resemble the condition of ischemic muscle pain.[110,117] On the other hand, eccentric muscle work may cause delayed onset of muscle soreness with peak soreness after 24 to 48 h.[110,118–121] The mechanism underlying delayed onset muscle soreness is probably related to ultrastructural damage resulting in the release of algogenic substances.[119,120] This may produce an inflammatory reaction as nonsteroidal antiinflammatory drugs (NSAID) appear to have an effect on this type of jaw muscle soreness.[122,123] Howell et al.[124] were, however, unable to demonstrate an NSAID effect on delayed soreness in limb muscles. Another feature of delayed onset of muscle soreness is that there is no pain at rest, but pain is evoked by muscle action and during palpation. This is in contrast to spontaneous pain induced by the exogenous experimental techniques. In general, the endogenous experimental techniques induce a widespread deep pain in muscles and other somatic structures, which may be used in studies that require an unspecific deep pain stimulus.

14.2.2.3 Electrical

In human studies intramuscular (i.m.) electrical stimulation can be used to assess the sensitivity of muscles,[125–131] to study basic aspects of deep pain,[132–134] and to investigate electrophysiological properties of muscle afferents by microneurography.[135–137] Electrical stimulation is a reliable model to study sensory manifestations of muscle pain such as referred pain[138,139] and temporal summation,[132] but is confounded by concurrently activated muscle twitches. Thus, the effect of electrically induced muscle pain cannot be assessed during the muscle activity.[140] Intraneural stimulation of muscle afferents[135–137] more selectively elicits muscle pain accompanied by referred pain, with pain areas that increase with increasing pain intensities.

14.2.2.4 Mechanical

Mechanical painful stimulation can be achieved with pressure algometers. Recently, a computer-controlled pressure algometer was designed in which the rate and peak pressure can be predefined and automatically controlled.[141] It is important to recognize that pressure stimulates both skin and muscle. Anesthetizing the skin can, however, reduce the skin pain contribution during pressure stimulation.[142,143]

14.2.2.5 Chemical

Intramuscular injections of algogenic substances such as capsaicin,[55,137,144,145] bradykinin, serotonin,[146–148] calcitonin gene-related peptide, neurokinin A, and substance P,[147,149] potassium chloride,[150] levo-ascorbic acid,[151] and hypertonic saline have been used to induce human muscle pain. Recently, infusion of an acid phosphate buffered to pH 5.2 has also been shown to induce muscle pain.[116] The experimental method that has been used extensively is i.m. injection of hypertonic saline, as the quality of the induced pain is comparable to acute clinical muscle pain with localized and referred pain.[105,152–157] The work of Kellgren[105] and Lewis[152] in the late 1930s initiated the method of saline-induced muscle pain,[152] and the safety of the technique is supported by the absence of reports of side effects after more than 1000 i.m. infusions.[154,158] The general procedure for this type of experimental model is to inject a small volume (e.g., 0.5 ml) or apply continuous infusion of hypertonic saline into the muscle.[143,158–182] In most of the earlier studies manual bolus infusions of hypertonic saline have been used; however, standardization of the infusion of small volumes is easier to accomplish by computer-controlled infusion pumps.[158,159] A systematic evaluation of the infusion parameters (infusion concentration, volume, rate, and tissue) has been carried out on pain intensity, quality, and local and referred pain patterns.[160] A major advantage of the hypertonic saline model is that a detailed description of sensory and motor effects can be obtained. Further, it seems to be the most suitable, experimental model to study the phenomenon of referred pain from musculoskeletal structures.

Referred pain is probably a combination of central processing and peripheral input as it is possible to induce referred pain to limbs with complete sensory loss due to spinal injury[183] or anaesthetic block.[153,184] The involvement of peripheral input from the referred pain area is not clear, as anaesthetizing this area shows inhibitory or no effects on the referred pain intensity.[106,138,183,185,186] Central sensitization may be involved in the generation of referred pain. Animal studies show a development of new receptive fields by noxious muscle stimuli.[187,188] Recordings from a dorsal horn neuron with a receptive field located in the biceps femoris muscle show new receptive fields in the tibialis anterior muscle and at the foot after i.m. injection of bradykinin into the tibialis anterior muscle.[188] In the context of referred pain, the unmasking of new receptive fields due to central sensitization could mediate referred pain.[189] This has been suggested as the phenomenon of secondary hyperalgesia in deep tissue. A number of studies have found that the area of the referred pain correlated with the intensity of the muscle pain,[135,138,161,185,190,191] which parallels the observations for cutaneous secondary hyperalgesia where the hyperalgesic area is related to the capsaicin-induced pain intensity.[90]

Hypertonic saline will probably give an unspecific (non-nociceptive and nociceptive) excitation of muscle afferents.[192] In animal studies, hypertonic saline is shown to excite group III (Aδ)[193,194] and group IV (C)[194–196] muscle afferents. Hypertonic saline may less consistently affect other afferents, as afferents responding to muscle stretch show increased,[195] decreased, or unchanged[193,194] activity after hypertonic saline. Recent animal studies have indicated that stretch receptors are not excited nor sensitized by hypertonic saline.[197] Nevertheless, i.m. injections of hypertonic saline might produce an afferent barrage in several types of nerve fibers, but most notably in nociceptive fibers. Furthermore, i.m. injections could also produce excitation (indirectly or directly) of muscle fibers. It has, however, not been possible to show spontaneous electromyographic activity from muscle tissue infiltrated with hypertonic saline compared with isotonic saline.[163,167,198] Intramuscular pressure has been discussed as another mechanism underlying saline-induced muscle nociception. After injection of hypertonic saline, however, it has not been possible to detect an i.m. pressure that is sufficient to induce muscle pain.[163] Moreover, the saline infusions result in an i.m. pressure (< 120 mmHg),[163] which is only slightly higher than the i.m. pressure found during normal exercise.[199,200] In keeping with this, i.m. infusion of isotonic saline with an infusion pressure of 400 mmHg did not produce muscle pain.[201]

14.2.2.6 Summation

Intraneural microstimulation of muscle nociceptive afferents causes muscle pain and a referred pain sensation (area and intensity) that is dependent on the stimulation

TABLE 14.3
The Various Stimulus Modalities
Available for the Induction of
Controlled Experimental Visceral
Pain in Humans

Modality	Stimulus
Electrical	Transmucosal
Thermal	
Heat	Water
Cold	Water
Mechanical	Distension
Chemical	Capsaicin

duration (temporal summation)[137] and the number of stimulated afferents (spatial summation).[136] Few experimental studies have studied the summation of muscle pain. Temporal summation to i.m. electrical stimulation has been demonstrated.[132] Moreover, repeated infusions of hypertonic saline were given in another study.[161] A manifold device was used for bolus infusion at spatially separated sites. If the infusions were separated by 90 s compared with 360 s, increased pain perception and increased referred pain areas were observed (temporal summation). The mechanism behind temporal summation is another key parameter when accumulating information about the human nociceptive system.

14.2.3 VISCERAL PAIN

Investigations of visceral pain have lagged behind those of other types of pain. One reason is that the viscera are more difficult to access than skin and muscle. It has been a puzzle for many years why cutting viscera was not painful when a small ulcer could cause severe pain. Experimental studies related to visceral pain have been concentrated on the gastrointestinal (GI) tract, the structure involved in the largest number of clinical complaints. The GI tract transducts multiple sensory modalities: electrical-, mechanical-, thermal-, and chemical-sensitive stimuli (Table 14.3). Several reviews[202–204] have been published on this issue, and it is covered in Chapter 5 of the present book.[10]

14.2.4 SUMMARY

Many standardized techniques are available for painful stimulation of skin, muscle, and viscera but the optimal stimulus method, which exclusively activates nociceptors, does not exist. As different stimulus modalities activate different afferent pathways and hence central pathways, it is essential to construct an experimental model on different stimulation techniques to obtain differentiated information about the human nociceptive process.

TABLE 14.4
The Techniques Available for Assessment of Experimental Pain in Humans

Type	Method
Psychophysical	
Response-dependent	Intensity
	Visual analogue scale (VAS)
	Verbal descriptor scale
	Numerical scale
	Cross-modality matching
	Stimulus-response function
	Quality
	McGill Pain Questionnaire
	Unpleasantness
	VAS
	Distribution
	Drawings (area, location)
Stimulus-dependent	Pain detection and tolerance thresholds
Electrophysiological	Evoked potentials (brain and brainstem)
	Microneurography
	Electroencephalogram
	Withdrawal/jaw reflexes
Imaging Techniques	Functional magnetic resonance imaging (fMRI)
	Positron emission tomography (PET)
	Single-photon emission computed tomography (SPECT)
	Magnetoencephalography (MEG)
Autonomic Reactions	Temperature/thermography/laser doppler
	Blood flow
	Blood pressure
	Heart rate and variability
	Pupil dilatation
	Galvanic skin resistance

14.3 ASSESSMENT OF EXPERIMENTAL PAIN

Methods of assessing pain are based on psychophysical, electrophysiological, and imaging techniques (Table 14.4). In general, these assessment techniques apply to phasic (stimulus with short duration) painful stimulation of skin, muscles, and viscera. Experimental muscle pain is a relatively unexplored area compared with experimental superficial pain. Muscle pain is, however, of extreme clinical importance and thus, a detailed discussion on assessment of experimental muscle pain is presented below. Examples of how tonic pain (several minutes) can be assessed are given also. The involvement of the autonomic nervous system during experimental pain also can be assessed (Table 14.4), but will not be discussed in detail. Extensive reviews of the role of electroencephalography and imaging techniques for assessment of pain have been presented,[205–208] and are covered in Chapter 12 in the present volume.[209]

14.3.1 PSYCHOPHYSICAL ASSESSMENT

Psychophysical determinations can be divided into response-dependent and stimulus-dependent methods.[5,210] The response-dependent methods are constructed by a series of fixed stimulus intensities with a score given to each stimulus. The score can be a visual analogue scale, verbal descriptor scale, magnitude estimation, or cross-modality match (Table 14.4). Visual analogue scales (VAS), verbal descriptor scales (VDS), McGill Pain Questionnaire (MPQ), and similar scales and questionnaires may be very helpful for the assessment of perceived intensity and quality.[5] One important advantage of the VAS compared to ordinal scales (e.g., numerical scales 0 to 10) is that the VAS have ratio scale properties,[211] i.e., for two different stimuli intensities, where one of them is perceived as the double amount of the other, the high VAS score should be double that of the low VAS score.

The stimulus-dependent methods are based on adjustment of the stimulus intensity until a pre-defined response, typically a threshold (e.g., detection, pain, or tolerance), is reached. The stimulus intensity required to reach the threshold is in physical units, and therefore, the use of subjective scales is avoided. In general, two methods to adjust the stimulus intensity exist. In the method of limits, the stimulus intensity is gradually increased until the threshold is reached and then decreased again until the stimulus intensity is just below threshold. The threshold is defined as the mean of the two limits. Ascending and descending trials can be repeated to improve reliability. In the method of constant stimuli, the stimulus intensity is incremented in fixed steps and presented to the subject several times in contrast to the gradual increase as in the method of limits. The threshold is defined as the stimulus intensity that is just above the threshold in 50% of the stimulus trials.

Stimulus-response functions are more informative than a threshold determination as suprathreshold response characteristics can be derived from the data. Nevertheless, the stimulus-response function may be established often with stimuli intensities near the pain threshold and, therefore, both assessment methods are valuable.

14.3.2 ELECTROPHYSIOLOGICAL ASSESSMENT TECHNIQUES

14.3.2.1 Evoked Potentials

Many studies on electrophysiological assessment of pain are based on evoked potentials. In 1883, Fleischl von Marxow[212] observed that peripheral stimuli applied to animals produced electrical brain potentials, and Berger[213] assumed that this could also be found in the human electroencephalogram. To his surprise, acute sensory stimuli, such as painful pinpricks, always caused a depression in the electroencephalogram lasting for some seconds. One of the interesting characteristics of the vertex potential is the relation between amplitude, stimulus intensity, and hence magnitude rating, and the potential has therefore been widely used in perceptual studies. This relation was established in 1953, and the vertex potential was suggested to be an "objective correlate for the auditory perception."[214] Therefore, it was obvious to ask whether this potential could also be used as "an objective correlate of pain."[215] Vertex potentials elicited by pinpricks[216] and strong nociceptive electrical stimuli[217] were first shown in the human electroencephalogram in the 1950s. Spreng and Ichioka[218] and Schmidt[219] were among the first to show that the amplitude of the human vertex potential, evoked by nociceptive electrical dental stimuli, increased with increasing stimulus intensity and hence pain intensity. The usefulness of vertex potentials evoked by nociceptive stimuli has been discredited due to the large intra- and interindividual variability of the responses.[220] The variability has been related to artifacts or to poor stimulus control, but it is now known that this variability is a general feature of the vertex potential.

The shape of the vertex potential does not change when the intensity of the electrical stimulus exceeds the pain threshold, but the amplitude saturates.[221–223] This has led to the suggestion that vertex potentials evoked by nociceptive electrical stimuli are not reliable correlates of changes within the nociceptive system.[224] For strange reasons, these observations have not hampered the use of electrically evoked vertex potentials in experimental pain research, and modulation of this potential has been widely accepted as an indicator of the excitability of the nociceptive system (e.g., Bromm et al.[225]). Clearly, the non-nociceptive-related components of the potential must be identified and isolated before the potential can be used to measure nociceptive processes.[226] Another neglected aspect in drug testing is controlling sedation, as attention clearly modulates the amplitude of vertex potentials.[224,227]

Many methodological studies have shown that potentials evoked by noxious laser stimuli correlate with the pain magnitude rating,[29,228–231] and provide an attractive alternative to electrically evoked potentials.[232] Despite the relationship between rating and amplitude for the laser-evoked potentials, many factors can disturb this relationship.[233] Central habituation[234,235] and variation in the inter-stimulus interval[236] change the amplitude of vertex potentials without concurrent changes in pain intensity. Thus, it seems as if vertex potentials and pain intensity ratings reflect different phenomena or different sensitivities, and therefore, both should be recorded simultaneously. If they reflected the same phenomenon, it could be argued that it would be redundant to measure the brain potentials. As vertex potentials are affected by many factors, it is mandatory to have control recordings. Such control recordings could be vertex potentials to nonpainful stimuli (e.g., visual[237]), potentials evoked from nonaffected areas (e.g., patients with unilateral affections), or from conditions without interaction (e.g., placebo drug).

14.3.2.2 Nociceptive Withdrawal Reflex

The nociceptive withdrawal reflex may be used as an additional supplement to psychophysical methods to elucidate aspects of spinal nociceptive processing. Sherrington[238] carried out early extensive studies of limb reflexes in animal preparations and observed characteristic ipsilateral flexion and extension in the contralateral limb to preserve balance. The generation of a withdrawal reflex is initiated by nociceptive input, but extensive processing takes place within the spinal cord. The neural connection from primary sensory neurons to motor neurons is a polysynaptic pathway; other afferent input, descending activity, and the excitability of neurons in this pathway modulate the generation of the spinal nociceptive reflex.

The reflex can be used for assessment in two ways: as a reflex threshold, or as an amplitude to a fixed suprathreshold stimulus intensity. Kugelberg[239] found that stimulus strengths sufficient to depolarize Aδ-fibers were most effective in eliciting the first reflex component though large fibers may contribute. Hugon[240] separated the A-fiber-mediated reflex response further into a component mediated by tactile (group II) afferents with a latency of 40 to 60 ms, and a component mediated by group III afferents with a latency of 85 to 120 ms. He denoted these two reflex components the RII and RIII reflex, respectively. Hagbarth[241] systematically investigated the reflex response in a number of extensor and flexor muscles as a function of the stimulation site, and Shahani and Young[242] studied the influence of stimulus intensity on the latency and size of the reflex. In an experimental situation, the nociceptive withdrawal reflex can be elicited by heat[243,244] or electrical stimulation delivered to a sensory nerve[245] or to a cutaneous area.[245,246] The reflex has been claimed to be a very robust measure with a causal relationship to pain intensity.[245,247,248] However, a number of studies have recently observed discrepancies and found poor correlation between reflex size and pain rating.[73,88,244,249] A new interesting approach is mapping of receptive fields related to the withdrawal reflex.[250] Several spatially different sites are stimulated in a sequential manner, and the reflex amplitude for each stimulus can be used to estimate the receptive field.[250] This is a very interesting technique as it allows a noninvasive mapping of the withdrawal reflex organization, e.g., before and during central sensitization.

Temporal summation can be assessed by summation pain detection/tolerance thresholds or by the nociceptive withdrawal reflex. Dimitrijevic and Nathan[251] stimulated patients with complete transection of the spinal cord by four strong electrical pulses per second and found that the amplitude and duration of the evoked nociceptive reflexes progressively increase. This technique was later refined,[78] and automated procedures have been developed to assess the threshold when healthy reflex builds up.[252] The reflex build-up correlates with the threshold when the subjects report that the pain intensity increases during the stimulus series (psychophysiocal summation pain threshold).

14.3.3 ASSESSMENT OF EXPERIMENTAL MUSCLE PAIN

Many of the general techniques discussed in Table 14.4 also can be used to characterize the sensory manifestations in muscle pain. Assessments of the sensory aspects involve evaluations of muscle pain (local pain) and of the somatic structures related to the referred pain area; i.e., the ongoing pain intensity and the sensitivity must be described for both. The effect of muscle pain on motor function can also be assessed (Table 14.5).

Verbal assessments of the muscle pain intensity experienced and other subjective characteristics of the pain are obviously needed in any clinical and experimental muscle pain studies. The pain intensity is usually scored in a continuous mode on an electronic VAS to characterize the time profile of experimental muscle pain. Studies of experimental muscle pain have found that the most frequently used word descriptors to characterize muscle pain are "drilling", "aching", "boring", and

TABLE 14.5

Assessment Techniques of Experimental Muscle Pain in Humans

Effects	Type	Methods
Sensory (local and referred pain areas, respectively)	Sensitivity	Stimulus-response function
		Pain detection and tolerance thresholds
	Ongoing pain	Distribution of pain (area)
		VAS intensity
		McGill Pain Questionnaire
Motor	Electromyography	Surface electrodes
		Needle electrodes
		Indwelling wire electrodes
	Kinetics and kinematics	Jaw tracking devices
		Optoelectronic devices
	Force	Dynamometers
		Force platforms
		Bite force meters

"taut."[160,253] The intensity of muscle pain is easily measured using VAS; however, this is only a one-dimensional aspect of the experienced pain, and additional VAS could be applied to monitor, e.g., unpleasantness and soreness.

In addition to verbal assessments, psychophysical tests are valuable adjuncts for the examination of sensitivity in the local and referred muscle pain areas. Both deep and superficial sensitivity should be assessed as differential effects on the skin and deep structures might be observed.[51,160] Many studies have used pressure algometry for quantitative analysis of muscle pain and tenderness.[254] The pressure pain thresholds vary substantially between regions, and methodological studies are necessary for each new location examined.[255,256] In some of the commercially available pressure algometers, the pressure application rate can be monitored, which is important for the reliability of the results. The computer-controlled pressure algometer solves the pressure-rate problem because the rate can be automatically controlled.[141] In addition to pressure algometry, i.m. electrical stimulation may be used to assess sensitivity in the local and referred pain areas[257]; Graven-Nielsen, 1997;[51] Arendt-Nielsen, 1997.[132] Electrical stimulation also gives the opportunity for the assessment of temporal summation.

Svensson et al.[156] performed an experimental study in which increased tenderness, assessed by pressure algometry, was observed after the jaw muscle had been exposed to experimental muscle pain (hypertonic saline). This corresponds to the finding that thresholds to intramuscular electrical stimulation are significantly lower in muscles 24 h after they have been exposed to hypertonic saline.[128] These findings on pressure and intramuscular electrical pain thresholds were also seen after infusion of isotonic saline in the leg muscles.[51]

Experimentally induced muscle pain by hypertonic saline also results in referred pain with cutaneous[128,153,181,258] and muscular[128] sensitivity changes; however, Kellgren[105] could not detect sensitivity changes in referred areas with this method whereas others have reported both hypo- and hyperaesthesia.[153,258] Recently, tactile hypoaesthesia in the referred pain area has been reported.[181] Hypoalgesia to pressure in the referred pain area was detected as a generalized phenomenon and not related to the referred pain area per se.[143] Moreover, hyperalgesia to pressure distal to the referred pain area has been found.[259] The controversies concerning hypo- and hyperaesthesia in the referred area might be due to an imbalance between inhibitory and excitatory neural mechanisms involved in referred pain and descending inhibitory control.

14.3.3.1 Sensory–Motor Interaction during Muscle Pain

The interaction between muscle pain and motor control is dependent on the motor task (rest, static, and dynamic contractions).[260] This differentiated effect of muscle pain illustrates the importance of assessing electromyographic, kinetic, kinematic, and force parameters (Table 14.5). A previous hypothesis has been put forward concerning the accompaniment of muscle pain by muscle hyperactivity,[261–263] but this so-called hyperactivity or vicious cycle theory has not gained much scientific support in the literature.[264] In addition, saline-induced muscle pain causes no increase in EMG activity at rest.[164,167] Muscle pain reduces maximal voluntary contraction and endurance time during submaximal contractions.[164,166,265] Moreover, muscle pain causes changes in coordination during dynamic tasks with increased antagonistic and decreased agonistic muscle activity.[164–166,175,180,253,266] This pattern is in accord with the so-called pain-adaptation model[264] predicting a decrease in movement amplitude and velocity by muscle pain. The neural mechanisms underlying pain-evoked changes in dynamic motor function can be related best to alternating inhibition and excitation of motoneurons, but the exact pathways responsible for the modulation are still not known.

The pain-adaptation model describes only modulation in muscle activity due to pain and not the reason for the persistence of muscle pain. From the experimental findings, the reason for muscle pain cannot be explained fully by altered muscle activity, and might also be due to mechanisms in the central nervous system. For instance, peripheral sensitization of muscle nociceptors, neurogenic inflammation, or facilitated central processing of peripheral input (central sensitization) could be involved in the pathophysiological mechanisms of muscle pain. Furthermore, contracture-induced pain (no EMG findings) and biochemically maintained muscle pain (peripheral hyperalgesia) may be other reasons for muscle pain. The pain-adaptation model explains the effects of muscle pain on motor function by reflex mediated circuits; i.e., changes in the voluntary drive in conscious humans are not an essential part of the model. Recent studies have clearly shown that experimental muscle pain is capable of changing reflex pathways. The nature of these complicated changes is dependent on the specific type of afferent fiber activated and whether the reflex is inhibitory or excitatory.[169–172,176,177,179,180,182] Nevertheless, these reflex findings suggest that muscle pain has a strong influence on reflex pathways without involving supraspinal mechanisms.

14.3.4 SUMMARY

Various techniques within psychophysical and electrophysiological methods can be used to assess the effects of phasic and tonic pain stimuli. Using a variety of methods to obtain the most detailed description of the perception and electrophysiological effects of the experimental stimuli is recommended.

14.4 EXAMPLES OF EXPERIMENTAL PAIN USED IN CLINICAL AND PHARMACOLOGICAL STUDIES

The aim of this section is to provide examples of how experimental pain models can be used as quantitative methods in clinical and pharmacological research.

14.4.1 CLINICAL STUDIES

The sensitivity of the nociceptive system in "pain patients" can be assessed by experimental pain techniques. This is important in the light of mechanism-based classification of pain.[267]

14.4.1.1 Quantitative Sensory Testing

Differentiated sensory testing has been used extensively to characterize sensory deficits in patients with neurogenic disorders involving hypo-/hyperaesthesia and hypo-/hyperalgesia (for a review, see Boivie et al.[8]). Modality-specific thresholds are determined on the healthy side vs. affected side or compared with control values, which give valuable information for the clinical examination. To differentiate the dysfunction related to various disorders of the sensory system it is mandatory to compose a sensory test series of different stimulus modalities activating different afferent neurons. Assessment of the somatosensory system can be based on psychophysical and electrophysiological techniques. The customary electrophysiological technique is electrically evoked somatosensory brain potentials, but just as with the compound sensory action potentials, this only assesses thick myelinated nerve function. The thick fibers assessed in daily clinical, neurophysiological investigations constitute, however, only a minor fraction of the total population of afferents and may not always be involved in the pathophysiology of pain.

Many disorders, e.g., diabetic neuropathy, affect the thin fibers before the thick fibers, and a clinical diagnosis concerning nerve impairment can be obtained only when the thick fibers start to show signs of dysfunction. At that time, the thin fibers may be severely affected with the possibility of developing severe neurogenic chronic pain. The need for methods to assess early impairment of thin fiber function is evident. Assessment of neuropathic pain patients for heat and cold functions might solve this problem, but here also the quality of the stimulus-induced sensation must be assessed because paradoxical sensations (i.e., cold perceived as heat) are a typical problem in these patients.[268,269] Moreover, hyper- or hypoalgesic response to touch, pinprick, warmth, or cold was characteristic compared with asymptomatic sites in neurogenic pain patients, although not reflected in the pain thresholds.[269] The stimulus-response function to mechanical stimuli was similarly affected, with left and right shifts of the curve relating the intensity to the pain response.[269]

Temporal summation can also be affected in neuropathic pain patients. In patients with a complex regional pain syndrome, temporal summation of burning pain induced by gentle mechanical stimuli or by electrical stimulation at rates of 0.3 Hz, can be blocked by sympathetic ganglia block.[270] Similarly, Eide and Rabben[271] found facilitated temporal summation to repetitive mechanical stimuli within symptomatic vs. contralateral sites in trigeminal neuropathic pain patients; however, a combination of facilitated temporal summation to mechanical stimuli but unchanged temporal summation to heat stimuli in the affected areas in neuropathic pain patients also has been reported.[46] An explanation for the finding of abnormal temporal summation of pain probably involves central sensitization.

14.4.1.2 Evoked Potentials

In routine neurophysiological practice, sensory-evoked potentials are derived from peripheral electrical stimulation. Unfortunately, some clinical papers assume that the evoked potentials to painful electrical stimulation represent aspects of nociceptive transmission. It is obvious that the electrically evoked barrage projects to the dorsal column tract and hence assesses neural transmission of sensory qualities such as light touch, vibration, and pressure.[226] On the other hand, sensory-evoked potentials following CO_2 laser stimulation relate to pain and nociceptive impulses projected in the spinothalamic tract.[272] To substantiate this statement further, strong painful tactile, electrical and laser pulses were applied to the skin before and after epidural analgesia.[232] No changes were found in the overall shape of the electrically evoked potential although the potentials evoked by laser were completely abolished after analgesia.

The large inter-individual variation in the amplitude of the laser-evoked potentials suggests that they may not be suitable for routine examinations in clinical practice. A large set of normative data based on laser-evoked potentials from normal healthy, age/sex-adjusted controls is essential. In studies where the patient and control groups serve as their own controls, e.g., comparing the differences in amplitude for potentials evoked from two areas or follow-up after surgery, laser-evoked potentials are suitable for monitoring. The laser-evoked potentials can provide useful information on the integrity of the nociceptive system that is not accessible by conventional electrophysiological techniques.

Laser-evoked potentials have been shown to be of value in assessing impairment of pain and temperature sensation in patients with peripheral neuropathies.[273] Correlation between pain/temperature impairment and changes in the laser potentials has been found in patients with syringomyelia,[274] multiple sclerosis,[275] and in neurological patients with various dissociated sensory deficits.[276] Dissociated sensory impairment in brainstem disorders suggests a lateral lesion involving the spinothalamic tract. Laser-evoked potentials have been used to follow (0.5 to 4 years) the recovery of the pain pathways in patients with isolated lateral brainstem lesions. Laser-evoked potentials recovered in these patients in parallel with normal mechanosensitivity.[277]

Sensory testing and clinical neurophysiology studies have indicated that patients with central pain syndromes occasionally have impairment of pain and temperature sensation. Central pain syndromes could be caused by disinhibition of spinothalamic excitability or by the reduction of spinothalamic function due to other central changes or disease in the brain. Casey et al.[278] found that central pain patients (cerebral or brainstem infarctions) with normal tactile sensation had significantly lower laser-evoked potentials on the effected side compared with the non-effected side. This study supports a deficit in spinothalamic tract function, but does not suggest excessive central responses to the activation of cutaneous nociceptive pathways.

In other chronic pain patients, a relative insensitivity to pain has been found. In patients with intractable pain, minor superficial burn lesions were inflicted by the laser without being too painful.[279] These patients showed a decrease in amplitude as the intensity of the painful CO_2 laser stimuli increased.[279] For electrical pain stimulation a similar tendency has been found[280,281] and has been interpreted as a way to cope with pain. In contrast, fibromyalgia patients show increased amplitudes of evoked potentials to cutaneous laser stimulation compared with controls.[282,283] The major augmentation of laser-evoked potentials resides only in the late components (N170–P390). These effects suggest the presence of exogenous factors such as reduced cortical and subcortical inhibition or central hypervigilance to the nociception, probably involving the limbic mid-cingulate generator. It has been shown that hypnotically induced hyperalgesia can also increase the laser-evoked vertex potentials.[284] Furthermore, a significant positive correlation was found between an increase in the laser potential and an increase in the subjective intensity of hypnotically induced emotion of depression.[285]

14.4.1.3 Experimental Deep and Referred Pain

Experimental techniques for cutaneous pain have been used extensively in clinical settings. Other than pressure algometry, experimental muscle pain has been used only rarely to assess the sensitivity of the pain system in patients. In general, reduced pressure pain thresholds have been found in patients with musculoskeletal disorders as compared to healthy controls.[131,156,256,286–290] Notably, stimulus-response functions have been found to shift to the left in musculoskeletal pain patients, which has been interpreted as additional involvement of low-threshold mechanoreceptors caused by central sensitization.[291,292] In line with pressure algometry, i.m. electrical stimulation has been used to determine muscular hyperalgesia in musculoskeletal pain patients.[131,257]

Characteristically, referred pain and muscle hyperalgesia are often found in musculoskeletal pain disorders. Chronic musculoskeletal pain has been shown to respond better to NMDA-antagonist (ketamine) treatment than to conventional morphine management,[293,294] indicating a role of central sensitization in these patients. It is, therefore, obvious to propose that muscle pain conditions[192,295] can involve central sensitization. From studies on cutaneous hyperalgesia, central summation of nociceptive input from muscles and referred pain areas is expected to be exaggerated in musculoskeletal pain conditions if central sensitization is involved. Experimental muscle pain can be used to quantify the central changes due to muscle pain (e.g., muscle hyperalgesia and referred pain) in clinical studies. In fibromyalgia patients[131] and whiplash-injury patients,[296] saline-induced muscle pain was used to quantify the pattern of referred pain. The area and intensity of referred pain were found to be exaggerated in patients compared with control subjects. Moreover, in fibromyalgia patients, i.m. electrical stimulation was used to assess the efficacy of temporal summation of painful muscle stimuli, and temporal summation was found more potent in the patients compared with control subjects.[131] Increased referred pain areas and efficacy of temporal summation in patients suggest that the gain of central processing is increased (central sensitization) in these patients. In addition, it was found that referred pain expansion and exaggerated temporal summation were partly inhibited by administration of an NMDA-antagonist.[297]

As seen in musculoskeletal disorders, an atypical referred pain pattern can be a sign of changed central excitability, as in irritable bowel syndrome (IBS)[298] or functional dyspepsia[299] patients. IBS patients have been extensively investigated, and hypersensitivity to gut distension is a general finding[300] despite normal somatic pain thresholds.[301] This increased referred pain as a result of visceral hypersensitivity is in accordance with experimental studies.[302] It has been proposed that in various visceral[303] pain conditions the central nervous system may be sensitized. Thus, summation of nociceptive input from viscera is expected to play an important role in visceral pain conditions.

14.4.1.4 Summary

Various stimulus modalities and assessment techniques should be combined to get sufficiently advanced and differential information about the human nociceptive system under pathophysiological conditions. This is essential in mechanism-based classification of pain.[267] Moreover, the experimental techniques for deep pain provide unique and clinically relevant information regarding mechanisms (referred pain and temporal summation) involved and affected in musculoskeletal and visceral pain patients.

14.4.2 PHARMACOLOGICAL STUDIES

Human experimental studies are the intermediate step between animal and clinical pharmacological trials. The advantage of experimental studies is that the final clinical trials can provide evidence for which specific pain modalities and mechanisms should be targeted. Clinical studies also should include many patients due to the heterogeneity of patient populations. Today, experimental studies are so advanced that they can provide knowledge of exactly which mechanisms a given drug affects. This will provide information on which clinical conditions the drug might be effective (e.g., neuropathic pain, musculoskeletal pain, etc.). To optimize this approach it is essential to describe the pre-clinical drugs with an outcome profile after a multimodal testing approach (Fig. 14.1). Such analgesic profiles can be used to plan clinical trials as specific profiles predict efficacy for specific pain conditions. The following sections illustrate how experimental pain can be used in pharmacological studies.

14.4.2.1 Introduction to Techniques and Study Design

The actual hypothesis of the experimental model determines the study design. A crossover, randomized double-blind design is the "gold standard." An important factor is that each subject should always be examined by the same investigator. Changing investigators can increase the inter-session variability.[304] All drugs,

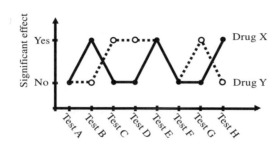

FIGURE 14.1 Schematic illustration of the analgesic profile for two different analgesic drugs (X and Y). The tests (A–H) illustrate different experimental testing modalities, e.g., heat pain tolerance, pressure pain threshold, temporal summation threshold, cold pressure pain VAS, saline-induced muscle, and referred pain VAS, etc. The two drugs affect the individual modalities differentially.

TABLE 14.6
Important Pharmacokinetic Aspects in Studies Involving Pharmacological Modulation of Experimental Pain in Humans

- The bioavailability of the drug in the mode of administration to be applied: i.v., orally, spinally.
- Dose-effect relationships, minimally effective doses.
- Establishment of a steady-state level of drug, which needs to be time-locked to the pain stimulus. This could be achieved by target controlled infusion (TCI).[327]
- Are genetic differences important for metabolization of the study drug? If so, the study can be confined to rapid or slow metabolizers of the compound.
- Adverse effects.
- Appropriate time for drug washout.

---

irrespective of mode and rate of administration, require a placebo control and if possible an active control. The inclusion of healthy subjects should match what is clinically relevant and both sexes should be included. In psychophysical studies, the difference in outcome parameters between gender is small and not conclusive. Previous studies have shown either decreased or increased pain sensitivity to painful skin stimulation between males and females.[305–307] However, inter-individual variation is greater than inter-sex variations, and thus a stratification of the material does not seem to be compulsory.

Pharmacodynamics is crucial and the actual drug should target the pain mechanisms involved in the experimental model. In line with this, different pharmacokinetic aspects need to be considered (Table 14.6).

Pain originating from different tissues, such as cutaneous, musculoskeletal, and visceral pain, has its specific characteristics in neuroanatomy and physiology, which demand differentiation in treatment and thereby also in the experimental models utilized for pharmacological testing.

The experimental design should include a multi-modal approach to obtain information on which pathways are affected by the drug. In addition, mechanism-based investigations (e.g., temporal summation or central sensitization) are important. An example of the multi-modal and mechanism-based design is quantitative sensory testing before and after capsaicin-induced secondary hyperalgesia (central sensitization). The multi-modal approach is obtained by evaluation of the tactile perception threshold (Aβ-fiber mediated), tactile pain threshold (Aδ-fiber mediated), and pain evoked by single (Aδ-fiber mediated) and suprathreshold tactile (C-fiber mediated) stimulation. Tactile threshold measurements are performed using von Frey filaments. Thresholds for cold perception and cold pain (Aδ-fiber function), perception of warmth and heat pain, and pain at suprathreshold stimulation (C-fiber mediated) are assessed. Cutaneous electrical stimulation is also used for the evaluation of threshold determinations and suprathreshold pain intensity, activating Aβ-, Aδ- and C-fibers unselectively. The quantitative sensory test should be applied pre-drug and post-drug (pre- and post-capsaicin) during drug administration. In this example, the experimental design can provide information on which pathways are modulated by the drug, with and without central sensitization.

When findings from pharmacological modulation of experimental pain are interpreted in healthy volunteers, it is important to realize that the experimental pain models only evoke spontaneous pain for a limited time. What can actually be investigated is the impact of drugs on pain mechanisms as it appears in the normal nervous system. Knowledge of what is true for the pathologically altered sensory nervous system must still be acquired from patient studies, but experimental models provide useful methods and instruments. The ability of analgesics to attenuate responses to experimental pain is of both mechanistic and clinical interest.[308,309]

14.4.2.2 Effect of Analgesic Drugs on Cutaneous Pain Stimuli in Normal Tissue

The many quantitative stimulation techniques available make studying differential effects of drugs possible. In previous studies, e.g., dose-response effect of epidural analgesia,[45] a highly specific blocking pattern was found for 12 different psychophysical measures to electrical, heat, laser, cold, tactile, and pressure stimuli of different durations. A better effect was found on pain evoked by short-lasting stimuli than on pain evoked by long-lasting stimuli of the same modality, indicating the importance of temporal summation. The modality-related blocking "order" was heat stimuli blocked before mechanical stimuli and again before electrical stimuli. The modality-related blocking sequence is probably a reflection of which pathways are activated by the different stimulus modalities.

Adenosine, but not morphine or the NMDA-receptor antagonists ketamine and dextromethorphan, increases the heat pain threshold.[310,311] The same drugs do not reduce pain at suprathreshold heat stimulation,[103,310] although dextromethorphan does have this effect.[311] Thus, results are somewhat contradictory in terms of response to NMDA-receptor blocking agents. The explanation for this may be due to pharmacokinetic reasons, but individual genetic differences in drug metabolism also may be of importance.[312]

Temporal summation is a potent mechanism which is not blocked by conventional anesthetic procedures such as epidural analgesia[313] or isoflurane anesthesia,[252] suggesting that additional drugs should be added to inhibit summation selectively. Furthermore, ketamine and dextromethorphan attenuate temporal summation to electrical stimulation, but not single impulse stimulation.[71,102]

14.4.2.3 Effect of Analgesic Drugs on Cutaneous Pain Caused by Central Sensitization

In experimental models involving central sensitization (e.g., secondary hyperalgesia), the primary efficacy variable is typically the pharmacological reduction of central sensitization, mainly expressed as areas of allodynia and hyperalgesia, or temporal summation to repetitive stimuli.

In the model employing intradermal injection of capsaicin, i.v. ketamine, alfentanil, and peripherally injected morphine reduce central sensitization, whereas i.v. midazolam and orally administered dextromethorphan do not.[73,93,94,314] From studies on topical capsaicin-induced cutaneous hyperalgesia, temporal summation to electrical stimuli in the secondary hyperalgesic area was facilitated compared with baseline measurements, and this facilitation was reversed by the application of ketamine.[73] In the model applying mustard oil to the skin, adenosine administered i.v. or intrathecally attenuates the development of secondary hyperalgesia.[56,57,97]

Due to its stability, the skin burn model is frequently used for pharmacological intervention. Reduction of central sensitization is demonstrated by oral dextromethorphan, oral ibuprofen, i.v. ketamine, i.v. adenosine, epidural morphine,[45,57,99,311,315] but not by s.c. or local ketamine or i.v. morphine.[99,103] Reduction of pain induced by thermal, electrical, tactile, or pressure stimulation seems to be parallel to the efficacy of central sensitization on temporal and spatial components. It may be speculated that the discrepancy between the negative results during i.v. administration of morphine, compared with the expressed reduction in central sensitization after epidural administration, is due to proximity of morphine to its site of action during epidural administration, which would result in a higher drug concentration at the effect site. A parallel situation could be the case when ketamine is administrated by different routes. The first passage elimination is marked, and the concentration at the effect site could be insufficient.

The reduction in the areas of secondary allodynia or hyperalgesia can be used as an estimation of the ability of the study treatment to reduce central sensitization, with respect to both dynamic (Aβ) and static (Aδ and C) components. Relative changes in tactile and thermal threshold differ in magnitude between studies. Nevertheless, C-fiber function is generally more compromised in the capsaicin and mustard oil models.

Stimulus intensity varies between different studies, and there seems to be an increased pharmacological modulation for high stimulus intensity for most commonly used pain modalities. This detail may account for many discrepancies in results, and thus also in the interpretation of psychophysical measurements and drug effects. The mode of drug administration is also of importance for the efficacy of the drug reaching its site of action.

14.4.2.4 Effect of Analgesic Drugs on Experimental Muscle and Referred Pain

Since the techniques for experimental muscle pain models are less developed than those for cutaneous pain models, the number of studies with pharmacological modulation of experimental muscle and referred pain is still sparse. However, referred pain mechanisms might be as relevant as secondary hyperalgesia has been in previous pharmacological studies, at least when looking at the clinical implications. In addition, temporal summation of muscle stimuli also should be assessed.

Systemic administration of opioids, ketamine, adenosine, theophylline, caffeine, and naproxen/paracetamol (acetaminophen) in combination are demonstrated to reduce ischemic pain in the tourniquet model,[316-323] while dextromethorphan, nonsteroidal antiinflammatory drugs in monotherapy, diazepam, and lidocaine are demonstrated not to do so.[312,317,324]

This also applies to the effect of ketamine on central sensitization and temporal summation in cutaneous models; ketamine has been shown to reduce experimentally induced referred pain areas and to attenuate temporal summation of muscle stimuli.[297] In addition, i.v. remifentanil reduces the slope of stimulus-response curves for electrical stimulation, more for i.m. stimulation compared with cutaneous stimulation.[325]

Laursen et al.[326] studied the effect of a differential tourniquet block followed by a regional block with lidocaine on muscle pain induced by either hypertonic saline or i.m. electrical stimulation. The main finding was that electrical painful stimulation primarily was mediated via Aδ-fibers (group III) and saline-induced muscle pain through C-fibers (group IV). In this type of study a known pharmacological effect was used to infer new knowledge of neural mechanisms involved in muscle pain. The above examples are used as illustrations of how experimental pain models can be used in drug-effect studies.

14.4.2.5 Summary

The multi-modal and mechanism-based experimental pain models are valuable tools for evaluating and documenting analgesic properties of drugs. The outcome from pharmacological modulation of experimental pain can predict aspects related to the clinical efficacy of a specific drug as the differentiated action on the different pathways and mechanisms can be teased out. The combination of mechanism-based classification of pain patients with specific knowledge of which pain mechanisms a drug affects will provide better opportunities for the development of new pain treatment regimes. More studies are needed to characterize the pharmacological mechanisms involved in experimental deep and referred pain derived from muscle and viscera.

14.5 CONCLUSION

The availability of methods for induction and assessment of experimental, superficial, and deep pain has been discussed. This chapter has, however, been restricted to models of skin, muscle, and visceral pain. Other models of experimental pain, from joints, ligaments, oral muscosa, nasal muscosa, and teeth deserve further consideration in the future development of experimental models. Nevertheless, the models of skin, muscle, and visceral pain provide a suitable basis to study: (1) basic pain mechanisms, (2) the nociceptive system under normal and pathophysiological conditions, and (3) drug efficacy in specific pain mechanisms. The majority of previous studies have concentrated on skin pain models. The clinical importance of pain from deep structures is evident. Thus, experimental models of deep pain in humans will probably be especially valuable in future experimental, clinical, and pharmacological studies.

The multi-modal experimental test paradigm is essential for obtaining sufficiently advanced and differential information about the human nociceptive system in normal and pathophysiological conditions: a requirement for mechanism-based classification and treatment of pain. Similarly, the multi-modal approach is important in pre-clinical pharmacological studies to provide evidence for which specific pain modalities and mechanisms should be targeted in clinical studies.

ACKNOWLEDGMENT

The Danish National Research Foundation supported the time spent preparing this chapter.

REFERENCES

1. Hardy, J.D., Wolff, H. G., and Goodell, H., Methods for the study of pain thresholds, in *Pain Sensations and Reactions,* Hardy, J.D., Wolff, H.G., and Goodell, H., Eds., Williams & Wilkins, Baltimore, 1952, 52.
2. Wolff, B.B., The role of laboratory pain induction methods in the systematic study of human pain, *Acupunct. Electro.-Therapeut. Res. Int. J.,* 2, 271, 1977.
3. Procacci, P., Zoppi, M., and Maresca, M., Experimental pain in man, *Pain,* 6, 123, 1979.
4. Handwerker, H.O. and Kobal, G., Psychophysiology of experimentally induced pain, *Physiol. Rev.,* 73, 639, 1993.
5. Gracely, R.H., Studies of pain in human subjects, in *Textbook of Pain,* Wall, P.D. and Melzack, R., Eds., Churchill Livingstone, Edinburgh, 1999, 385.
6. Bromm, B., *Pain Measurement in Man,* Elsevier, Amsterdam, 1984.
7. Chapman, R.C. and Loeser, J.D., *Issues in Pain Measurements,* Raven Press, New York, 1989.
8. Boivie, J., Hansson, P., and Lindblom, U., *Touch, Temperature, and Pain in Health and Disease: Mechanisms and Assessments,* IASP Press, Seattle, 1994.
9. Arendt-Nielsen, L., Induction and assessment of experimental pain from human skin, muscle, and viscera, in *Proc. 8th World Congress on Pain,* Jensen, T.S., Turner, J.A., and Wiesenfeld-Hallin, Z., Eds., IASP Press, Seattle, 1997, 393.
10. Ness, T.J. and Gebhart, G.F., Methods in visceral pain research, in *Methods in Pain Research,* Kruger, L., Ed., CRC Press, Boca Raton, 2001.
11. Beecher, H.K., *Measurement of Subjective Responses,* Oxford University Press, New York, 1959.
12. von Frey, M., *Untersuchungen über die menschliche Haut. Des XXIII Bandes der Abhandlungen der matematich-physischen Classe der Königl,* Sächsischen Gesellschaft der Wissenschaften, No. III, Bei S. Hirzel, Leipzig, 1896.

13. Goldscheider, A., Die spezifische Energie der gefülsnerven der Haut, *Monatschr. Prakt. Dermatol.*, 3, 283, 1884.

14. Lewis, T., *Pain*, Macmillan, New York, 1942.

15. von Helmholtz, H., Über die Dauer und den Verlauf der durch Strömesschwankungen inducirten Electrischen Ströme, *Poggendorf's Anna. Physik Chemie*, 83, 505, 1851.

16. Tursky, B., Watson, P.D., and O'Connell, D.N., A concentric shock electrode for pain stimulation, *Psychophysiology*, 1, 296, 1965.

17. Marchettini, P., Cline, M., and Ochoa, J.L., Innervation territories for touch and pain afferents of single fascicles of the human ulnar nerve. Mapping through intraneural microrecording and microstimulation, *Brain*, 113, 1491, 1990.

18. Torebjörk, H.E., Human microneurography and intraneural microstimulation in the study of neuropathic pain, *Muscle Nerve*, 16, 1063, 1993.

19. Serra, J. et al., Activity-dependent slowing of conduction differentiates functional subtypes of C fibres innervating human skin, *J. Physiol. (Lond.)*, 515, 799, 1999.

20. Elo, J. and Nikula, A., Zur Topographie des Wärmesinnes, *Skand. Arch. Physiol.*, 24, 226, 1910.

21. Alrutz, S., Studien auf dem Gebiete der Temperatursinne, *Skand. Arch. Physiol.*, 7, 321, 1897.

22. Hardy, J.D., Wolff, H.G., and Goodell, H., A new method for measuring pain threshold: observations on spatial summation of pain, *J. Clin. Invest.*, 19, 649, 1940.

23. Mor, J. and Carmon, A., Laser emitted radiant heat for pain research, *Pain*, 1, 233, 1975.

24. Arendt-Nielsen, L. and Bjerring, P., Sensory and pain threshold characteristics to laser stimuli, *J. Neurol. Neurosurg. Psychiatry*, 51, 35, 1988.

25. Svensson, P. et al., Comparison of four laser types for experimental pain stimulation on oral mucosa and hairy skin, *Lasers. Surg. Med.*, 11, 313, 1991.

26. Bromm, B. and Chen, A.C., Brain electrical source analysis of laser evoked potentials in response to painful trigeminal nerve stimulation, *Electroencephalogr. Clin. Neurophysiol.*, 95, 14, 1995.

27. Nielsen, J. et al., Laser diode: pain threshold and temperature distribution in human skin, *Proc. 9th World Congress on Pain, Vienna, Austria*, Abstr. 400, 1999.

28. Andersen, O.K. et al., Evidence for central summation of C and A delta nociceptive activity in man, *Pain*, 59, 273, 1994.

29. Svensson, P. et al., Quantitative determinations of sensory and pain thresholds on human oral mucosa by argon laser stimulation, *Pain*, 49, 233, 1992.

30. Svensson, P. et al., Vertex potentials evoked by nociceptive laser stimulation of oral mucosa: a comparison of four stimulation paradigms, *Anesth. Pain Control. Dent.*, 4, 222, 1992.

31. Meyer, R.A. and Campbell, J.N., Myelinated nociceptive afferents account for the hyperalgesia that follows a burn to the hand, *Science*, 213, 1527, 1981.

32. Treede, R.-D. et al., Evidence for two different heat transduction mechanisms in nociceptive primary afferents innervating monkey skin, *J. Physiol. (Lond.)*, 483, 747, 1995.

33. Lindblom, U., Analysis of abnormal touch, pain, and temperature sensation in patients, in *Touch, Temperature, and Pain in Health and Disease: Mechanisms and Assessments,* Boivie, J., Hansson, P., and Lindblom, U., Eds., IASP Press, Seattle, 1994, 63.

34. Yeomans, D.C., Pirec, V., and Proudfit, H.K., Nociceptive responses to high and low rates of noxious cutaneous heating are mediated by different nociceptors in the rat: behavioral evidence, *Pain*, 68, 133, 1996.

35. Nielsen, J. and Arendt-Nielsen, L., The importance of stimulus configuration for temporal summation of first and second pain to repeated heat stimuli, *Eur. J. Pain*, 2, 329, 1998.

36. Nielsen, J. and Arendt-Nielsen, L., The influence of rate of temperature change and peak stimulus duration on pain intensity and quality, *Somatosens. Mot. Res.*, 15, 220, 1998.

37. Hines, E.A. and Brown, G.E., A standard stimulus for measuring vasomotor reactions: its application in the study of hypertension, *Proc. Staff Meetings Mayo Clin.*, 7, 332, 1932.

38. Wolff, B.B. et al., Response of experimental pain to analgesic drugs: II: Codeine and placebo, *Clin. Pharmacol. Ther.*, 7, 323, 1966.

39. Morin, C. and Bushnell, M.C., Temporal and qualitative properties of cold pain and heat pain: a psychophysical study, *Pain*, 74, 67, 1998.

40. Davis, K.D., Cold-induced pain and prickle in the glabrous and hairy skin, *Pain*, 75, 47, 1998.

41. Beise, R.D., Carstens, E., and Kohlloffel, L.U., Psychophysical study of stinging pain evoked by brief freezing of superficial skin and ensuing short-lasting changes in sensations of cool and cold pain, *Pain*, 74, 275, 1998.

42. Harrison, J.L. and Davis, K.D., Cold-evoked pain varies with skin type and cooling rate: a psychophysical study in humans, *Pain*, 83, 123, 1999.

43. Della Corte, M. et al., Study on the cutaneous pricking pain threshold in normal man, *Arch. Fisiol.*, 64, 141, 1965.

44. Bromm, B. and Scharein, E., Principal component analysis of pain-related cerebral potentials to mechanical and electrical stimulation in man, *Electroencephalogr. Clin. Neurophysiol.*, 53, 94, 1982.

45. Brennum, J. et al., Quantitative sensory examination of epidural anaesthesia and analgesia in man: dose-response effect of bupivacaine, *Pain*, 56, 315, 1994.

46. Gottrup, H. et al., The relationship between sensory thresholds and mechanical hyperalgesia in nerve injury, *Pain*, 75, 321, 1998.

47. Levin, S., Pearsall, G., and Ruderman, R.J., Von Frey's method of measuring pressure sensibility in the hand: an engineering analysis of the Weinstein-Semmes pressure aesthesiometer, *J. Hand Surg.*, 3, 211, 1978.

48. Kohlloffel, L.U., Koltzenburg, M., and Handwerker, H.O., A novel technique for the evaluation of mechanical pain and hyperalgesia, *Pain*, 46, 81, 1991.

49. Kilo, S. et al., Different patterns of hyperalgesia induced by experimental inflammation in human skin, *Brain*, 117, 385, 1994.

50. Arendt-Nielsen, L., et al., Magnetoencephalographic responses to painful impact stimulation, *Brain Res.*, 839, 203, 1999.

51. Graven-Nielsen, T. et al., Quantification of deep and superficial sensibility in saline-induced muscle pain — a psychophysical study, *Somatosens. Mot. Res.*, 15, 46, 1998.

52. Koltzenburg, M., Lundberg, L.E.R., and Torebjörk, H.E., Dynamic and static components of mechanical hyperalgesia in human hairy skin, *Pain*, 51, 207, 1992.

53. LaMotte, R.H. et al., Neurogenic hyperalgesia: psychophysical studies of the underlying mechanisms, *J. Neurophysiol.*, 66, 190, 1991.

54. Eisenach, J.C., Hood, D.D., and Curry, R., Intrathecal, but not intravenous, clonidine reduces experimental thermal or capsaicin-induced pain and hyperalgesia in normal volunteers, *Anesth. Analg.*, 87, 591, 1998.

55. Witting, N. et al., Intramuscular and intradermal injection of capsaicin: a comparison of local and referred pain, *Pain*, 84, 407, 2000.

56. Rane, K. et al., Intrathecal adenosine administration: a phase 1 clinical safety study in healthy volunteers, with additional evaluation of its influence on sensory thresholds and experimental pain, *Anesthesiology*, 89, 1108, 1998.

57. Sjölund, K.F., Segerdahl, M., and Sollevi, A., Adenosine reduces secondary hyperalgesia in two human models of cutaneous inflammatory pain, *Anesth. Analg.*, 88, 605, 1999.
58. Olausson, B., Recordings of human polymodal single C-fiber afferents following mechanical and argon-laser heat stimulation of inflamed skin, *Exp. Brain Res.*, 122, 55, 1998.
59. Culp, W.J. et al., Heat and mechanical hyperalgesia induced by capsaicin. Cross modality threshold modulation in human C nociceptors, *Brain*, 112, 1317, 1989.
60. Torebjörk, H.E., Lundberg, L.E.R., and LaMotte, R.H., Central changes in processing of mechanoreceptive input in capsaicin-induced secondary hyperalgesia, *J. Physiol. (Lond.)*, 448, 765, 1992.
61. Koltzenburg, M., Torebjörk, H.E., and Wahren, L.K., Nociceptor modulated central sensitization causes mechanical hyperalgesia in acute chemogenic and chronic neuropathic pain, *Brain*, 117, 579, 1994.
62. Keele, C.A. and Armstrong, D., *Substances Producing Pain and Itch*, Arnold, London, 1964.
63. Pedersen-Bjergaard, U. et al., Algesia and local responses induced by neurokinin A and substance P in human skin and temporal muscle, *Peptides*, 10, 1147, 1989.
64. Manning, D.C. et al., Pain and hyperalgesia after intradermal injection of bradykinin in humans, *Clin. Pharmacol. Ther.*, 50, 721, 1991.
65. Steen, K.H. et al., Inflammatory mediators potentiate pain induced by experimental tissue acidosis, *Pain*, 66, 163, 1996.
66. Steen, K.H. and Reeh, P.W., Sustained graded pain and hyperalgesia from harmless experimental tissue acidosis in human skin, *Neurosci. Lett.*, 154, 113, 1993.
67. Ren, K., Wind-up and the NMDA receptor: from animal studies to humans, *Pain*, 59, 157, 1994.
68. Dickenson, A.H. and Sullivan, A.F., Evidence for a role of the NMDA receptor in the frequency dependent potentiation of deep rat dorsal horn nociceptive neurones following C fibre stimulation, *Neuropharmacology*, 26, 1235, 1987.
69. Wall, P.D. and Woolf, C.J., Muscle but not cutaneous C-afferent input produces prolonged increase in the excitability of the flexion reflex in the rat, *J. Physiol. (Lond.)*, 356, 443, 1984.
70. Arendt-Nielsen, L., Sonnenborg, F.A., and Andersen, O.K., Facilitation of the withdrawal reflex by repeated transcutaneous electrical stimulation: an experimental study on central integration in humans, *Eur. J. Appl. Physiol.*, 81, 165, 2000.
71. Price, D.D. et al., The N-methyl-D-aspartate receptor antagonist dextromethorphan selectively reduces temporal summation of second pain in man, *Pain*, 59, 165, 1994.
72. Arendt-Nielsen, L. et al., The effect of N-methyl-D-aspartate antagonist (Ketamine) on single and repeated nociceptive stimuli: a placebo-controlled experimental human study, *Anesth. Analg.*, 81, 63, 1995.
73. Andersen, O.K. et al., The effect of Ketamine on stimulation of primary and secondary hyperalgesic areas induced by capsaicin — a double-blind, placebo-controlled, human experimental study, *Pain*, 66, 51, 1996.
74. Arendt-Nielsen, L. et al., Effect of racemic mixture and the (S+)-isomer of ketamine on temporal and spatial summation of pain, *Br. J. Anaesth.*, 77, 625, 1996.
75. Price, D.D. et al., Spatial and temporal transformations of input to spinothalamic tract neurons and their relation to somatic sensations, *J. Neurophysiol.*, 41, 933, 1978.
76. Wagman, I.H. and Price, D.D., Responses of dorsal horn cells of M. mulatta to cutaneous and sural nerve A and C fiber stimuli, *J. Neurophysiol.*, 32, 803, 1969.
77. Price, D.D. et al., Peripheral suppression of first pain and central summation of second pain evoked by noxious heat pulses, *Pain*, 3, 57, 1977.

78. Arendt-Nielsen, L. et al., Electrophysiological and psychophysical quantification of temporal summation in the human nociceptive system, *Eur. J. Appl. Physiol.*, 68, 266, 1994.

79. Kenshalo, D.R., Decker, T., and Hamilton, A., Spatial summation on the forehead, forearm, and back produced by radiant and conducted heat, *J. Comp. Physiol. Psychol.*, 63, 510, 1967.

80. Stevens, J.C. and Marks, L.E., Spatial summation and the dynamics of warmth sensation, *Perceptual Psychophys.*, 9, 291, 1971.

81. Machet-Pietropaoli, H. and Chery-Croze, S., Spatial summation of thermal pain in human beings, *Sens. Processes*, 3, 183, 1979.

82. Kojo, I. and Pertovaara, A., The effects of stimulus area and adaptation temperature on warm and heat pain thresholds in man, *Int. J. Neurosci.*, 32, 875, 1987.

83. Price, D.D., McHaffie, J.G., and Larson, M.A., Spatial summation of heat-induced pain: influence of stimulus area and spatial separation of stimuli on perceived pain sensation intensity and unpleasantness, *J. Neurophysiol.*, 62, 1270, 1989.

84. Douglass, D.K., Carstens, E., and Watkins, L.R., Spatial summation in human thermal pain perception: comparison within and between dermatomes, *Pain*, 50, 197, 1992.

85. Greene, L.C. and Hardy, J.D., Spatial summation of pain, *J. Appl. Physiol.*, 13, 457, 1958.

86. Bouhassira, D. et al., Dorsal horn convergent neurones: negative feedback triggered by spatial summation of nociceptive afferents, *Pain*, 62, 195, 1995.

87. Nielsen, J. and Arendt-Nielsen, L., Spatial summation of heat induced pain within and between dermatomes, *Somatosens. Mot. Res.*, 14, 119, 1997.

88. Andersen, O.K., Gracely, R.H., and Arendt-Nielsen, L., Facilitation of the human nociceptive reflex by stimulation of A beta-fibres in a secondary hyperalgesic area sustained by nociceptive input from the primary hyperalgesic area, *Acta Physiol. Scand.*, 155, 87, 1995.

89. LaMotte, R.H., Lundberg, L.E.R., and Torebjörk, H.E., Pain, hyperalgesia and activity in nociceptive C units in humans after intradermal injection of capsaicin, *J. Physiol. (Lond.)*, 448, 749, 1992.

90. Simone, D.A., Baumann, T.K., and LaMotte, R.H., Dose-dependent pain and mechanical hyperalgesia in humans after intradermal injection of capsaicin, *Pain*, 38, 99, 1989.

91. Magerl, W., Wilk, S.H., and Treede, R.D., Secondary hyperalgesia and perceptual wind-up following intradermal injection of capsaicin in humans, *Pain*, 74, 257, 1998.

92. Wallace, M.S. et al., Concentration-effect relations for intravenous lidocaine infusions in human volunteers: effects on acute sensory thresholds and capsaicin-evoked hyperpathia, *Anesthesiology*, 86, 1262, 1997.

93. Kinnman, E., Nygards, E.B., and Hansson, P., Effects of dextromethorphan in clinical doses on capsaicin-induced ongoing pain and mechanical hypersensitivity, *J. Pain Symptom. Manage.*, 14, 195, 1997.

94. Kinnman, E., Nygards, E.B., and Hansson, P., Peripherally administered morphine attenuates capsaicin-induced mechanical hypersensitivity in humans, *Anesth. Analg.*, 84, 595, 1997.

95. Witting, N. et al., Repetitive intradermal capsaicin: differential effect on pain and areas of allodynia and punctate hyperalgesia, *Somatosens. Mot. Res.*, 17, 5, 2000.

96. Ali, Z., Meyer, R.A., and Campbell, J.N., Secondary hyperalgesia to mechanical but not heat stimuli following a capsaicin injection in hairy skin, *Pain*, 68, 401, 1996.

97. Segerdahl, M. et al., Systemic adenosine attenuates touch evoked allodynia induced by mustard oil in humans, *Neuroreport*, 6, 753, 1995.

98. Möiniche, S., Dahl, J.B., and Kehlet, H., Time course of primary and secondary hyperalgesia after heat injury to the skin, *Br. J. Anaesth.*, 71, 201, 1993.

99. Warncke, T., Stubhaug, A., and Jörum, E., Ketamine, an NMDA receptor antagonist, suppresses spatial and temporal properties of burn-induced secondary hyperalgesia in man: a double-blind, cross-over comparison with morphine and placebo, *Pain*, 72, 99, 1997.

100. Pedersen, J.L. and Kehlet, H., Secondary hyperalgesia to heat stimuli after burn injury in man, *Pain*, 76, 377, 1998.

101. Pedersen, J.L. and Kehlet, H., Hyperalgesia in a human model of acute inflammatory pain: a methodological study, *Pain*, 74, 139, 1998.

102. Pedersen, J.L. et al., Hyperalgesia and temporal summation of pain after heat injury in man, *Pain*, 74, 189, 1998.

103. Pedersen, J.L., Galle, T.S., and Kehlet, H., Peripheral analgesic effects of ketamine in acute inflammatory pain, *Anesthesiology*, 89, 58, 1998.

104. Petersen, K.L. and Rowbotham, M.C., A new human experimental pain model: the heat/capsaicin sensitization model, *Neuroreport*, 10, 1511, 1999.

105. Kellgren, J.H., Observations on referred pain arising from muscle, *Clin. Sci.*, 3, 175, 1938.

106. Hockaday, J.M. and Whitty, C.W.M., Patterns of referred pain in the normal subject, *Brain*, 90, 481, 1967.

107. Bonica, J.J., General considerations of acute pain, in *The Management of Pain,* Bonica, J.J., Ed., Lea & Febiger, Philadelphia, 1990, 159.

108. Svensson, P. and Arendt-Nielsen, L., Induction and assessment of experimental muscle pain, *J. Electromyogr. Kinesiol.*, 5, 131, 1995.

109. Lewis, T., Pain in muscular ischemia, *Arch. Int. Med.*, 49, 713, 1932.

110. Vecchiet, L., Giamberardino, M.A., and Marini, I., Immediate muscular pain from physical activity, in *Pain and Mobility,* Tiengo, M. et al., Eds., Raven Press, New York, 1987, 193.

111. Smith, G.M. et al., Experimental pain produced by the submaximum effort tourniquet technique: further evidence of validity, *J. Pharmacol. Exp. Ther.*, 163, 468, 1968.

112. Smith, G.M. et al., An experimental pain method sensitive to morphine in man: the submaximum effort tourniquet technique, *J. Pharmacol. Exp. Ther.*, 154, 324, 1966.

113. Moore, R.M., Moore, R.E., and Singleton, A.O., Experiments on the chemical stimulation of pain-endings associated with small blood-vessels, *Am. J. Physiol.*, 107, 594, 1934.

114. Harpuder, K. and Stein, I.D., Studies on the nature of pain arising from an ischemic limb, *Am. Heart J.*, 25, 429, 1943.

115. MacIver, M.B. and Tanelian, D.L., Activation of C fibers by metabolic perturbations associated with tourniquet ischemia, *Anesthesiology*, 76, 617, 1992.

116. Issberner, U., Reeh, P.W., and Steen, K.H., Pain due to tissue acidosis: a mechanism for inflammatory and ischemic myalgia?, *Neurosci. Lett.*, 208, 191, 1996.

117. Newham, D.J. and Mills, K.R., Muscles, tendons and ligaments, in *Textbook of Pain,* Wall, P.D. and Melzack, R., Eds., Churchill Livingstone, Edinburgh, 1999, 517.

118. Asmussen, E., Observations on experimental muscular soreness, *Acta Rheumatol. Scand.*, 2, 109, 1956.

119. Newham, D.J., The consequences of eccentric contractions and their relation to delayed onset muscle pain, *Eur. J. Appl. Physiol.*, 57, 353, 1988.

120. Howell, J.N., Chleboun, G., and Conatser, R., Muscle stiffness, strength loss, swelling and soreness following exercise-induced injury in humans, *J. Physiol. (Lond.)*, 464, 183, 1993.

121. Bajaj, P., Graven-Nielsen, T., Wright, A., Davies, I., and Arendt-Nielsen, L., Muscle hyperalgesia in post exercise muscle soreness assessed by single and repetitive ultrasound stimuli, *J. Pain,* 1, 111, 2000.

122. Svensson, P. and Arendt-Nielsen, L., Effect of topical NSAID on post-exercise jaw muscle soreness: a placebo-controlled experimental study, *J. Musculoskel. Pain*, 3, 41, 1995.

123. Svensson, P., Houe, L., and Arendt-Nielsen, L., Effect of systemic versus topical nonsteroidal anti-inflammatory drugs on postexercise jaw-muscle soreness: a placebo-controlled study, *J. Orofac. Pain*, 11, 353, 1997.

124. Howell, J.N. et al., The effect of nonsteroidal anti-inflammatory drugs on recovery from exercise-induced muscle injury. 2. Ibuprofen, *J. Musculoskel. Pain*, 6, 69, 1998.

125. Meadows, J.C., Observations on muscle pain in man, with particular reference to pain during needle electromyography, *J. Neurol. Neurosurg. Psychiatry*, 33, 519, 1970.

126. Brucini, M. et al., Pain thresholds and electromyographic features of periarticular muscles in patients with osteoarthritis of the knee, *Pain*, 10, 57, 1981.

127. Duranti, R., Galletti, R., and Pantaleo, T., Relationships between characteristics of electrical stimulation, muscle pain and blink responses in man, *Electroencephalogr. Clin. Neurophysiol.*, 55, 637, 1983.

128. Vecchiet, L. et al., Modifications of cutaneous, subcutaneous and muscular sensory and pain thresholds after the induction of an experimental algogenic focus in the skeletal muscle, *Clin. J. Pain*, 4, 55, 1988.

129. Kawakita, K., Miura, T., and Iwase, Y., Deep pain measurement at tender points by pulse algometry with insulated needle electrodes, *Pain*, 44, 235, 1991.

130. Ishimaru, K., Kawakita, K., and Sakita, M., Analgesic effects induced by TENS and electroacupuncture with different types of stimulating electrodes on deep tissues in human subjects, *Pain*, 63, 181, 1995.

131. Sörensen, J. et al., Hyperexcitability in fibromyalgia, *J. Rheumatol.*, 25, 152, 1998.

132. Arendt-Nielsen, L. et al., Temporal summation in muscles and referred pain areas: an experimental human study, *Muscle Nerve*, 20, 1311, 1997.

133. Svensson, P. et al., Human intramuscular and cutaneous pain: psychophysical comparisons, *Exp. Brain Res.*, 114, 390, 1997.

134. Svensson, P. et al., Non-painful and painful stimulation of human skin and muscle: analysis of cerebral evoked potentials, *Electroencephalogr. Clin. Neurophysiol.*, 104, 343, 1997.

135. Torebjörk, H.E., Ochoa, J.L., and Schady, W., Referred pain from intraneural stimulation of muscle fascicles in the median nerve, *Pain*, 18, 145, 1984.

136. Simone, D.A. et al., Identification of muscle afferents subserving sensation of deep pain in humans, *J. Neurophysiol.*, 72, 883, 1994.

137. Marchettini, P. et al., Pain from excitation of identified muscle nociceptors in humans, *Brain Res.*, 740, 109, 1996.

138. Laursen, R.J. et al., Referred pain is dependent on sensory input from the peripheral — a human experimental study, *Eur. J. Pain*, 1, 261, 1997.

139. Laursen, R.J. et al., The effect of differential and complete nerve block on referred pain — a psychophysical study, *Pain*, 80, 257, 1998.

140. Stohler, C.S., Zhang, X., and Ashton-Miller, J.A., An experimental model of jaw muscle pain in man, in *The Biological Mechanisms of Tooth Movement and Craniofacial Adaptation,* Davidovitch, Z., Ed., The Ohio State University College of Dentistry, Columbus, 1992, 503.

141. Jensen, M.P., Computer controlled pressure algometry in human joints, Ph.D. thesis, Aarhus University, Denmark, 1999.

142. Kosek, E. and Ekholm, J., Modulation of pressure pain thresholds during and following isometric contraction, *Pain*, 61, 481, 1995.

143. Graven-Nielsen, T. et al., Experimentally induced muscle pain induces hypoalgesia in heterotopic deep tissues, but not in homotopic deep tissues, *Brain Res.*, 787, 203, 1998.

144. Sohn, M.K. Graven-Nielsen, T., Arendt-Nielsen, L., and Svensson, P., Inhibition of motor unit firing during experimental muscle pain in humans, *Muscle Nerve,* 23, 1219, 2000.

145. Arima, T., Svensson, P., and Arendt-Nielsen, L., Capsaicin-induced muscle hyperalgesia in exercised and non-exercised human masseter, *J. Orofac. Pain,* 14, 213, 2000.

146. Jensen, K. et al., Pain and tenderness in human temporal muscle induced by brady-kinin and 5-hydroxytryptamine, *Peptides*, 11, 1127, 1990.

147. Babenko, V. et al., Experimental human muscle pain induced by intramuscular injec-tions of bradykinin, serotonin, and substance P, *Eur. J. Pain*, 3, 93, 1999.

148. Babenko, V. et al., Experimental human muscle pain and muscular hyperalgesia induced by combinations of serotonin and bradykinin, *Pain*, 82, 1, 1999.

149. Pedersen-Bjergaard, U. et al., Calcitonin gene-related peptide, neurokinin A and substance P: Effects on nociception and neurogenic inflammation in human skin and temporal muscle, *Peptides*, 12, 333, 1991.

150. Jensen, K. and Norup, M., Experimental pain in human temporal muscle induced by hypertonic saline, potassium and acidity, *Cephalalgia*, 12, 101, 1992.

151. Rossi, A. and Decchi, B., Changes in Ib heteronymous inhibition to soleus motoneurones during cutaneous and muscle nociceptive stimulation in humans, *Brain Res.*, 774, 55, 1997.

152. Lewis, T., Suggestions relating to the study of somatic pain, *Br. Med. J.*, 1, 1321, 1938.

153. Feinstein, B. et al., Experiments on pain referred from deep tissues, *J. Bone Joint Surg.*, 36, 981, 1954.

154. Stohler, C.S. and Lund, J.P., Effects of noxious stimulation of the jaw muscles on the sensory experience of volunteer human subjects, in *Biological & Psychological Aspects of Orofacial Pain,* Stohler, C.S. and Carlson, D.S., Eds., Center for Human Growth & Development, The University of Michigan, Ann Arbor, 1994, 55.

155. Stohler, C.S. and Lund, J.P., Psychophysical and orofacial motor response to muscle pain — validation and utility of an experimental model, in *Brain and Oral Functions,* Morimoto, T., Matsuya, T., and Takada, K., Eds., Elsevier, 1995, 227.

156. Svensson, P. et al., Effect of chronic and experimental jaw muscle pain on pain-pressure thresholds and stimulus-response curves, *J. Orofac. Pain*, 9, 347, 1995.

157. Stohler, C.S. and Kowalski, C.J., Spatial and temporal summation of sensory and affective dimensions of deep somatic pain, *Pain*, 79, 165, 1999.

158. Graven-Nielsen, T., Sensory manifestations and sensory-motor interactions during experimental muscle pain in man, Ph.D. thesis, Aalborg University, Denmark, 1997.

159. Zhang, X., Ashton-Miller, J.A., and Stohler, C.S., A closed-loop system for maintaining constant experimental muscle pain in man, *IEEE Trans. Bio. Eng.*, 40, 344, 1993.

160. Graven-Nielsen, T. et al., Experimental muscle pain: a quantitative study of local and referred pain in humans following injection of hypertonic saline, *J. Musculoskel. Pain*, 5, 49, 1997.

161. Graven-Nielsen, T. et al., Quantification of local and referred muscle pain in humans after sequential i.m. injections of hypertonic saline, *Pain*, 69, 111, 1997.

162. Graven-Nielsen, T. et al., Stimulus-response functions in areas with experimentally induced referred muscle pain — A psychophysical study, *Brain Res.*, 744, 121, 1997.

163. Graven-Nielsen, T. et al., In vivo model of muscle pain: quantification of intra-muscular chemical, electrical, and pressure changes associated with saline-induced muscle pain in humans, *Pain*, 69, 137, 1997.

164. Graven-Nielsen, T., Svensson, P., and Arendt-Nielsen, L., Effects of experimental muscle pain on muscle activity and co-ordination during static and dynamic motor function, *Electroencephalogr. Clin. Neurophysiol.*, 105, 156, 1997.

165. Svensson, P., Houe, L., and Arendt-Nielsen, L., Bilateral experimental muscle pain changes electromyographic activity of human jaw-closing muscles during mastication, *Exp. Brain Res.*, 116, 182, 1997.

166. Svensson, P., Arendt-Nielsen, L., and Houe, L., Muscle pain modulates mastication: An experimental study in humans, *J. Orofac. Pain*, 12, 7, 1998.

167. Svensson, P. et al., Experimental muscle pain does not cause long-lasting increases in resting EMG activity, *Muscle Nerve*, 21, 1382, 1998.

168. Svensson, P., Graven-Nielsen, T., and Arendt-Nielsen, L., Mechanical hyperesthesia of human facial skin induced by tonic painful stimulation of jaw muscles, *Pain*, 74, 93, 1998.

169. Svensson, P. et al., Experimental jaw-muscle pain does not change heteronymous H-reflexes in the human temporalis muscle, *Exp. Brain Res.*, 121, 311, 1998.

170. Svensson, P. et al., Modulation of inhibitory reflexes in single motor units in human masseter by tonic painful stimulation, *Pain*, 83, 441, 1999.

171. Svensson, P. et al., Modulation of stretch-evoked reflexes in single motor units in human masseter muscle by experimental pain, *Exp. Brain Res.*, 132, 65, 2000.

172. Andersen, O.K. et al., Interaction between cutaneous and muscle afferent activity in polysynaptic reflex pathways: a human experimental study, *Pain*, 84, 29, 2000.

173. Madeleine, P. et al., Sensory manifestations in experimental and work-related chronic neck-shoulder pain, *Eur. J. Pain*, 2, 251, 1998.

174. Madeleine, P., Voigt, M., and Arendt-Nielsen, L., Reorganisation of human step initiation during acute experimental muscle pain, *Gait. Posture*, 10, 240, 1999.

175. Madeleine, P. et al., Shoulder muscle co-ordination during chronic and acute experimental neck-shoulder pain. An occupational pain study, *Eur. J. Appl. Physiol.*, 79, 127, 1999.

176. Matre, D.A. et al., Experimental muscle pain increases the human stretch reflex, *Pain*, 75, 331, 1998.

177. Matre, D.A. et al., The influence of experimental muscle pain on the human soleus stretch reflex during sitting and walking, *Clin. Neurophysiol.*, 110, 2033, 1999.

178. Moe-Nilssen, R., Ljunggren, A.E., and Torebjörk, E., Dynamic adjustments of walking behavior dependent on noxious input in experimental low back pain, *Pain*, 83, 477, 1999.

179. Wang, K., Svensson, P., and Arendt-Nielsen, L., Modulation of exteroceptive suppression periods in human jaw-closing muscles by local and remote experimental muscle pain, *Pain*, 82, 253, 1999.

180. Zedka, M. et al., Voluntary and reflex control of human back muscles during induced pain, *J. Physiol. (Lond.)*, 520, 591, 1999.

181. Leffler, A.S., Kosek, E., and Hansson, P., Injection of hypertonic saline into musculus infraspinatus resulted in referred pain and sensory disturbances in the ipsilateral upper arm, *Eur. J. Pain*, 4, 73, 2000.

182. Wang, K., Svensson, P., and Arendt-Nielsen, L., Effects of tonic muscle pain on short-latency jaw-stretch reflexes in humans, *Pain*, 88, 189, 2000.

183. Whitty, C.W.M. and Willison, R.G., Some aspects of referred pain, *Lancet*, 2, 226, 1958.

184. Laursen, R.J. et al., The effect of compression and regional anaesthetic block on referred pain intensity in humans, *Pain*, 80, 257, 1999.

185. Sinclair, D.C., Wenddell, G., and Feindel, W.H., Referred pain and associated phenomena, *Brain*, 71, 184, 1948.

186. Klingon, G.H. and Jeffreys, W.H., Distribution of cutaneous hyperalgesia, *Neurology*, 8, 272, 1958.

187. Cook, A.J. et al., Dynamic receptive field plasticity in rat spinal cord dorsal horn following C-primary afferent input, *Nature*, 325, 151, 1987.

188. Hoheisel, U. et al., Appearance of new receptive fields in rat dorsal horn neurons following noxious stimulation of skeletal muscle: a model for referral of muscle pain?, *Neurosci. Lett.*, 153, 9, 1993.

189. Mense, S., Referral of muscle pain. New aspects, *APS*, 3, 1, 1994.

190. Lewis, T. and Kellgren, J.H., Observations relating to referred pain, viscero-motor reflexes and other associated phenomena, *Clin. Sci.*, 4, 47, 1939.

191. Inman, V.T. and Saunders, J.B.C.M., Referred pain from skeletal structures, *J. Nerv. Ment. Dis.*, 99, 660, 1944.

192. Mense, S., Nociception from skeletal muscle in relation to clinical muscle pain, *Pain*, 54, 241, 1993.

193. Paintal, A.S., Functional analysis of group III afferent fibres of mammalian muscles, *J. Physiol. (Lond.)*, 152, 250, 1960.

194. Kumazawa, T. and Mizumura, K., Thin-fibre receptors responding to mechanical, chemical and thermal stimulation in the skeletal muscle of the dog, *J. Physiol. (Lond.)*, 273, 179, 1977.

195. Iggo, A., Non-myelinated afferent fibres from mammalian skeletal muscle, *J. Physiol. (Lond.)*, 155, 52, 1961.

196. Serratrice, G. et al., Cutaneous, muscular and visceral unmyelinated afferent fibres: comparative study, in *Peripheral Neuropathies*, Canal, N. and Pozza, G., Eds., Elsevier, Amsterdam, 1978, 69.

197. Capra, N.F. and Ro, J.Y., Experimental muscle pain produces central modulation of proprioceptive signals arising from jaw muscle spindles, *Pain*, 86, 151, 2000.

198. Stohler, C.S., Zhang, X., and Lund, J.P., The effect of experimental jaw muscle pain on postural muscle activity, *Pain*, 66, 215, 1996.

199. Crenshaw, A.G. et al., Knee extension torque and intramuscular pressure of the vastus lateralis muscle during eccentric and concentric activities, *Eur. J. Appl. Physiol.*, 70, 13, 1995.

200. Jensen, B.R. et al., Soft tissue architecture and intramuscular pressure in the shoulder region, *Eur. J. Morphol.*, 33, 205, 1995.

201. Wolff, B.B. and Jarvik, M.E., Quantitative measures of deep somatic pain: further studies with hypertonic saline, *Clin. Sci.*, 28, 43, 1965.

202. Ness, T.J. and Gebhart, G.F., Visceral pain: a review of experimental studies, *Pain*, 41, 167, 1990.

203. Gebhart, G.F., *Visceral Pain*, IASP Press, Seattle, 1995.

204. Drewes, A.M., Experimentally evoked pain in the human gastrointestinal tract, Ph.D. thesis, Aalborg University, Denmark, 1998.

205. Chen, A.C., Human brain measures of clinical pain: a review. I. Topographic mappings, *Pain*, 54, 115, 1993.

206. Chen, A.C., Human brain measures of clinical pain: a review. II. Tomographic imagings, *Pain*, 54, 133, 1993.

207. Jones, A.K.P. and Derbyshire, S.W.G., Position emmission topography as a tool for understanding the cerebral processing of pain, in *Touch, Temperature and Pain in Health and Disease: Mechanisms and Assessments,* Boivie, J., Hansson, P., and Lindblom, B., Eds., IASP Press, Seattle, 1994, 491.

208. Casey, K.L., Forebrain mechanisms of nociception and pain: analysis through imaging, *Proc. Natl. Acad. Sci. USA*, 96, 7668, 1999.

209. Davis, K., Functional brain imaging in humans: methodology and issues, in *Methods in Pain Research,* Kruger, L., Ed., CRC Press, Boca Raton, 2001.

210. Chapman, C.R. et al., Pain measurement: an overview, *Pain*, 22, 1, 1985.

211. Price, D.D., *Psychological Mechanisms of Pain and Analgesia*, IASP Press, Seattle, 1999.

212. Fleischl von Marxow, E., Mittheilung betreffend die physiologie der hirnrinde, *Centralblatt Physiologie*, 4, 538, 1890.

213. Berger, H., Über das elektrenkephalogramm das menschen. Dritte mitteilung, *Arch. Psychiat. Nervenkr.*, 97, 6, 1932.

214. Perl, E.R., Galambos, R., and Glorig, A., The estimation of hearing threshold by electroencephalography, *Electroencephalogr. Clin. Neurophysiol.*, 5, 501, 1953.

215. Chatrian, G.E. et al., Cerebral responses to electrical tooth pulp stimulation in man. An objective correlate of acute experimental pain, *Neurology*, 25, 745, 1975.

216. Roth, M., Shaw, J., and Green, J., The form, voltage distribution and physiological significance of the K-complex, *Electroencephalogr. Clin. Neurophysiol.*, 8, 385, 1956.

217. Bancaud, J., Bloch, V., and Paillard, J., Contribution E.E.G. à l'étude des potentials évoqués chez l'homme au niveau du vertex, *Rev. Neurolog.*, 89, 399, 1953.

218. Spreng, M. and Ichioka, M., Langsame rindenpotentiale bei schmerzreizung am menschen, *Pflügers Arch.*, 279, 121, 1964.

219. Schmidt, J., Die beeinflussung der langsamen hirnrindenpotentiale des menschen nach elektrischer zahnreizung durch analgetika, *Acta Biol. Med. Ger.*, 24, 361, 1970.

220. Davis, H. et al., The slow response of the human cortex to auditory stimuli: recovery process, *Electroencephalogr. Clin. Neurophysiol.*, 21, 105, 1966.

221. Uttal, W.R. and Cook, L., Systematics of the evoked somatosensory cortical potential: a psychophysical-electrophysiological comparison, *Ann. N.Y. Acad. Sci.*, 112, 60, 1964.

222. De Broucker, T. and Willer, J.C., Etude comparative du réflexe nociceptif et des composantes tardives du potentiel évoqué somésthésique lors de stimulations du nerf sural chez l'homme normal, *Rev. Electroencephalogr. Neurophysiol. Clin.*, 15, 149, 1985.

223. Brennum, J. and Jensen, T.S., Relationship between vertex potentials and magnitude of pre-pain and pain sensations evoked by electrical skin stimuli, *Electroencephalogr. Clin. Neurophysiol.*, 82, 387, 1992.

224. Leandri, M., Campbell, J.A., and Lahuerta, J., The effect of attention on tooth-pulp evoked potentials, in *Advances in Pain Research and Therapy,* Fields, H.L. et al., Eds., Raven Press, New York, 1985, 331.

225. Bromm, B., Meier, W., and Scharein, E., Antagonism between tilidine and naloxone on cerebral potentials and pain ratings in man, *Eur. J. Pharmacol.*, 87, 431, 1983.

226. Dowman, R., Spinal and supraspinal correlates of nociception in man, *Pain*, 45, 269, 1991.

227. Miltner, W. et al., Somatosensory event-related potentials to painful and non-painful stimuli: effects of attention, *Pain*, 38, 303, 1989.

228. Carmon, A., Dotan, Y., and Sarne, Y., Correlation of subjective pain experience with cerebral evoked responses to noxious thermal stimulations, *Exp. Brain Res.*, 33, 445, 1978.

229. Bromm, B. and Treede, R.D., Human cerebral potentials evoked by CO_2 laser stimuli causing pain, *Exp. Brain Res.*, 67, 153, 1987.

230. Arendt-Nielsen, L., First pain event related potentials to argon laser stimuli: recording and quantification, *J. Neurol. Neurosurg. Psychiatry*, 53, 398, 1990.

231. Svensson, P. et al., Vertex potentials evoked by nociceptive laser stimulation of oral mucosa: relationship to stimulus intensity, *Anesth. Pain Control. Dent.*, 2, 27, 1993.

232. Arendt-Nielsen, L. and Chen, A.C.N., Evoked potentials to painful laser stimulation, in *Clinical Neurophysiology: From Receptor to Perception,* Comi, G. et al., Eds., Elsevier, 1999, 311.

233. Arendt-Nielsen, L., Characteristics, detection, and modulation of laser-evoked vertex potentials, *Acta Anaesthesiol. Scand. Suppl.*, 101, 7, 1994.

234. Kobal, G. and Raab, W., The effects of analgesics on pain-related somatosensory evoked potentials, *Agents Actions Suppl.*, 19, 75, 1986.

235. Miltner, W., Larbig, W., and Braun, C., Biofeedback of somatosensory event-related potentials: can individual pain sensations be modified by biofeedback-induced self-control of event-related potentials?, *Pain*, 35, 205, 1988.

236. Jacobson, R.C., Chapman, C.R., and Gerlach, R., Stimulus intensity and inter-stimulus interval effects on pain-related cerebral potentials, *Electroencephalogr. Clin. Neurophysiol.*, 62, 352, 1985.

237. Svensson, P. et al., Oral mucosal analgesia quantitatively assessed by argon laser-induced thresholds and single-evoked vertex potentials, *Anesth. Pain Control. Dent.*, 2, 154, 1993.

238. Sherrington, C.S., Flexion-reflex of the limb, crossed extension-reflex and reflex stepping and standing, *J. Physiol. (Lond.)*, 40, 28, 1910.

239. Kugelberg, E., Demonstration of A and C fibre components in the Babinski plantar response and the pathological flexion reflex, *Brain*, 71, 304, 1948.

240. Hugon, M., Exteroceptive reflexes to stimulation of the sural nerve in man, in *New Developments in Electromyography and Clinical Neurophysiology*, Desmedt, J.E., Ed., Karger, Basel, 1973, 713.

241. Hagbarth, K.E., Spinal withdrawal reflexes in human lower limbs, *J. Neurol. Neurosurg. Psychiatry*, 23, 222, 1960.

242. Shahani, B.T. and Young, R.R., Human flexor reflexes, *J. Neurol. Neurosurg. Psychiatry*, 34, 616, 1971.

243. Willer, J.C., Boureau, F., and Berny, J., Nociceptive flexion reflexes elicited by noxious laser radiant heat in man, *Pain*, 7, 15, 1979.

244. Campbell, I.G., Carstens, E., and Watkins, L.R., Comparison of human pain sensation and flexion withdrawal evoked by noxious radiant heat, *Pain*, 45, 259, 1991.

245. Willer, J.C., Comparative study of perceived pain and nociceptive flexion reflex in man, *Pain*, 3, 69, 1977.

246. Kugelberg, E., Eklund, K., and Grimby, L., An electromyographic study of the nociceptive reflexes of the lower limb. Mechanism of the plantar responses, *Brain*, 83, 394, 1960.

247. Chan, C.W. and Dallaire, M., Subjective pain sensation is linearly correlated with the flexion reflex in man, *Brain Res.*, 479, 145, 1989.

248. De Broucker, T., Willer, J.C., and Bergeret, S., The nociceptive flexion reflex in humans: a specific and objective correlate of experimental pain, in *Issues in Pain Measurement*, Chapman, C.R. and Loeser, J.D., Eds., Raven Press, New York, 1989, 337.

249. Garcia-Larrea, L. et al., Flexion reflexes following anterolateral cordotomy in man: dissociation between pain sensation and nociceptive reflex RIII, *Pain*, 55, 139, 1993.

250. Andersen, O.K., Sonnenborg, F.A., and Arendt-Nielsen, L., Modular organization of human leg withdrawal reflexes elicited by electrical stimulation of the foot sole, *Muscle Nerve*, 22, 1520, 1999.

251. Dimitrijevic, M.R. and Nathan, P.W., Studies of spasticity in man. 4. Changes in flexion reflex with repetitive cutaneous stimulation in spinal man, *Brain*, 93, 743, 1970.

252. Petersen-Felix, S. et al., The effects of isoflurane on repeated nociceptive stimuli (central temporal summation), *Pain*, 64, 277, 1996.

253. Svensson, P., Arendt-Nielsen, L., and Houe, L., Sensory-motor interactions of human experimental unilateral jaw muscle pain: a quantitative analysis, *Pain*, 64, 241, 1996.

254. Jensen, K., Quantification of tenderness by palpation and use of pressure algometers, in *Advances in Pain Research and Therapy,* Fricton, J. R. and Awad, E., Eds., Raven Press, New York, 1990, 165.

255. Brennum, J. et al., Measurements of human pressure-pain thresholds on fingers and toes, *Pain,* 38, 211, 1989.

256. Ohrbach, R. and Gale, E.N., Pressure pain thresholds, clinical assessment, and differential diagnosis: reliability and validity in patients with myogenic pain, *Pain,* 39, 157, 1989.

257. Vecchiet, L. et al., Comparative sensory evaluation of parietal tissues in painful and nonpainful areas in fibromyalgia and myofascial pain syndrome, in *Proc. 7th World Congress on Pain,* Gebhart, G. F., Hammond, D. L., and Jensen, T. S., Eds., IASP Press, Seattle, 1994, 177.

258. Steinbrocker, O. et al., Observations on pain produced by injections of hypertonic saline into muscles and other supportive tissues, *J. Clin. Inv.,* 32, 1045, 1953.

259. Graven-Nielsen, T. et al., Sensitivity to pressure stimuli in capsaicin-induced referred pain areas, *European Federation of IASP Chapters (EFIC), Nice, France,* p. 213, 2000 (Abstract).

260. Graven-Nielsen, T., Svensson, P., and Arendt-Nielsen, L., Effect of muscle pain on motor control — a human experimental approach, *Adv. Physiother.,* 2, 26, 1999.

261. Travell, J.G., Rinzler, S., and Herman, M., Pain and disability of the shoulder and arm, *J.A.M.A.,* 120, 417, 1942.

262. Schmidt, R.F., Kniffki, K.-D., and Schomburg, E.D., Der einfluss kleinkalibriger muskelafferenzen auf den muskeltonus, in *Therapie der Spastic,* Bauer, H., Koella, W.P., and Struppler, H., Eds., Verlag, München, 1981, 71.

263. Johansson, H. and Sojka, P., Pathophysiological mechanisms involved in genesis and spread of muscular tension in occupational muscle pain and in chronic musculoskeletal pain syndromes: a hypothesis, *Med. Hypoth.,* 35, 196, 1991.

264. Lund, J.P., Widmer, C.G., and Schwartz, G., What is the link between myofascial pain and dysfunction?, in *EMG of Jaw Reflexes in Man,* van Steenberghe, D. and De Laat, A., Eds., Leuven University Press, Leuven 1989, 427.

265. Wang, K. et al., EMG-force relationships are influenced by experimental jaw-muscle pain, *J. Oral Rehabil.,* 27, 394, 2000.

266. Arendt-Nielsen, L. et al., The influence of low back pain on muscle activity and coordination during gait: a clinical and experimental study, *Pain,* 64, 231, 1996.

267. Woolf, C.J. et al., Towards a mechanism-based classification of pain?, *Pain,* 77, 227, 1998.

268. Rowbotham, M.C. and Fields, H.L., The relationship of pain, allodynia and thermal sensation in post-herpetic neuralgia, *Brain,* 119, 347, 1996.

269. Berglund, B., Harju, E.-L., and Lindblom, U., Psychophysical testing of somatosensory functions in neuropathic pain patients, in *Proc. 8th World Congress on Pain,* Jensen, T. S., Turner, J. A., and Wiesenfeld-Hallin, Z., Eds., IASP Press, Seattle, 1997, 435.

270. Price, D.D., Bennett, G.J., and Rafii, A., Psychophysical observations on patients with neuropathic pain relieved by a sympathetic block, *Pain,* 36, 273, 1989.

271. Eide, P.K. and Rabben, T., Trigeminal neuropathic pain: pathophysiological mechanisms examined by quantitative assessment of abnormal pain and sensory perception, *Neurosurgery,* 43, 1103, 1998.

272. Treede, R.-D. and Bromm, B., Neurophysiological approaches to study of spinothalamic tract functions in humans, in *Pain and Central Nervous System Diseases: The Central Pain Syndrome,* Casey, K.L., Ed., Raven Press, New York, 1991, 117.

273. Kakigi, R. et al., CO_2 laser-induced pain-related somatosensory evoked potentials in peripheral neuropathies: correlation between electrophysiological and histopathological findings, *Muscle Nerve*, 14, 441, 1991.

274. Kakigi, R. et al., Pain-related somatosensory evoked potentials in syringomyelia, *Brain*, 114, 1871, 1991.

275. Kakigi, R. et al., Physiological study of the spinothalamic tract conduction in multiple sclerosis, *J. Neurol. Sci.*, 107, 205, 1992.

276. Bromm, B. and Treede, R.D., Laser-evoked cerebral potentials in the assessment of cutaneous pain sensitivity in normal subjects and patients, *Rev. Neurol. (Paris)*, 147, 625, 1991.

277. Hansen, H.C. et al., Recovery from brain-stem lesions involving the nociceptive pathways: comparison of clinical findings with laser-evoked potentials, *J. Clin. Neurophysiol.*, 13, 330, 1996.

278. Casey, K.L. et al., Laser-evoked cerebral potentials and sensory function in patients with central pain, *Pain*, 64, 485, 1996.

279. Coger, R.W. et al., Somatosensory evoked potentials and noxious stimulation in patients with intractable, noncancer pain syndromes, *Psychiatry Res.*, 2, 279, 1980.

280. Buchsbaum, M. and Silverman, J., Stimulus intensity control and the cortical evoked response, *Psychosom. Med.*, 30, 12, 1968.

281. Buchsbaum, M.S., Average evoked response augmenting reducing in schizophrenia and affective disorders, in *The Biology of the Major Psychoses: A Comparative Analysis*, Freedman, D.X., Ed., Raven Press, New York, 1975, 129.

282. Gibson, S.J. et al., Altered heat pain thresholds and cerebral event-related potentials following painful CO_2 laser stimulation in subjects with fibromyalgia syndrome, *Pain*, 58, 185, 1994.

283. Lorenz, J., Grasedyck, K., and Bromm, B., Middle and long latency somatosensory evoked potentials after painful laser stimulation in patients with fibromyalgia syndrome, *Electroencephalogr. Clin. Neurophysiol.*, 100, 165, 1996.

284. Arendt-Nielsen, L., Zachariae, R., and Bjerring, P., Quantitative evaluation of hypnotically suggested hyperaesthesia and analgesia by painful laser stimulation, *Pain*, 42, 243, 1990.

285. Zachariae, R. et al., The effect of hypnotically induced emotional states on brain potentials evoked by painful argon laser stimulation, *Clin. J. Pain*, 7, 130, 1991.

286. Reeves, J.L., Jaeger, B., and Graff-Radford, S.B., Reliability of the pressure algometer as a measure of myofascial trigger point sensitivity, *Pain*, 24, 313, 1986.

287. Schiffman, E. et al., A pressure algometer for myofascial pain syndrome: reliability and validity testing, in *Proc. Vth World Congress on Pain*, Dubner, R., Gebhart, G.F., and Bond, M.R., Eds., Elsevier, Amsterdam, 1988, 407.

288. Lautenbacher, S., Rollman, G.B., and McCain, G.A., Multi-method assessment of experimental and clinical pain in patients with fibromyalgia, *Pain*, 59, 45, 1994.

289. Reid, K.I., Gracely, R.H., and Dubner, R.A., The influence of time, facial side, and location on pain-pressure thresholds in chronic myogenous temporomandibular disorder, *J. Orofac. Pain*, 8, 258, 1994.

290. Kosek, E., Ekholm, J., and Hansson, P., Modulation of pressure pain thresholds during and following isometric contraction in patients with fibromyalgia and in healthy controls, *Pain*, 64, 415, 1996.

291. Bendtsen, L., Jensen, R., and Olesen, J., Qualitatively altered nociception in chronic myofascial pain, *Pain*, 65, 259, 1996.

292. Bendtsen, L. et al., Evidence of qualitatively altered nociception in patients with fibromyalgia, *Arthritis Rheum.*, 40, 98, 1997.

293. Sörensen, J. et al., Pain analysis in patients with fibromyalgia: effects of intravenous morphine, lidocaine and ketamine, *Scand. J. Rheumatol.*, 24, 360, 1995.

294. Sörensen, J. et al., Fibromyalgia — are there different mechanisms in the processing of pain? A double blind crossover comparison of analgesic drugs, *J. Rheumatol.*, 24, 1615, 1997.

295. Henriksson, K.G. and Mense, S., Pain and nociception in fibromyalgia: clinical and neurobiological considerations on aetiology and pathogenesis, *Pain Rev.*, 1, 245, 1994.

296. Johansen, M.K. et al., Generalised muscular hyperalgesia in chronic whiplash syndrome, *Pain*, 83, 229, 1999.

297. Graven-Nielsen, T. et al., Ketamine reduces muscle pain, temporal summation, and referred pain in fibromyalgia patients, *Pain*, 85, 483, 2000.

298. Swarbrick, E.T. et al., Site of pain from the irritable bowel, *Lancet*, 2, 443, 1980.

299. Mertz, H. et al., Evidence for altered spinal processing of gastric afferent information in non-ulcer dyspepsia, *Gastroenterology*, 104, A551, 1993.

300. Mayer, E.A. et al., Visceral hyperalgesia and irritable bowel syndrome, in *Visceral Pain*, Gebhart, G.F., Ed., IASP Press, Seattle, 1995, 429.

301. Whitehead, W.E. et al., Tolerance for rectosigmoid distention in irritable bowel syndrome, *Gastroenterology*, 98, 1187, 1990.

302. Ness, T.J., Metcalf, A.M., and Gebhart, G.F., A psychophysiological study in humans using phasic colonic distension as a noxious visceral stimulus, *Pain*, 43, 377, 1990.

303. Mayer, E.A. and Raybould, H.E., Role of visceral afferent mechanisms in functional bowel disorders, *Gastroenterology*, 99, 1688, 1990.

304. Antonaci, F., Sand, T., and Lucas, G.A., Pressure algometry in healthy subjects: interexaminer variability, *Scand. J. Rehabil. Med.*, 30, 3, 1998.

305. Feine, J.S. et al., Sex differences in the perception of noxious heat stimuli, *Pain*, 44, 255, 1991.

306. Fillingim, R.B. et al., Sex differences in temporal summation but not sensory-discriminative processing of thermal pain, *Pain*, 75, 121, 1998.

307. Walker, J.S. and Carmody, J.J., Experimental pain in healthy human subjects: gender differences in nociception and in response to ibuprofen, *Anesth. Analg.*, 86, 1257, 1998.

308. Chapman, C.R., Experimental pain models and analgesic efficacy, in *The Design of Analgesic Clinical Trials*, Max, M., Portenoy, R., and Laska, E., Eds., Raven Press, New York, 1991, 49.

309. Gracely, R.H., Experimental pain models, in *The Design of Analgesic Clinical Trials*, Max, M.B., Posner, J., and Laska, E.M., Eds., Raven Press, New York, 1991, 33.

310. Ekblom, A., Segerdahl, M., and Sollevi, A., Adenosine increases the cutaneous heat pain threshold in healthy volunteers, *Acta Anaesthesiol. Scand.*, 39, 717, 1995.

311. Ilkjaer, S. et al., Effect of systemic N-methyl-D-aspartate receptor antagonist (dextromethorphan) on primary and secondary hyperalgesia in humans, *Br. J. Anaesth.*, 79, 600, 1997.

312. Plesan, A., Sollevi, A., and Segerdahl, M., The NMDA-receptor antagonist dextromethorphan lacks analgesic effect in a human experimental ischemic pain model, *Acta Anaesthesiol. Scand.*, 44, 924, 2000.

313. Curatolo, M. et al., Temporal summation during extradural anaesthesia, *Br. J. Anaesth.*, 75, 634, 1995.

314. Park, K.M. et al., Effects of intravenous ketamine, alfentanil, or placebo on pain, pinprick hyperalgesia, and allodynia produced by intradermal capsaicin in human subjects, *Pain*, 63, 163, 1995.

315. Petersen, K.L., Brennum, J., and Dahl, J.B., Experimental evaluation of the analgesic effect of ibuprofen on primary and secondary hyperalgesia, *Pain*, 70, 167, 1997.

316. Clements, J.A. and Nimmo, W.S., Pharmacokinetics and analgesic effect of ketamine in man, *Br. J. Anaesth.*, 53, 27, 1981.

317. Posner, J., A modified submaximal effort tourniquet test for evaluation of analgesics in healthy volunteers, *Pain*, 19, 143, 1984.

318. Jonzon, B., Sylvén, C., and Kaijser, L., Theophylline decreases pain in the ischaemic forearm test, *Cardiovasc. Res.*, 23, 807, 1989.

319. Maurset, A. et al., A new version of the ischemic tourniquet pain test, *Meth. Find. Exp. Clin. Pharmacol.*, 13, 643, 1991.

320. López-Fiesco, A. et al., Evaluation of the analgesic efficacy of a combination of naproxen and paracetamol versus dipyrone on the tourniquet test in healthy volunteers, *Proc. West. Pharmacol. Soc.*, 35, 207, 1992.

321. Segerdahl, M., Ekblom, A., and Sollevi, A., The influence of adenosine, ketamine, and morphine on experimentally induced ischemic pain in healthy volunteers, *Anesth. Analg.*, 79, 787, 1994.

322. Myers, D.E., Shaikh, Z., and Zullo, T.G., Hypoalgesic effect of caffeine in experimental ischemic muscle contraction pain, *Headache*, 37, 654, 1997.

323. Rae, C.P. et al., Analgesic effect of adenosine on ischaemic pain in human volunteers, *Br. J. Anaesth.*, 82, 427, 1999.

324. Rowlingson, J.C. et al., Lidocaine as an analgesic for experimental pain, *Anesthesiology*, 52, 20, 1980.

325. Curatolo, M. et al., Remifentanil inhibits muscular more than cutaneous pain in humans, *Br. J. Anaesth.*, 85, 529, 2000.

326. Laursen, R.J. et al., The effect of differential and complete nerve block on experimental muscle pain in humans, *Muscle Nerve*, 22, 1564, 1999.

327. Shafer, S.L. and Gregg, K.M., Algorithms to rapidly achieve and maintain stable drug concentrations at the site of drug effect with a computer-controlled infusion pump, *J. Pharmacokinet. Biopharm.*, 20, 147, 1992.

Index